# Management for Professionals

More information about this series at http://www.springer.com/series/10101

Carsten Herbes · Christian Friege

Editors

# Marketing Renewable Energy

Concepts, Business Models and Cases

 Springer

*Editors*
Carsten Herbes
University of Nuertingen-Geislingen
Nuertingen, Germany

Christian Friege
Friege-Consulting
Stuttgart, Germany

ISSN 2192-8096             ISSN 2192-810X   (electronic)
Management for Professionals
ISBN 978-3-319-46426-8     ISBN 978-3-319-46427-5   (eBook)
DOI 10.1007/978-3-319-46427-5

Library of Congress Control Number: 2017936674

Cover illustration: © Grüne Landschaft I steffi-bartels@gmx.de I 2014

Printed on acid-free paper

This Springer imprint is published by Springer Nature
The registered company is Springer International Publishing AG
The registered company address is: Gewerbestrasse 11, 6330 Cham, Switzerland

# Preface

Fighting climate change in the world is a priority today for many people. Transforming our energy system from a heavy reliance on fossil fuels towards sustainability is deemed a major contributor to this goal. Therefore, many countries around the world have set ambitious targets for the growth of renewable energy (RE). In Germany, the "Energiewende" (energy transition) has been pursued for almost 20 years now and has resulted in significant change of the energy system. Often, the project is considered the most comprehensive change project of the German society in the last decades. This development is carefully observed in many countries, given that Germany is one of the largest economies in the world. Ultimately, there is no role model for such a fundamental new direction in an economy. While the case study of the Energiewende is frequently referred to in this book, many countries all over the world have taken steps towards decarbonization of their energy systems. After more conceptual contributions, examples from different countries are also analyzed.

Technically, the production of energy from renewable sources is constantly evolving. Meanwhile, however, other barriers appear. In countries where a particularly strong expansion of renewables takes place, increasing acceptance problems can be observed. This is mainly due to the local effects of wind mills or photovoltaic systems on residents and even more fundamentally due to perceived negative effects, for example, regarding the use of energy crops for bioenergy production or the extra costs for taxpayers and consumers. One consequence of these acceptance problems among the population is a declining willingness of policy makers to further accelerate the expansion of RE by subsidies. In Germany, the example of the significant cut-backs in the promotion of biogas underlines this. Another, often more visible barrier is the insufficient integration of the generation of renewable energy into the market and—as for green electricity—into grid management. These issues become ever more limiting with increasing production of RE. These last two factors, that is, the modified political support and the increasing need for the integration of RE into existing and future structures of power distribution, require a real marketing for RE.

For the German RE industry the change from long-term government guaranteed feed-in tariffs (based on the Renewables Energy Act (REA) ) to an increasingly market-based system is nothing less than a paradigm shift. It was precisely the system of the REA, eliminating virtually all risks in the sales market, which led to the strong expansion of renewables. Today, marketing to customers not only entails substantial additional risks on the sales side, but requires completely new skills and strategies in the RE industry.

This book is intended to support the adaption of marketing for RE. It has its roots in a German language volume (Herbes, C./Friege, C. (Eds.): Marketing Erneuerbarer Energien; Wiesbaden: Springer Gabler 2015), which originated at the interface between universities and the energy industry. The positive uptake of this book and many requests from international colleagues led to this English edition. Many of the German contributions have been adapted and updated for this book—and a number of new chapters were specifically added for this English edition. The book is aimed at specialists and managers of utility companies, be they large suppliers, municipal utilities, green power providers or renewable energy cooperatives, and at teachers and students. The contributions are meant to enhance the understanding of basal concepts that underlie the commercialization of RE. At the same time, practical tools and strategies for the marketing of RE are explained. The contributions often focus on green electricity, but also go beyond and expand on other aspects of RE.

The first part of the book "Foundations of Renewable Energy Marketing" collects contributions, which are relevant for RE marketing in general. *Friege and Herbes* identify in their introductory chapter the basic marketing characteristics of RE and develop an exemplary marketing mix for RE. *Bloche-Daub et al.* display the global potential of RE explaining how big the RE markets may be in perspective. The contribution of *Menges and Beyer* analyzes consumer preferences comprehensively and systematically, thereby laying the foundation for a RE marketing concept targeted at private consumers. RE products also open up special opportunities for a direct sales model as the chapter by *Friege* shows. *Tabi, Hille and Wüstenhagen* explain, in which way target group segmentation can be achieved in a green electricity market and a second contribution from the University of St. Gallen by *Chassot et al.* reviews, whether providing green electricity as a default product is an option. RE is a credence good that cannot be experienced directly in its environmentally friendly properties by the consumer. Therefore, considerations about certificates play a particularly important role in the elaboration of marketing strategies. These certificates, their mechanism and function are the topic of the contribution by *Leprich, Hoffmann and Luxenburger*.

Part two of the book looks at "Special Markets and New Business Models". Marketing strategies for biomethane are particularly complex, because there are four parallel paths to the market. *Herbes* analyzes these paths in his essay within their respective frameworks. *Klöpfer and Kliemczak* review the ever-growing contracting market, where using RE leads to very specific considerations. *Gervers* illustrates how RE can be used in the marketing of tourism enterprises. This is an interesting market with travelers being increasingly aware of their environmental responsibilities, who at the same time want to escape issues such as climate change for a while. The fourth paper in this section comes from *Schlemmermeier and Drechsler*. They present innovative business models, that originate in the development of RE and in a fundamentally decentralized energy system. Marc *Ringel* discusses in his essay the relationship between RE and electric mobility leading to a new business model "Green Mobility". Equally, *Schott and Koch* present their view on the marketing of batteries as an enabler of marketing RE.

The concluding part of the book looks at "Marketing of Renewables in Regional Markets". *Claudia Kemfert* asks in her essay, whether the Energiewende in Germany serves as a role model or leads into a dead end. *Graichen, Redl and Steigenberger* from the German think tank Agora Energiewende explain successes and challenges in the German Energiewende. Michel *Cruciani* allows for a view at marketing green energy in France and *Spataru and Arcuri* focus on the situation in the United Kingdom. *Bigerna, Bollino and Polinori* contribute an essay on marketing RE in Italy and Jörg *Raupach-Sumiya* portrays in detail the market in Japan. Thus a multitude of differing regulatory backgrounds are taken to discuss, in which way RE can be marketed to customers, what the specifics of each market and product are, and which challenges and opportunities stem from this. And all of this is presented with the intention to allow for learning across borders and to contribute to the further development of marketing of RE.

We would like to thank all authors, who made this book possible with their conceptual and extensive practical knowledge and with admirable commitment. They all aim to contribute to the successful commercialization of renewable energies. Our thanks also go to Prashanth Mahagaonkar and Ruth Milewski, our partners at Springer for the excellent support and to Nürtingen-Geislingen University for the provision of resources for the production of this book.

Nürtingen, Stuttgart                                                        Carsten Herbes
                                                                            Christian Friege

# Contents

# List of Contributors

**Bruno Arcuri**  UCL Energy Institute, London, UK

**Gregor Beyer**  Clausthal University of Technology, Department of Macroeconomics, Clausthal-Zellerfeld, Germany

**Simona Bigerna**  Department of Economics, University of Perugia, Perugia, Italy

**Karina Bloche-Daub**  Deutsches Biomasseforschungszentrum gGmbH, Leipzig, Germany

**Carlo Andrea Bollino**  Department of Economics, University of Perugia, Perugia, Italy

**Sylviane Chassot**  Institute for Economy and the Environment, Good Energies Chair for Management of Renewable Energies, University of St. Gallen, St. Gallen, Switzerland

**Michel Cruciani**  CGEMP, Université Paris-Dauphine, Paris, France

**Björn Drechsler**  LBD-Beratungsgesellschaft mbH, Berlin, Germany

**Nicole Fahr**  Institute for Economy and the Environment, Good Energies Chair for Management of Renewable Energies, University of St. Gallen, St. Gallen, Switzerland

**Christian Friege**  Stuttgart, Germany

**Susanne Gervers**  Hochschule für Wirtschaft und Umwelt Nürtingen-Geislingen, Nürtingen, Germany

**Peter Graf**  Institute for Economy and the Environment, Good Energies Chair for Management of Renewable Energies, University of St. Gallen, St. Gallen, Switzerland

**Patrick Graichen**  Agora Energiewende, Berlin, Germany

**Carsten Herbes**  Institute for International Research on Sustainable Management and Renewable Energy, Nuertingen-Geislingen University, Nuertingen, Germany

**Stefanie Lena Hille** Institute for Economy and the Environment (IWÖ-HSG), Good Energies Chair for Management of Renewable Energies, University of St. Gallen, St. Gallen, Switzerland

**Patrick Hoffmann** IZES gGmbH, Saarbrücken, Germany

**Claudia Kemfert** German Institute for Economic Research (DIW), Berlin, Germany
Hertie School of Governance, Berlin, Germany

**Ralf Klöpfer** MVV Energie AG, Mannheim, Germany

**Ulrich Kliemczak** MVV Energy Solutions GmbH, Mannheim, Germany

**Oliver Koch** sonnen GmbH, Wildpoldsried, Germany

**Volker Lenz** Deutsches Biomasseforschungszentrum gGmbH, Leipzig, Germany

**Uwe Leprich** IZES gGmbH, Saarbrücken, Germany

**Martin Luxenburger** IZES gGmbH, Saarbrücken, Germany

**Roland Menges** Clausthal University of Technology, Department of Macroeconomics, Clausthal-Zellerfeld, Germany

**Michael Nelles** Deutsches Biomasseforschungszentrum gGmbH, Leipzig, Germany
Waste Management and Material Flow of the Faculty of Agricultural and Environmental Sciences of the University of Rostock, Rostock, Germany

**Paolo Polinori** Department of Economics, University of Perugia, Perugia, Italy

**Jörg Raupach-Sumiya** College of Business Administration, Ritsumeikan University, Osaka, Japan

**Christian Redl** Agora Energiewende, Berlin, Germany

**Marc Ringel** Faculty of Economics and Law, Energy Economics, Nuertingen Geislingen University, Geislingen, Germany

**Ben Schlemmermeier** LBD-Beratungsgesellschaft mbH, Berlin, Germany

**Benjamin Schott** sonnen GmbH, Wildpoldsried, Germany

**Catalina Spataru** UCL Energy Institute, London, UK

**Markus Steigenberger** Agora Energiewende, Berlin, Germany

**Andrea Tabi** Institute for Economy and the Environment (IWÖ-HSG), Good Energies Chair for Management of Renewable Energies, University of St. Gallen, St. Gallen, Switzerland

**Rolf Wüstenhagen** Institute for Economy and the Environment, Good Energies Chair for Management of Renewable Energies, University of St. Gallen, St. Gallen, Switzerland

**Janet Witt** Deutsches Biomasseforschungszentrum gGmbH, Leipzig, Germany

Faculty of Landscape Architecture, Horticulture and Forestry, University of Applied Science Erfurt, Erfurt, Germany

# Part I

# Foundations of Renewable Energy Marketing

# Some Basic Concepts for Marketing Renewable Energy

Christian Friege and Carsten Herbes

**Abstract**

Against the background of a modern understanding of marketing, which stresses value orientation and the interactive web, the attributes of renewable energy (commodity, low-involvement product, credence good, partially public good, product that needs explanation in two dimensions, and prosumer good), as well as the aims of the consumers of renewable energy, a marketing mix for green energy is developed. Policies on the product, pricing, distribution, and communication are analyzed in detail and presented with a particular focus on the specifics of regenerative energy.

**Keywords**

Renewable energy • Green energy • Marketing

## 1    Introduction

Why does a separate analysis of the marketing of renewable energy not only enhance our understanding but in fact serve as a necessary supplement to our marketing knowledge? It is of course true that the fundamentals of marketing are also applicable to the marketing of renewable energy. Additionally, the challenges of differentiating commodities, turning low-involvement products into branded

A previous version of this chapter has been published in Herbes, C.; Friege, Chr. (Hrsg): Marketing Erneuerbarer Energien. Grundlagen, Geschäftsmodelle, Fallbeispiele, 2015, Springer Gabler.

C. Friege (✉)
Stuttgart, Germany
e-mail: cf@friege-consulting.de

C. Herbes
Institute for International Research on Sustainable Management and Renewable Energy,
Nuertingen-Geislingen University, Neckarsteige 6-10, 72622 Nuertingen, Germany

© Springer International Publishing AG 2017
C. Herbes, C. Friege (eds.), *Marketing Renewable Energy*, Management for Professionals, DOI 10.1007/978-3-319-46427-5_1

products, and marketing new technologies that are still caught between subsidies, testing, and commercial viability are also known per se. However, the marketing of renewable energy is a significantly more complex matter. Its singular importance is revealed by the comprehensive public debate about the energy revolution and the opportunity to turn renewable energy into an engine of growth for the economy of the twenty-first century. This puts the marketing of renewable energy into a unique societal context, and the tasks of marketing experts are challenging, multifaceted, and without any obvious examples from other industries or situations. Against this backdrop, a number of conceptual considerations about the marketing of renewable energy will be developed. The following issues will be addressed:

- Which understanding of marketing is suitable as a framework for the marketing of renewable energy? Which societal changes must be taken into consideration as the basis for distribution and marketing activities—in addition to a legal framework and subsidies?
- Which aims of consumers are most important for the marketing of renewable energy? Renewable energy grew out of the environmental movement and plays a major role in the debate about global climate change. This aspect largely influences the marketing of renewable energy.
- What are specific attributes of renewable energy and what are the effects of these attributes on marketing strategies?
- What is the relevance of renewable energy as an input factor for the provision of other goods and services? How can this input be utilized in their marketing?

In what follows, each of these questions is discussed in a separate section and a summary with a focus on concrete action is provided.

For all these questions, we define *renewable energy* as follows:

Renewable energy includes those types of energy sources that are available in virtually unlimited amounts (such as energy from the sun, wind or rivers), immediately grow again or are steadily available (biomass or bio-waste), and all forms of energy (such as green electricity or heating power) that are fully derived from these energy sources.

## 2    Marketing 3.0

Let us first of all address the issue of identifying a marketing concept that is suitable as a framework for the marketing of renewable energy. Kotler et al. (2010) clearly stress that the field of marketing has witnessed significant development over the years and is currently characterized as marketing 3.0. It defines a modern form of marketing as one that is embedded in social media, is based on many-to-many communication, and actively accepts the idea of social responsibility of corporations and integrates it into a marketing strategy (Fig. 1).

This value-based understanding of marketing is also stressed by the American Marketing Association, which in 2013 defined marketing as "the activity, set of

| | Marketing 1.0<br>Product-centric<br>Marketing | Marketing 2.0<br>Consumer-oriented<br>Marketing | Marketing 3.0<br>Values-driven<br>Marketing |
|---|---|---|---|
| Objective | Sell products | Satisfy and retain the consumers | Make the world a better place |
| Enabling forces | Industrial revolution | Information technology | New wave technology / Web 2.0+* |
| How companies see the market | Mass buyers with physical needs | Smarter consumer with mind and heart | Whole human with mind, heart, and spirit |
| Key marketing concept | Product development | Differentiation | Values |
| Company marketing guidelines | Product specification | Corporate and product positioning | Corporate mission, vision and values |
| Value propositions | Functional | Functional and emotional | Functional, emotional, and spiritual |
| Interaction with consumers | One-to-many transaction | One-to-one relationship | Many-to-many collaboration |

* Social media propelled by (1) cheap computers and mobile phones etc., (2) inexpensive internet access and (3) open source technology.

**Fig. 1** Comparison of the marketing approaches 1.0, 2.0 and 3.0 (source: adapted from Kotler et al. (2010), p. 6)

institutions, and processes for creating, communicating, delivering, and exchanging offerings that have value for customers, clients, partners, **and society at large**" (AMA 2013, emphasis added). The arguments by Kotler et al. (2010) are grouped around two focal points, value orientation and the interactive Web, that shape our society in the twenty-first century. Both of these focal points are helpful in developing conceptual foundations for the marketing of renewable energy (Fig. 1).

1. Value orientation

Kotler et al. (2010) identify world improvement as the purpose of their understanding of marketing and summarize as follows: "Instead of treating people simply as consumers, marketers approach them as whole human beings with minds, hearts, and spirits. Increasingly, consumers are looking for solutions to their anxieties about making the globalized world a better place. In a world full of confusion, they search for companies that address their deepest needs for social, economic and environmental justice in their mission, vision and values. They look for not only functional and emotional fulfilment but also human spirit fulfilment in the products and services they choose. . . .Supplying meaning is the future value proposition in marketing" (pp. 4, 20). This includes a holistic perspective on the consumer as a market perspective ["Marketing 3.0 lifts the concept of marketing into the arena of human aspirations, values and spirit" (p. 4)] and a business, which submits to its own explicit value framework in addition to mission and vision.

2. Interactive Web

So-called new wave technology (Kotler et al. 2010, p. 6, Fig. 1) is interpreted in Fig. 1 as Web 2.0+ since it refers not only to the Internet as a platform and technology but also to its rapid global spread based on the availability of inexpensive hardware (particularly smartphones and tablet computers that make mobile Internet access possible), simple and inexpensive Internet access, and open-source technologies, which allow easy access to software and its quick updating. Of similar importance is the developing culture of online collaboration, either via quick communication in social media (many-to-many) or via online cooperation (co-creation, crowdsourcing, and so forth). Value orientation and an interactive web characterize social realities as well as modern marketing 3.0. For that reason, they are also appropriate measures by which to classify the particular features of renewable energy and are suitable for use as a conceptual framework for further considerations.

## 3    Attributes of Renewable Energies and Their Effects on the Marketing of Renewable Energy

Renewable energies are characterized by a number of common attributes: they are (1) commodities, (2) low-involvement products and at the same time (3) credence goods. They are (4) partially public goods and (5) require explanation in two dimensions. Finally, they increasingly become (6) prosumer goods. What are the causes and effects of these attributes of renewable energy? What is the role of the two elements of marketing 3.0 mentioned earlier, namely the web and value orientation?

(1) Commodities

Commodities are goods with a quality that is subject to clearly defined criteria. Therefore, there is no differentiation and they are fully fungible, in other words interchangeable. Prices of commodities are typically determined at trading places or exchanges. Electricity and gas are typical commodities, as are fuels; the heating market, for example, is derived from the commodity market for energy.

*Value orientation* means that it is not only the defined quality that matters for renewable energy, but rather the energy source, which becomes the decisive criterion for customers' purchase decision. At issue is not so much the commodity of electricity or gas, for example, but rather the differentiation resulting from its generation from renewable primary energy. This differentiation is supported by the *interactive Web,* which not only provides detailed information about energy production but also facilitates discussion in the relevant communities. In addition, the Internet also allows for the comprehensive presentation of the different types of certifications that guarantee the origin of the renewable energy [on this issue see Leprich et al. (2017)].

(2) Low-involvement products

The extent of involvement with a product prior to the buying decision is defined as "a person's perceived relevance of the object based on inherent needs, values and interests" (Zaichkowsky 1985, p. 342). Electricity, gas, and fuel are generally considered to be low-involvement products (e.g., Busch et al. (2009); Lohse and Künzel (2011)).

Thanks to the value orientation, the offering can be differentiated not alone via brand development (as is the case for mineral oil fuels), but in addition via the societal positioning of the product as a contribution to slowing down climate change, using resources in a sustainable manner, and so forth. "Making the world a better place" as an aim of marketing 3.0 becomes particularly apparent in this regard. The activating effect of the *interactive web* further strengthens this positioning. As an example, the platform utopia.de was established with the explicit aim of "contributing towards sustainable change in the consumer behavior and lifestyle of millions of people" (Utopia 2014). Forums on a wide range of topics, especially concerning energy and energy consumption, can be found there.

(3) Credence goods

Green electricity and biomethane are credence goods. Meffert et. al. (2015, pp. 38–39) define the credence qualities of a service or good as follows: "In this case the consumer is unable to ascertain certain attributes or qualities either before or after the purchase, even though these attributes are important to him and he is willing to pay an appropriate price for them." This is apparently also the case for renewable energy, since the consumer—apart from certifications (Leprich et al. 2017)—has no way of ascertaining whether the stated energy sources were in fact utilized and what ecological value added is indeed achieved when purchasing the product. The same is true for the use of renewable energy in transportation services and in the offerings of tourism companies (Gervers 2017) or in the case of e-mobility (Ringel 2017). In every instance, the consumer ultimately depends on the promise that a "green" contribution is indeed being made.

In a study about credence goods, Dulleck et al. (2011) found that a significant share of the providers are in fact honest. In comparison, the subsequent verification of the attributes of a credence good—even where this would be possible—has no particular significance. Of much greater importance is the liability for the promised services—in the case of renewable energy, this is replaced mainly by reputation.[1]

---

[1]However, Dulleck et al. (2011) only assigned high relevance to reputation in cases where liability and competition were not strongly developed. In the case of renewable energy, reputation largely replaces liability.

In principle, *value-oriented marketing* is more suitable for creating the necessary trust. Corporations with a focus on value will be able to construct their orientation and positioning in such a way that mission, vision, and values are highly consistent and integrated and that any deviation from them would result in a massive loss of reputation, providing incentives to adhere to the self-imposed set of values. The necessary transparency is assured in this case via the social control of the *interactive web*, which also enhances the reputational risk as guarantor of appropriate behavior.

(4) Partially public goods

The aspect of climate protection turns renewable energy into a partially public good. Assuming that any increase in the demand for renewable energy also results in increased production and thus in a reduction of $CO_2$ emissions, which in turn contributes to a slowing of climate change, everybody benefits from the decision of an individual to purchase renewable energy products, including all persons who remain inactive. This supports free rider behavior and ultimately results in a suboptimal market outcome (e.g., Menges and Beyer 2017).

*Value orientation* in marketing and by consumers works against this free rider problem, and the motivation to reach conscious and socially responsible purchasing decisions is promoted if corporate values are stressed, especially if they are in line with the values of the target customer. Furthermore, the psychological benefit of the "warm glow", in other words the positive feeling of superiority which comes with a good deed, is value based.

The *interactive web* significantly facilitates the generation of an additional psychological benefit for the consumer via demonstrative consumption: "self-expressive benefits from conspicuous environmentally sound consumption" (Hartmann and Apaolaza-Ibáñez 2012, p. 1254). Thus it is not merely the social control but also the self-portrayal of the environmentally conscious purchase that is facilitated significantly by the Internet and the social media (on renewable energy as input factors, see the following section "Relevance of Renewable Energy as an Input Factor in Marketing Other Goods and Services").

(5) Goods that require an explanation in two dimensions

It may appear surprising at first glance that renewable energy is characterized both as a commodity and a good requiring explanation. It is its origin that leads to differentiation (see earlier discussion) and thus also calls for explanations. This is true at two different levels: initially, the basic attributes of the product need to be explained, especially its origin, but additional questions directly related to the product (for example the connection to a local heat network) must also be answered. In addition, questions about the effect of the purchasing decision on overarching goals, such as the implementation of the energy revolution or the containment of climate change, must be asked at a second level.

This second level of explanation very directly requires the *value orientation* postulated by marketing 3.0, which allows for differentiation and specifically provides entry points for product and communication policies. This mechanism is again intensified by the *interactive web,* which not only facilitates the transfer of knowledge among target customers but additionally transports the—in parts certainly not trivial—explanations of the providers.

(6) Prosumer goods

Especially renewable energy is increasingly consumed and produced simultaneously by the end customers of the utilities and is thus a typical example of a prosumer good. This will fundamentally change the energy sector (Schlemmermeier and Drechsler 2017) but at the same time pose new challenges for the marketing of renewable energy.

Especially during the transitional period between high subsidies for renewable energy, which was formative in Germany until the Renewable Energy Act 2014, and the clearly apparent movement towards more market-based solutions and new market models—for example in the new Renewable Energy Act, but also in the discussion about the best electricity market design (Schlemmermeier and Drechsler 2017)—many decisions, and especially buying decisions, concerning renewable energy are driven by the corresponding *value orientation.* Why? Compared to the situation a few years ago, investments of private households in photovoltaic systems only generate rather small returns. Thus financial considerations are less and less sufficient as an investment motive, and value-oriented targets such as environmental and climate protection, as well as increased independence from major utilities, gain in importance.

In this situation, marketing requires a meeting of similar values shared by the supply and demand sides. And again the interactive web serves as a platform that amplifies communication and interaction. But above all, it serves as the technological platform for the informational connection of the energy distribution networks, which is a decisive component of smart grids and the corresponding management of the demand and supply sides.

As a first conclusion of the arguments presented, it can be stated that the six highlighted attributes of renewable energy are characteristic and relevant for marketing. In addition, numerous aspects confirm that the description of marketing that Kotler et al. (2010) have labeled marketing 3.0, especially the value orientation and the interactive web, are particularly suitable for the marketing of renewable energy. It is thus not surprising that a modern understanding of marketing, influenced by our societal reality, and the challenges stemming from that reality, namely the need to market a fundamentally new product category, come together in multiple ways.

## 4    Aims of Consumers When Buying Renewable Energy

The societal framework for the marketing of renewable energy developed in the previous section would remain incomplete if it did not incorporate the specific motives and motivations for supply and demand in the market for renewable energy.

The starting point for the diffusion of renewable energy is the concern about the long-term environmental effects of fossil fuels and nuclear power and the increasing awareness about climate change and its effects. Environmental awareness is one of the most important psychographic drivers of purchasing decisions and willingness to pay for renewable energy (Rowlands et al. 2002; MacPherson and Lange 2013; Herbes et al. 2015). Thus the primary issue is not a new business idea, but rather a value-driven innovation in the sense of Kotler et al. (2010).

A first approach to understanding the motivation of consumers who purchase renewable energy and who frequently are willing to pay more than for products that do not incorporate renewable energy is provided by the motivation of private households to invest their savings in renewable energy projects, and it differs from the motives of institutional investors. An analysis of the motivation for investments in renewable energy by institutional investors shows that financial goals dominate (e.g., Taylor Wessing 2012, p. 12). In the case of private investors, who are decisive in the spread of renewable energy and are responsible for increasingly decentralized production, the motive of sustainability clearly plays an important role (see overview of Friege and Voss 2015): investments by private investors are thus driven both by considerations of sustainability and values as well as by financial incentives. A similar motivation can also be assumed when it comes to the consumption patterns of these private investors.

This is also confirmed in detail by a comprehensive analysis of the literature (see overview by Herbes and Ramme [2014] in Fig. 2). The motivation of consumers can fundamentally be organized in two groups: first, the purchase is expected to lead to actual change, such as a limitation on climate change. Second, the purchase helps the purchaser to feel better or to achieve a status gain.

(1) Actual change/utilitarian benefits

Consumers want to accomplish aims or maximize benefits by targeting real change, which is beneficial not only to themselves but to others as well, so-called utilitarian benefits. In this case altruism plays a role as a basic value (Litvine and Wüstenhagen 2011). These utilitarian benefits include, for example, environmental and climate protection, regional production or the support of specific ways to generate renewable energy. At the same time, consumers want to assure that the renewable energy contracts they select actually contribute towards achieving these targets, so that their purchasing decision has a real effect ("perceived consumer effectiveness"). This is related to the attributes of a credence good, discussed earlier. Consumers can assess the effects of their purchasing decisions with the help of information about the renewable energy product or with reference to labels.

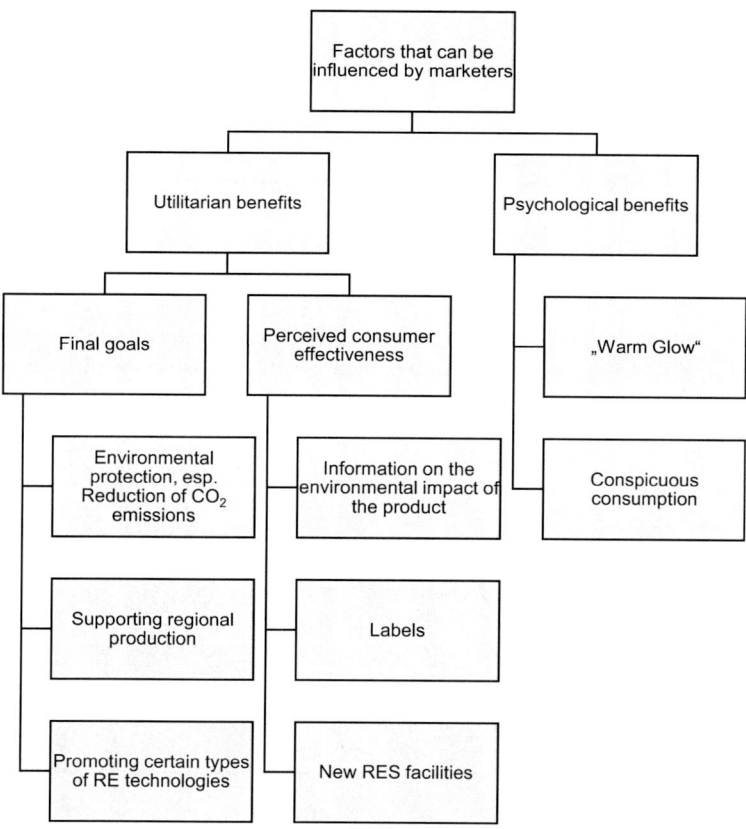

**Fig. 2** Overview of the aims of buyers of renewable energy contracts (adapted from Herbes and Ramme (2014), p. 259)

(2) Psychological benefit

In addition to the so-called utilitarian benefits, consumers can also derive so-called psychological benefits from their purchasing decision (Hartmann and Apaolaza-Ibáñez 2012). To some extent they are not as aware of these types of benefits compared to the previously mentioned goals of environmental or climate protection. Nonetheless, these types of benefits are important. One of them is the so-called warm glow, the pleasant feeling of moral superiority that results from a good deed. A second aspect is the gain in social distinction that can be derived from the demonstrative consumption of renewable energy. For the warm glow it is not necessary that the act of consumption be observed by a third party; however, it is decisive for demonstrative consumption. With all the focus on value orientation, it also needs to be pointed out that the average green electricity rate in Germany is only 2 % above the corresponding gray electricity rate and that consumers are in some cases even able to reduce their cost when switching to a green rate (Top agrar online 2012). Thus financial motives are likely to play a role as well.

Distributers of renewable energy always need to keep an eye on the entire target portfolio of their (potential) customers and use it as the basis of their marketing mix, particularly in their product and communication policies.

## 5    Marketing Mix for Renewable Energy

Now that marketing 3.0 has been revealed in principle as a suitable concept for marketing renewable energy and six major attributes of renewable energy have been proposed for the conceptual presentation of the marketing activities, the most important aspects of the so-called 4Ps that are common to all renewable energy products must be identified, including the motives for demanding renewable energy (Fig. 3).

### 5.1    Product Policy for Renewable Energy

The major importance of the origin of green electricity, biomethane, and heating power when configuring renewable energy products was already stressed earlier. The origin is mainly documented via certificates and detailed presentations of all power plants on the Internet.

(1) Certificates, or quality labels, serve several functions (Manta 2012; Leprich et al. 2017):
   a. Commodities such as electricity or gas are assigned specific attributes, which imply a sort of de-commoditization,
   b. Customers may select from among different product specifications based on their preferences,
   c. The certificate/quality label confirms the attributes specified,
   d. Providers use certificates to differentiate their products.
      An interrelationship exists between certificates and the degree of customer involvement. Manta (2012) demonstrated that the relevance of quality labels increases as customer degree of involvement is reduced. Quality labels are also more relevant than product attributes. However, earlier studies found a relatively minor relevance of certificates for the buying decision of customers in the field of green electricity (Kaenzig et al. 2013).
      Corporations and certifiers have recently presented the first quality label that goes beyond an assessment of the product and instead includes an overall evaluation of the provider. For "pioneers of the energy revolution" it has been confirmed that "goals and requirements of the energy revolution are not only firmly anchored in the corporate policy, but are also applied consistently in practice" (TÜV Süd 2014). The focus on differentiation becomes particularly

| | Product Policy | Price Policy | Distribution Policy | Communication Policy |
|---|---|---|---|---|
| **Commodities** | • Differentiated by origin<br>• Differentiation of services and customer retention | • High price transparency | • Differentiation via distribution channels | • Marketing is decisive for differentiation from commodities |
| **Low involvement products** | • Product components that increase involvement<br>• Certificates | • High price sensitivity | • Direct distribution or online distribution – many of the classical channels are problematic | • Decisive is the identification of communication opportunities |
| **Credence goods** | • Origin defines the product<br>• Certification<br>• Use of reputation | • Competitive price | • Direct distribution, online distribution or distribution via existing customers – the focus is always on trust | • Brand building<br>• Endorsements from environmental agencies<br>• Sponsoring of climate protection projects |
| **Partially public goods** | • Product must allow differentiation from free riders | • Willingness to pay is limited by free rider behavior | • Convincing customers to achieve differentiation from free riders | • Communicate usefulness for customers and society at large |
| **Explanations required in two dimensions** | • Increased complexity | • Differentiation via second level of usefulness | • Suitable for direct distribution | • Different perspectives allow additional communication |
| **Prosumer goods** | • Strongly dependent on regulation of renewable energy<br>• Complexity due to uncertainty about future developments | • Leads to significantly reduced price transparency | • Innovative distribution channels | • Complex communications challenges |

**Fig. 3** Marketing mix for renewable energy

apparent in this case. In addition, it has become increasingly clear that it is ultimately the reputation of the provider that ensures the attributes certified.

(2) Online presentation of all power plants

The offering of green electricity is occasionally supported by a detailed presentation of the facilities in use on the web pages of the provider. Transparency appears to be prima facie greater in the case of "pure providers of green electricity" in Germany compared to suppliers of green and grey electricity (Friege & Herbes, 2015, p. 15). Most providers rely only on the legally required proof of origin.

The potential importance of this transparency can be derived from the results of a survey by the federal environmental agency (UBA 2014). As an example, of 100 users of proofs of origin in Germany, only 27 obtain them in combination with physical delivery (p. 60). Of the 27, 2 stated that they obtain their electricity from the electricity exchange EEX (p. 64), which obviously makes the combined purchase impossible.

Decisive for the product combination is the question whether an additional ecological benefit should be offered and, if so, what kind (see comprehensive discussion in Leprich et al. 2017). Not all certificates demand the same standards. Most importantly, however, it is possible to achieve a high standing of certificates among consumers, even though the purchase of green electricity triggers no additional investments in power plants or does not support the energy revolution in any other way. This is the case, for example, if the green electricity is generated exclusively in old hydropower plants, which in some cases have already been in operation for more than 100 years. A different option for obtaining an ecological value added is, for example, the "sun cent" of the green utility EWS,[2] which may, however, in the meantime be obsolete owing to the effective subsidies from the Renewable Energy Act and the fact that grid parity has meanwhile been achieved. Against this backdrop, the certification of companies is an interesting approach, which makes it easier for consumers to identify those energy suppliers that are serious about the energy revolution.

Another example of product attributes that are suitable to support de-commoditization are—especially for commercial customers—certificates of origin that can be placed prominently around business premises (e.g. saying "Our goods are manufactured using solely (brand name) green energy."). They are also suitable for achieving a certain differentiation from free riders. Both the support contributions and the certificates of origin can at the same time be used to increase customer involvement.

All these measures in the field of product policy aim at differentiating the product. In addition, Enke et al. (2011) point out that de-commoditization can also be achieved via superior customer relations. Of course, this is also a possible approach for further product differentiation when offering renewable energy. For example,

---

[2]http://www.ews-schoenau.de/oekostrom/kundenfoerderung.html; accessed 14 December 2014.

LichtBlick, a provider of green energy, can document that it has implemented this approach. Thanks to its superior customer orientation, the company has held the leading position—in addition to many other awards—among all energy providers, including those companies that distribute electricity of unknown origin, in the German Customer Monitor (Deutscher Kundenmonitor) for 6 years in a row (LichtBlick 2014).

All these product attributes will ultimately imply higher complexity, which results in the need to provide explanations about renewable energy products in two different dimensions. Complexity grows further if renewable energy is increasingly established in the market as a prosumer good. This widens the range of relevant products to include contracting and the efficient generation of energy by customers (Klöpfer and Kliemczak 2017), in addition to the traditional matters of distribution of green electricity, sustainably generated heating power, or biomethane. It must be pointed out in this context that the uncertainty surrounding regulatory framework conditions and technological progress is currently huge. Nonetheless, even if these effects are uncertain, they need to be considered when structuring new products.

Overall, the product policy for renewable energy is dominated by the product component "energy source" and the relevant documentation for the customer, ideally in the form of a transparent additional ecological benefit. With the growing importance of prosumers as customers, the challenge will be to structure extremely complex products in such a way that they can be explained and, thus, placed successfully in the market.

## 5.2   Pricing Policy for Renewable Energy

Looking at the entries in the column "pricing policy" in Fig. 3, contradictions appear to be present. The high price transparency that follows from the fact that we are dealing with a commodity is reduced by the growing importance of prosumer goods in the field of renewable energy. This is basically plausible since there can be no doubt that the ability to differentiate among various combinations of generation and use—notwithstanding all the work concerning certificates and documentation of production units—is larger.

More interesting, however, are the effects concerning low-involvement products and credence goods, where price plays a larger role than most other attributes; for different reasons there is a specific focus on the price when the purchase decision is made. In the case of low-involvement products, the price is a simple indicator and replaces any examination of the brand or the product[3]; in the case of credence goods, it is predominantly competition that sets the price (Dulleck et al. 2011). In

---

[3]For example: "In contrast, the involvement in the case of utilities tends to be lower. Without involvement, true brand loyalty cannot emerge....The lower the involvement, the higher is the importance of market factors such as market presence or price." (TNS Infratest 2008).

this context, the issue of additional ecological benefits of renewable energy becomes a second area in need of explanation on top of the differentiation accomplished by the product policy.

Current research on the willingness to pay for renewable energy is mainly characterized by methodological debates. These debates are fueled by the fact that studies about the willingness to pay regularly present encouraging results, whereas the actual numbers of customers who switch to green electricity contracts of high quality are much lower (Rowlands et al. 2002; Kaenzig et al. 2013; Stigka et al. 2014).

The study by A.T. Kearney (2012) (Fig. 4) reveals scope for action for the years 2011 and 2012. Analyzed as "pure providers of green electricity" are LichtBlick, Tchibo,[4] Naturstrom, and EWS. Looking at the price charged, a pure provider of green electricity obtains a price premium of EUR 120 compared to municipal utilities and EUR 125 compared to discounters in 2012. The numbers are slightly lower for 2011, at EUR 80 and EUR 100, respectively.

Assuming the costs of certificates of origin were approximately EUR 4/MWh for the year 2012 (this relates to newer hydropower from Austria) (UBA 2014), it becomes apparent that the consumer who was taken as the basis of the calculation in Fig. 4 had to absorb a maximum cost difference of EUR 14. It can thus be shown that an ecological structure of the renewable energy product not only results in the availability of a price premium but is also relatively more profitable. On the other hand, the price differences between green electricity and grey electricity are so low that it can be safely assumed that old Norwegian hydropower with costs for the certificates of origin of less than EUR 1 was used as the basis of the calculations in Fig. 4.

And indeed, the federal environmental office (UBA) compares the costs of the certificates of origin indicatively with the costs for rebuilding the energy system in its study and reaches the following conclusion: "Given these conditions, final consumers cannot expect that the choice of a green electricity rate contributes towards the financing of the further development of renewable energies" (UBA 2014, p. 146).

On balance, the high price premiums of the past will not be sustainable over time. This is supported by the following arguments:

• Increased supply of renewable energy, which is increasingly competitive in the marketplace due to the passage of the Renewable Energy Act;
• Willingness to pay is limited by potential free riders—the higher the difference between the so-called market price without additional ecological benefit and price of renewable energy, the more relevant will be the attribute of renewable energy as a partially public good;

---

[4]Since 2010, the coffee shop and retailer have been offering green electricity, which is certified by ok-power and TÜV, respectively, however, without naming the power plants of origin. The "climate-friendly gas" is created through a complete compensation of $CO_2$ emissions via gold standard certificates (Tchibo 2014).

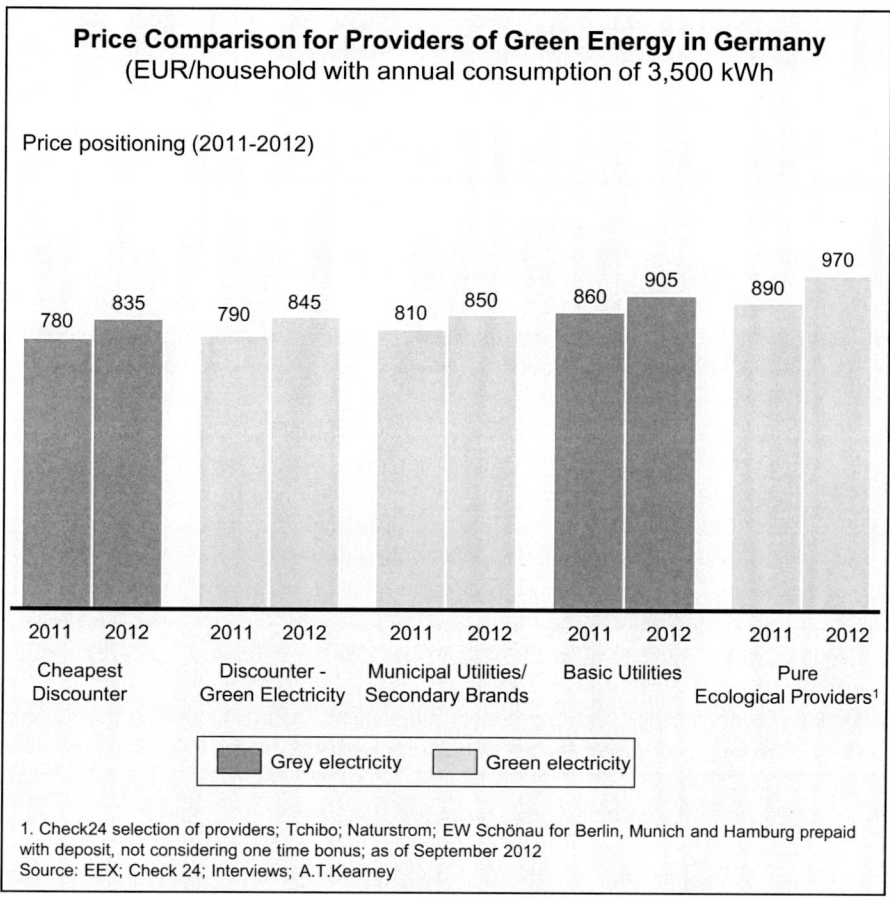

**Fig. 4** Comparison of prices of providers of green electricity in Germany (Source: A. T. Kearney 2012, p. 27)

- The continued high support for the energy revolution (Losse 2014) should not mask the fact that energy nonetheless remains a low-involvement product, and thus the willingness to pay is limited
- Finally, online portals, which are driven purely by price, continue to hold a major market share: "Already 80 % of households obtain information from a comparison portal and close to 50 % switch online" (A. T. Kearney 2012, p. 3).

In the case of biomethane, the setting of prices is more challenging for suppliers compared to electricity from renewable energy since, depending on the products used, the production and purchase costs for biomethane can be significantly higher than those for natural gas (Herbes et al. 2016). Unlike electricity from renewable energy, the additional costs are not single-digit percentages but can easily be twice as high in the case of 100 % biomethane products. For the pricing policy of renewable energy, this implies that it follows directly from the product policy: the more transparent the differentiation of the renewable energy product is, the more it contrasts with other renewable energy products, the more understandable is the ecological usefulness, and the more the involvement of customers can be developed, then the more likely there will be a price premium in the future. Support in this direction will also come from a trend towards prosumer goods.

## 5.3    Distribution Policy for Renewable Energy

In some sense, the distribution policy follows the product and pricing policies (Fig. 3): the stronger the product configuration supports a competitive—also vis-à-vis basic suppliers in electricity and gas—price, the easier it will be to distribute the renewable energy product online. Finalization of contracts and comprehensive explanations are not only easily done online these days, but such a distribution channel is even state of the art. This is the case not only for electricity or gas contracts but also, for example, for the distribution of roof solar panels in the USA, which to a large degree takes place online as well.

At the same time, the need to provide explanations in two dimensions about renewable energy products requires conversations with the customer, and for that reason renewable energy products are also very suitable for direct selling, both as prosumer good "rooftop solar panels for partial self-sufficiency" and as utility product green electricity and eco-gas. This distribution channel is particularly suitable in situations where the additional ecological benefit is pronounced, when the product is complex, or if the response is otherwise unsatisfactory due to the low-involvement attributes (on this issue see Friege 2016).

A. T. Kearney (2012) reports that in the first half of 2012, approximately half of the changes in providers and rates for electricity and gas contracts were implemented online, more than one fourth were initiated via direct distribution, while the share of the next largest distribution channel, telephone sales, is approximately 10 %. Because of the restrictive regulations in Germany, the latter channel is mainly limited to contract changes with the same provider.

However, in addition to the dominant channels online and direct distribution, an additional important distribution channel are referrals by existing customers. While this channel only leads to high growth rates in absolute terms if the initial level of customers is high, it is also a lasting form of advertising characterized by low costs. As the market research institute YouGov states: "Close to 20 % of the customers currently state that over the past two weeks they had conversations with others about EWS Schönau. In the case of other companies, this number is at most half as large" (Geißler 2014). It can also be assumed that this will have a positive effect on referrals by friends.

Compared to these three channels (online, direct selling, referrals by friends) all others are far less effective: telephone sales are strongly limited by legal rules, the response rates of direct mail are mostly too low, while advertisements, bill boards, and radio and TV spots rarely achieve satisfactory and measurable results.

Finally, it needs to be pointed out that innovative multiple distribution channels will develop for complex renewable energy products characterized as prosumer goods. While a broad-based configuration still needs to evolve, a framework for such a multichannel strategy is already established (Friege 2017).

And thus, the distribution policy for renewable energy turns out to be a perfect fit for marketing 3.0: distribution of renewable energy is value-driven, and this conviction is transported personally (direct distribution, referral by existing customers) or via the interactive web (online marketing, anchoring in social media). Trust in the sales representative, the customer who recommends the service, and the service offered takes center stage and is coupled with the assurance that a legitimate provider does not want to risk the loss of reputation.

## 5.4 Communication Policy for Renewable Energy

An important instrument for de-commoditization is brand management, which obviously needs to be implemented across all components of the marketing mix and is still included at this point in the context of communication policy. Why? Especially in the context of renewable energy, target-specific communication is decisive for creating awareness of the service differentiation. Wiedmann and Ludewig (2011), for example, develop the brand positioning for a municipal utility and summarize their resulting "corporate branding story" (p. 106) as follows: now the task is to communicate the developed brand positioning.

Concerning the communication aspects of brand management, the mineral oil industry may offer a few suggestions. Major providers, for example, have success-fully established premium gasoline brands, which are about EUR 0,05 - EUR 0,10 more expensive per liter compared to diesel or premium gasoline. Advertising, testimonials, sponsorships, and customer management were used (Möller and Roltsch 2011). As a rough sketch transferred to renewable energy, motorsport sponsorship will be replaced by support for climate protection projects and testimonials of Formula 1 pilots by the recommendations of environmental agencies. This was done, for example, by 17 environmental organizations that in the initiative "Quitting Nuclear Power—Let Us Do It Ourselves" support the switch to EWS, Greenpeace Energy, LichtBlick, or Naturstrom in Germany.[5] Marketing for renewable energy products will consist predominantly of transparent online information and active participation in the interactive web. The importance of customer management was already stressed when presenting the product policy for renewable energy. At this point, client-focused communication must be added.

In the context of the transparent presentation of renewable energy products and their suppliers, an effort will be made to build trust, which in the current environ-ment is more than ever the result of truthful, complete, and transparent communi-cation (Friege 2010). This also includes the need to point out the benefits for

---

[5]http://www.atomausstieg-selber-machen.de/ accessed 2 February 2015.

customers and society at large to reduce free rider behavior (Fig. 3). Unlike with many other products, any communication by providers is accompanied by a political and public media discussion, which can sometimes be very intense. In such discussions, storylines such as "plate or tank" respectively "maizification" (large quantities of maize being grown as an energy crop) are created (on this issue see Herbes et al. 2014), which run contrary to the interests of the suppliers of biomethane. The timely reaction of the providers is required in these instances.

It is also important to address all possible utility categories of renewable energy customers (Fig. 2) in the communication. For example, customers can be supported in their efforts to display conspicuous consumption if they are provided with information or stories by the producer that are suitable for communication in their own social network.

Overall, it must be the aim of the communication policy to continuously find new occasions to communicate with the target audience and to present the range of available services or at least some of its aspects. This is specifically relevant in the case of complex prosumer goods. And the greater the variety of perspectives used to enable communication with different subsegments of the target audience, the more sustainable will be the establishment of a comprehensive and holistic brand communication.

## 6  Relevance of Renewable Energy as an Input Factor in Marketing Other Goods and Services

In section four, the aims of consumers who order renewable energy products were discussed. These aims are of direct relevance for providers of renewable energy products, but at the same time, these products can be used by many different companies along the value chain, including, for example, the producers of consumer or investment goods. There are many reasons why these producers might decide to use renewable energy. Perhaps they are reacting to consumer preferences in favor of sustainably produced products and services and are using their decision to support renewable energy in the context of a marketing campaign focusing on consumers' sustainability targets. The most prominent example in this connection is the recently introduced Bahncard of Deutsche Bahn (German Railway), which allows rail travel at a discounted rate. The Bahncard is now issued in green, and Deutsche Bahn purchases large quantities of electricity from renewable sources to achieve $CO_2$ neutrality for distances traveled by owners of the card. Tour operators also use renewable energy in their marketing (Gervers 2017), and renewable energy is also used in the marketing of food. For example, the packaging of Milka chocolate includes the information that only renewable energy was used in its production. Such an approach is particularly promising in the marketing of products where sustainability generally plays a large role in their positioning, for example, bio food. What does this mean for the marketing of renewable energy? There is a need to support customers in the effort to communicate the advantages of using renewable energy to end users. This is exactly what many providers are already doing by offering communication manuals, word building blocks, or printable versions of eco-labels, which in turn can be used by their customers in their own marketing.

Consumer demand for sustainably produced products and services can also affect other participants in the value chain. Producers of consumer goods occasionally reformulate the preferences of the end customer as sustainability requirements for their suppliers, including the manufacturers of investment goods. These sustainability requirements can also include the use of renewable energy. In this way, the preferences of consumers can occasionally have an effect on several stages along the value chain.

In addition to the desire to address the concerns of consumers who demand the use of renewable energy, producers can also pursue further goals at all stages of the value chain when using renewable energy in the production process. First, there are financial goals. Until the Renewable Energy Act of 2014, generating electricity from biomethane was a way for manufacturing enterprises to achieve a reduction and long-term stabilization of heating costs. Second, the use of renewable energy can also originate from a comprehensive sustainability strategy, without any pressure by consumers or institutional clients.

However, such a use of renewable energy must always be apparent in the overall "brand substance" (Meffert et al. 2010, p. 30). It consists of three components (Meffert et al. 2010, pp. 30–31), always with regard to renewable energy:

- For the use of renewable energy, sustainability in aims and strategy means that credibility can only be achieved if it is consistent with the corporate strategy and positioning;
- Sustainability in the value chain with regard to renewable energy requires the development of processes and key figures that allow for the implementation and—if needed—communication of the promise "produced with renewable energy";
- Sustainability in service offerings means that the use of renewable energy as an input factor is meaningful only if it strengthens product differentiation or is appreciated by relevant stakeholders (e.g., consumers, media). The relevant metric to assess the decision is still sustainable profitability.

In summary, renewable energy can be used at different stages of the value chain, where it can also become a component of the branding strategy in line with the brand substance. Thus the distributors of renewable energy must not only address their direct customers in their marketing activities but also provide support in addressing the preferences of their final customers through the use of renewable energy (Fig. 5).

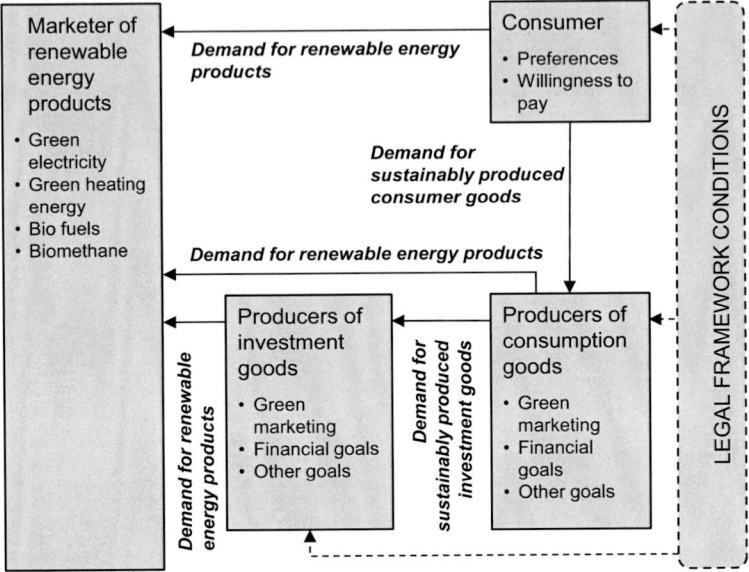

**Fig. 5** Demand effects for renewable energy products in the value chain

## 7 Summary: Most Important Steps for the Successful Marketing of Renewable Energy

The environment for marketing renewable energy is characterized by a value orientation and the interactive web. This framework reflects our times and the societal realities and is comprehensively suitable for the marketing of renewable energy. Within this framework, the goals for the consumption of renewable energy need to be considered. They are not limited to a cost/benefit analysis but in addition target actual change and pursue individual psychological aims. All attributes of renewable energy need to be considered as well since they frequently determine the scope of the distributor. The marketing of renewable energy takes place within this framework (Fig. 6).

In this process, the main steps of the marketing of renewable energy need to be stressed:

1. Of overarching relevance in product policy is transparency about energy generation. Here a distinction is made between sustainable products and products that are labelled as "greenwashing."[6] Ideally, the product provides an ecological value added, especially if it wants to satisfy consumers' motivation for actual change.

---

[6]The attributes of low-involvement and credence goods in fact imply that even products where greenwashing must be assumed can be successful in the marketplace, at least in the short run.

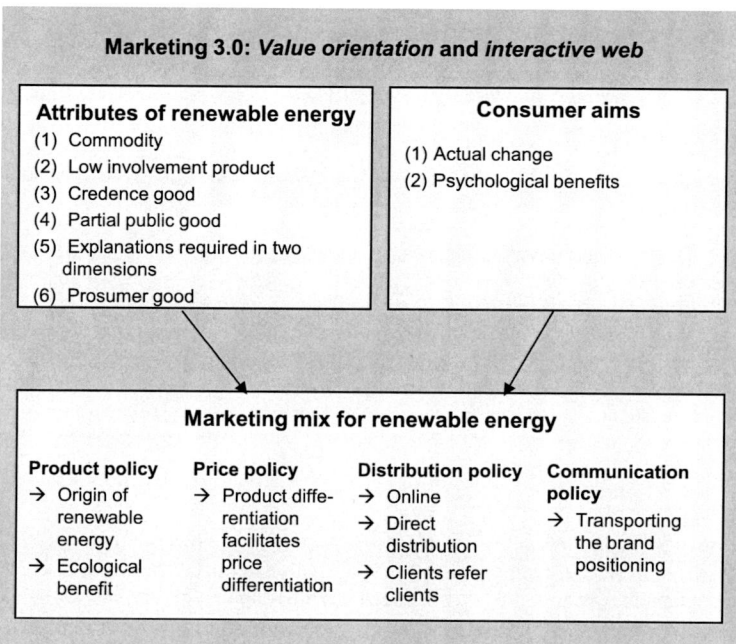

**Fig. 6** Factors influencing the marketing mix

2. This involves the potential for achieving additional revenue.
3. In addition to the largest distribution channel, the Internet, which, due to its high price transparency, limits the potential for additional revenue, direct selling and referral by existing customers are decisive for successful marketing.
4. The ultimate aim of any communication policy is to transport the brand positioning. This is not an easy task owing to the tension between low-involvement and credence goods on the one hand and the need to provide explanations along two dimensions on the other hand.

In summary, the marketing of renewable energy must be based on a number of clearly identifiable success factors and areas of activity that are derived from the attributes of renewable energy. In an environment of value orientation and interactive web, these success factors can be used in a targeted fashion. The renewable energy product is defined above all by its transparent origin, which therefore serves as the starting point for all marketing activities. However, these marketing efforts do not take place in a vacuum but are heavily influenced by legislators and regulators.

# References

AMA. (2013). *Definition of marketing*; approved July 2013. Accessed December 12, 2014, from https://www.ama.org/AboutAMA/Pages/Definition-of-Marketing.aspx.

Busch, H., Esch, F.-R., & Knörle, C. (2009). Integrierte Markenwertplanung der EnBW. In F.-R. Esch & W. Armbrecht (Eds.), *Best practice der Markenführung* (pp. 355–369). Wiesbaden: Gabler.

Dulleck, U., Kerschbamer, R., & Sutter, M. (2011). The economics of credence goods: An experiment on the role of liability, verifiability, reputation, and competition. *American Economic Review, 101*(2), 526–555.

Enge, M., Geigenmüller, A., & Leischnig, A. (2011). Commodity Marketing Eine Einführung. In M. Enke & A. Geigenmüller (Eds.), *Commodity marketing* (2nd ed., pp. 4–29). Wiesbaden: Gabler.

Friege, C. (2010). Kundenmanagement und Nachhaltigkeit—erfolgreiche Positionierung im Internetzeitalter. *Marketing Review St. Gallen, 27*(4), 42–46.

Friege, C. (2016). *Der Direktvertrieb in Mehrkanalstrategien.* Wiesbaden: Gabler.

Friege, C., & Herbes, C. (2015). Konzeptionelle Überlegungen zur Vermarktung von Erneuerbaren Energien. In C. Herbes & C. Friege (Eds.), *Marketing Erneuerbarer Energien* (pp. 3–28). Wiesbaden: Springer Gabler.

Friege, C., & Voss, H. (2015). Motive von Privatinvestoren bei Investitionen in EE-Projekte. In C. Herbes & C. Friege (Eds.), *Handbuch Finanzierung von Erneuerbare-Energien-Projekten* (pp. 89–105). München: UVK Lucius.

Friege, C. (2017). Direct Selling of renewable energy products. In C. Herbes & C. Friege (Eds.), *Marketing renewable energy.* Heidelberg: Springer.

Geißler, H. (2014). Verbraucher strafen Unternehmen für Strompreise nicht noch weiter ab. WirtschaftsWoche Online. Accessed December 14, 2014, from http://www.wiwo.de/ unternehmen/energie/brandindex-verbraucher-strafen-unternehmen-fuer-strompreise-nicht-noch-weiter-ab/9528742.html.

Gervers, S. (2017). Renewable energy in the marketing of tourism companies. In C. Herbes & C. Friege (Eds.), *Marketing renewable energy.* Heidelberg: Springer.

Hartmann, P., & Apaolaza-Ibáñez, V. (2012). Consumer attitude and purchase intention toward green energy brands: The roles of psychological benefits and environmental concern. *Journal of Business Research, 65*, 1254–1263.

Herbes, C., Braun, L., & Rube, D. (2016). Pricing of biomethane products targeted at private households in Germany—product attributes and providers' pricing strategies. *Energies, 9*(4), 252.

Herbes, C., Jirka, E., Braun, J.-P., & Pukall, K. (2014). Der gesellschaftliche Diskurs um den „Maisdeckel"vor und nach der Novelle des Erneuerbare-Energien-Gesetzes (EEG) 2012. *GAIA, 23*(2), 100–108.

Herbes, C., & Ramme, I. (2014). Online marketing of green electricity in Germany—A content analysis of providers' websites. *Energy Policy, 66*, 257–266.

Herbes, C., Friege, C., Baldo, D., & Mueller, K.-M. (2015). Willingness to pay lip service? Applying a neuroscience-based method to WTP for green electricity. *Energy Policy, 87*, 562–572.

Kaenzig, J., Heizle, S. L., & Wüstenhagen, R. (2013). Whatever the customer wants, the customer gets? Exploring the gap between consumer preferences and default electricity products in Germany. *Energy Policy, 53*, 311–322.

Kearney, A. T. (2012). Der Strom- und Gasvertrieb im Wandel. Accessed December 14, 2014, from http://www.atkearney.de/documents/856314/1214638/BIP_Der_Strom_und_Gasvertrieb_im_ Wandel.pdf/ee091e7c-9406-4b23-b5b3-608f936cbecc.

Klöpfer, R., & Kliemczak, U. (2017). Renewable energies in the contracting market. In C. Herbes & C. Friege (Eds.), *Marketing renewable energy.* Heidelberg: Springer.

Kotler, P., Kartaja, H., & Setiawan, I. (2010). *Marketing 3.0.* Hoboken, NJ: Wiley.

Leprich, U., Hoffmann, P., & Luxenburger, M. (2017). Certificates in the market for renewable energy in Germany. In C. Herbes & C. Friege (Eds.), *Marketing renewable energy.* Heidelberg: Springer.

LichtBlick. (2014). Vertrauen schaffen—mit ausgezeichneten Produkten und ausgezeichnetem Kundenservice. Accessed December 14, 2014, from http://www.lichtblick.de/privatkunden/strom/.

Litvine, D., & Wüstenhagen, R. (2011). Helping "light green" consumers walk the talk: Results of a behavioral intervention survey in the Swiss electricity market. *Ecological Economics, 70*(3), 462–474.

Lohse, L., & Künzel, M. (2011). Customer relationship management im Energiemarkt. In M. Enke & A. Geigenmüller (Eds.), *Commodity marketing* (2nd ed., pp. 382–400). Wiesbaden: Gabler.

Losse, B. (2014). Hohe Zustimmung für Energiewende; WirtschaftsWoche Online. Accessed December 14, 2014, from http://www.wiwo.de/politik/deutschland/allensbach-umfrage-hohe-zustimmung-fuer-energiewende/10037578.html.

MacPherson, R., & Lange, I. (2013). Determinants of green electricity tariff uptake in the UK. *Energy Policy, 62*, 920–933.

Manta, M. (2012). *Bedeutung von Gütesiegeln*. München: FGM.

Meffert, H., Burmann, C., & Kirchgeorg, M. (2015). *Marketing* (12th ed.). Wiesbaden: Springer Gabler.

Meffert, H., Rauch, C., & Lepp, H. L. (2010). Sustainable branding—mehr als ein neues Schlagwort?! *Marketing Review St. Gallen, 27*(5), 28–35.

Menges, R., & Beyer, G. (2017). Consumer preferences for renewable energy. In C. Herbes & C. Friege (Eds.), *Marketing renewable energy*. Heidelberg: Springer.

Möller, S., & Roltsch, S. (2011). Differenzierung von Commodities am Beispiel von Hochleistungskraftstoffen. In M. Enke & A. Geigenmüller (Eds.), *Commodity marketing* (2nd ed., pp. 458–477). Wiesbaden: Gabler.

Ringel, M. (2017). Driving renewables: Business models for the integration of renewable energy and e-mobility in Europe. In C. Herbes & C. Friege (Eds.), *Marketing renewable energy*. Heidelberg: Springer.

Rowlands, I. H., Parker, P., & Scott, D. (2002). Consumer perceptions of "green power". *Journal of Consumer Marketing, 19*(2), 112–129.

Schlemmermeier, B., & Drechsler, B. (2017). From energy supplier to capacity manager—New business models in green and decentralized energy markets. In C. Herbes & C. Friege (Eds.), *Marketing renewable energy*. Heidelberg: Springer.

Stigka, E. K., Paravantis, J. A., & Mihalakakou, G. K. (2014). Social acceptance of renewable energy sources: A review of contingent valuation applications. *Renewable & Sustainable Energy Reviews, 32*, 100–106.

Taylor Wessing, L. L. P. (2012). *Private capital and clean energy*. London: Taylor Wessing.

Tchibo. (2014). *Ökostrom & gas*. Accessed December 14, 2014, from http://www.tchibo.de/oekostrom-gas-nachhaltige-energie-zum-tchibo-tarif-c400001066.html#.

TNS Infratest. (2008). Markenwahl mit Herz und Verstand. Accessed December 21, 2014, from http://www.tns-infratest.com/presse/ftd-archiv/2008-01-21_ursachenforschung-2.asp.

Top agrar online. (2012). Teurer Ökostrom ist ein Irrglaube. Accessed April 12, 2013, from http://www.topagrar.com/news/Energie-Energienews-Oekostrom-nichtteurer-als-herkoemmliche-Tarife-909404.html.

TÜV Süd. (2014). Wegbereiter der Energiewende. Accessed December 14, 2014, from http://www.tuev-sued.de/anlagen-bau-industrietechnik/technikfelder/umwelttechnik/energie-zertifizierung/wegbereiter-der-energiewende.

UBA. (2014). Marktanalyse Ökostrom—Endbericht. Accessed December 14, 2014, from http://www.umweltbundesamt.de/sites/default/files/medien/376/publikationen/texte_04_2014_marktanalyse_oekostrom_0.pdf.

Utopia. (2014). Über Utopia. Accessed December 21, 2014, from http://www.utopia.de/utopia.

Wiedmann, K.-P., & Ludewig, D. (2011). Commodity branding. In M. Enke & A. Geigenmüller (Eds.), *Commodity marketing* (2nd ed., pp. 82–114). Wiesbaden: Gabler.

Zaichkowsky, J. L. (1985). Measuring the involvement construct. *Journal of Consumer Research, 12*(3), 341–352.

**Christian Friege** holds a doctorate in business administration from Catholic University Eichstaett. He was CEO of Lichtblick AG, Germany's leading green energy provider until 2012 and consulted in the field thereafter. Before joining Lichtblick he was Chief Customer Officer on the board of debitel AG and Chief Executive of BCA (a Bertelsmann subsidiary) in the UK. Currently, he is Chief Sales Officer on the board of CEWE Stiftung & Co KGaA. Throughout his managerial career he has always maintained a strong research interest in the field of marketing and sales. His academic publications have appeared in Journal of Service Research, Marketing Review St. Gallen and Energy Policy. Most recently he co-edited books in German on Marketing Renewable Energy and Financing Green Energy Projects together with Prof. Dr. Carsten Herbes.

**Carsten Herbes** is a professor of International Management and Renewable Energy at Nuertingen-Geislingen University (NGU) and Director of the ‚Institute for International Research on Sustainable Management and Renewable Energy'. Before joining NGU he worked for ten years with a leading management consulting firm in Europe and Asia. Subsequently he joined a bioenergy company where he finally became CFO. He obtained a Master's degree in business administration from Mannheim University and a Ph.D. from the University of Frankfurt (Oder). His research topics include marketing, acceptance and cost of renewable energy. He acts as advisor on sustainability issues to various companies and associations and is a frequent speaker at national and international conferences.

# Global Markets and Trends for Renewables

Karina Bloche-Daub, Janet Witt, Volker Lenz, and Michael Nelles

**Abstract**

Since the first oil crisis in the 1970s, technologies to use renewable energy (RE) have been developed and improved significantly, and the importance of these resources in the production of electricity, heat, and fuels has increased continuously. For each renewable energy source a unique set of technologies has developed bringing with it different forms of application. Although wind energy and solar energy have developed strongly in the past few years (especially in China, Germany, the USA, Brazil, India, and Japan), they still account for less than 1 % of primary energy consumption worldwide. So far, traditional biomass application and hydropower are the most commonly used REs, and they will retain their frontrunner position in the near future. Against this backdrop the following chapter aims to give an overview of the current usage of RE—worldwide, in the EU, North America, Asia, and the rest of the world. Furthermore, an outlook on the potential development of the different RE technologies up to 2020 is given.

A previous version of this chapter has been published in Herbes, C.; Friege, Chr. (Hrsg): Marketing Erneuerbarer Energien. Grundlagen, Geschäftsmodelle, Fallbeispiele, 2015, Springer Gabler.

K. Bloche-Daub (✉) • V. Lenz
Deutsches Biomasseforschungszentrum gGmbH, Torgauerstr. 116, 04347 Leipzig, Germany
e-mail: karina.bloche-daub@dbfz.de

J. Witt
Deutsches Biomasseforschungszentrum gGmbH, Torgauerstr. 116, 04347 Leipzig, Germany

Faculty of Landscape Architecture, Horticulture and Forestry, University of Applied Science Erfurt, Altonaer Straße 25, 99085 Erfurt, Germany

M. Nelles
Deutsches Biomasseforschungszentrum gGmbH, Torgauerstr. 116, 04347 Leipzig, Germany

Waste Management and Material Flow of the Faculty of Agricultural and Environmental Sciences of the University of Rostock, Justus-v.-Liebig-Weg 6, 18059 Rostock, Germany

**Keywords**
Renewable energy use • Renewable energy markets • Support schemes •
Renewable energy production • Renewable electricity • Heat and fuels

# 1    Support Schemes for Renewable Energies

The development of renewable energy (RE) markets is determined by national legal frameworks and especially by incentives for the development of RE technologies. According to regional availability and the purchasing power of the costumer, transregional and transcontinental trade is a result. Previously, this was limited to transportable solid and liquid biofuels, for example, wood chips, pellets, and biodiesel. However, public criticism—addressing especially direct and indirect land-use changes in Asia and South America (the so-called food versus fuel debate)—has shown that proof of origin and the sustainability of the traded products/resources are a necessity.

To achieve national RE targets, different national support mechanisms are used, and they can vary substantially from country to country. In what follows, an insight into commonly used political and economic incentives is given.

## 1.1    Support Schemes for Renewable Electricity

A list of typical models used to support renewable electricity is presented here (Renewable Energy Policy Network for the 21st Century 2016):

- FIT (Feed-in tariff or feed-in premium): This type of policy guarantees producers of REs specified payments per unit, for example, in US dollars per kilowatt hour (USD/kWh). The same policy may also regulate how producers can pool and sell power to the (public) grid. A variety of alternatives exist for defining the level of incentives. The two main options are the payment of a guaranteed minimum price (e.g., FIT) and payment floats on top of the wholesale electricity price (e.g., feed-in premium). A special form of the FIT is net metering. Under this policy, utility consumers with on-site electricity generators can receive credits for excess generation that can be applied to offset consumption in other billing periods.
- Renewable portfolio standard (RPS): This can be a governmental scheme requiring a utility company, group of companies, or consumers to provide or use a given minimum share of RE. This is also called renewable obligations or mandated market shares and is applied, for example, in Great Britain.
- Tendering: This is also called auction, reverse auction, or tender. The RE supply or capacity is auctioned by the seller and typically sells at the lowest price acceptable to them. The bidding may be assessed by price and nonprice factors. At the moment this approach is used especially in Central and South America,

but it is slated to be introduced in more countries, for example, in Germany in 2017.

- Trade with certificates: A limited number of certificates for the emission of greenhouse gases (GHGs) are issued. Ultimately the certificates should help trigger further GHG reductions, for example, by increasing the efficiency of existing plants or the installation of new RE plants to compensate fossil energy resources. Certificates will help this development by a growth in value and, if needed, a reduced availability of certificates on the market. This mechanism has been used, for example, in the European Union (EU) since 2003 as the EU Emissions Trading System (ETS) and is the first transnational trading system; so far, however, it has had limited success (European Commission 2016)
- Labeling: Currently there is no global framework (label) in place for the marketing of electricity from RE sources like, for example, so-called eco-electricity. Nevertheless, energy companies, energy platforms, and the electricity stock exchange, especially in Europe, offer a limited amount of eco-electricity for trading.

## 1.2    Support Schemes for Renewable Heat

Currently 45 countries have renewable heating and cooling targets; 31 of these countries are located on the European continent (Renewable Energy Policy Network for the 21st Century 2016). However, the promotion and support schemes for renewable heating and cooling technologies lag behind the implementation of policies in the renewable electricity and fuel in transport (Renewable Energy Policy Network for the 21st Century 2016). Measures that are often used to incentivize the generation of renewable heating/cooling include reduced taxes for renewable fuels (e.g., for wood pellets in Germany and Italy), national or regional subsidies for investment costs, and low-interest loans for the installation or redevelopment of regenerative heating/cooling systems. These approaches are prioritized in industrialized countries, while developing and emerging countries implement comparable approaches only partially. Generally, subsidies for the installation or reconstruction of renewable heating/cooling systems are temporary, and the magnitude of incentivizing is affected by economic fluctuations.

## 1.3    Support Schemes for Renewable Fuels

In the last few years many regulatory measures and fiscal incentives have been imposed on renewable fuels and electric vehicles. The vast majority of these policies targeted the production and use of biodiesel and ethanol. Typically approaches included installing targets, regulatory measures, and tax/financial incentives (Renewable Energy Policy Network for the 21st Century 2016).

Quota models are often used in the biofuels market. Quotas are set in reference to a pure biofuel (100 %) or a mixed biofuel (amount of biofuels mixed with fossil fuels). In Germany, the amendment of the law on biofuel quotas (BioKraftQuG) in 2015 changed from a quota defining an amount of biofuel on the market to a decarbonization quota. To evaluate the impact of the target quota, the GHG reduction must be calculated. Most policies, however, focus on road transport and do not include aviation, rail, or shipping. For example, no national support programs for the use of renewable fuels in the aviation sector currently exist. Nevertheless, some organizations try to introduce a target quota for the use of alternative fuels in the aviation sector, for example, the Aviation Initiative for Renewable Energy e.V. (Aireg), which wants to implement a quota of 10 % biofuels by 2025 (Aireg e.V. n.d.).

## 1.4    Policy Measures and Targets Worldwide

In Table 1 the relevant policies, fiscal incentives, and public financing are listed for three major regions of the world; they are also used subsequently in Sects. 2 and 3:

- Asia: The three leading countries in the RE sector are China, India, and Japan, while countries like South Korea and Indonesia may only play a marginal role at the moment.
- EU: The EU currently consists of 28 member states. In the statistics presented here, all EU countries are taken into consideration.
- North America: For the statistics analyzed here the North American continent is assumed to be the USA, Canada, and Mexico.
- Rest of the world: Important regions summarized under this category are Africa, Middle East, Central and South America (with Brazil and Argentina)

## 2    Primary Energy Consumption from Nonrenewable Sources

Nonrenewable primary energy sources are fossil fuels such as coal, natural gas, oil, and peat and mineral fuels (mainly natural uranium). Fossil fuels are based on organic decomposition products such as decomposed plants and animals. These substances have stored high amounts of solar energy from prehistoric times as part of the carbon lifecycle. The stored energy can be transferred into heat, electricity, and fuels applying thermochemical processes. Mineral sources refer to radioactive material that is used in nuclear power plants to produce thermal energy by neutron-induced nuclear fission.

## 2.1    Status Quo

*Global*  551 EJ of primary energy was consumed worldwide in 2015 (Fig. 1). Fossil as well as nuclear fuels are still the dominant sources for energy production. These

**Table 1** Number of countries with regulatory policies, renewable energy targets, and fiscal incentives/public financing [data from Renewable Energy Policy Network for the Twenty-First Century (2016)]

| | Regulatory policies | | | | | | | | Fiscal incentives and public financing | | | | |
|---|---|---|---|---|---|---|---|---|---|---|---|---|---|
| | Renewable energy target | Feed-in tariff/premium payment | Electric utility quota obligation | Net metering/net billing | Transport obligation/mandate | Heat obligation/mandate | Tradable REC | Tendering | Capital subsidy/grant or rebate | Investment or production tax credits | Reduction in sales, energy VAT, or other taxes | Energy production payment | Public investment, loans, or grants |
| Asia | 32 | 21 | 9 | 11 | 11 | 5 | 7 | 17 | 14 | 10 | 21 | 8 | 21 |
| EU | 28 | 24 | 7 | 10 | 25 | 7 | 15 | 13 | 20 | 13 | 19 | 5 | 19 |
| North America | 3 | 2 | 2 | 3 | 2 | 1 | 1 | 2 | 2 | 3 | 2 | | 3 |
| Rest of the world | 71 | 34 | 11 | 29 | 28 | 8 | 5 | 32 | 23 | 17 | 58 | 11 | 40 |
| World | 134 | 81 | 29 | 53 | 66 | 21 | 28 | 64 | 59 | 43 | 100 | 24 | 83 |

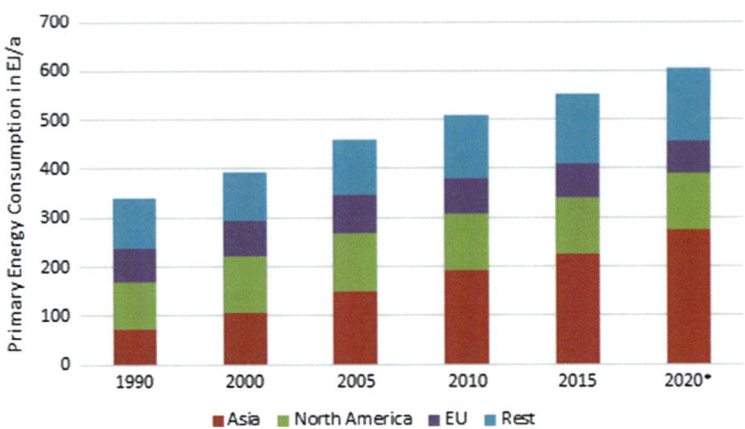

**Fig. 1** Primary energy consumption of selected regions [(*asterisk*) data estimated; illustration based on BP PLC (2016)]

nonrenewable sources provide 498 EJ of energy, which equals approximately 87 % of the total world primary energy demand. Oil has the highest market share, with a global consumption of 181 EJ (33 %). Coal (hard and soft) contributes 30 % to satisfy global energy demand. Furthermore, 131 EJ (24 %) are provided by natural gas and 24 EJ (4 %) by nuclear resources. Contrary to the relative constant global distribution of fossil energy sources, significant differences are observable on regional level. This is related to regional resource availability and differences in regional economies affecting primary energy demand. For example, US primary energy consumption declined between 2014 and 2015 by 0.9 %, while consumption increased in Germany (2.8 %) and China (1.5 %) (BP PLC 2016).

*EU* European primary energy consumption was approximately 69 EJ in 2015. With a share of 36 %, oil is the dominant fossil energy source, followed by natural gas with 22 %, coal (16 %), and nuclear (12 %). The latter is especially used in France, Germany, and the UK (BP PLC 2016).

*Asia* Coal is the most important energy source in Asia, where 115 EJ was consumed in 2015, more than half of the total Asian primary energy consumption in that year. Oil and natural gas are respectively the second and third most frequently used fossil energy sources. Oil contributes with 27 % and natural gas with 11 % to satisfy Asia's primary energy demand. At the moment nuclear energy is not as important in Asia as it is in Europe or the USA, but its use has increased in recent years. In East and South Asia, 40 new nuclear power plants are currently under construction and another 90 are planned. The highest growth in the use of this technology is expected in China, South Korea, and India. China is the world's largest consumer of primary energy from nuclear sources (126 EJ) (BP PLC 2016; World Nuclear Association 2016).

*North America*  Currently 117 EJ of primary energy is consumed in North America. The USA has the most dominant energy market on the North American continent. With a primary energy demand of 96 EJ in 2015, the country accounts for 65 % of North American and 17 % of the world's primary energy consumption (Fig. 1). The two most important fossil fuels are oil and natural gas, accounting for 37 and 31 % of primary energy consumption. Coal and nuclear-based energy are less important for the US energy market, with 17 and 8 % of primary energy consumption, respectively (BP PLC 2016).

*Rest of the world*  In 2015, 140 EJ of primary energy was consumed in the rest of the world. This accounts for 30 % of the world's primary energy consumption. In regional terms, the Middle East used the most primary energy sources (37 EJ) and accounts for 7 % of the world's primary energy consumption. South and Central America used 29 EJ (5 %) of primary energy. Here Brazil, with annual consumption of 12 EJ, is the biggest consumer. Primary energy consumption in Russia is at nearly the same level as South American energy consumption (28 EJ). With a share of 38 % natural gas is the most important fossil energy source in this part of the world, followed by oil with 37 %. Coal accounts for 11 % of primary energy consumption and nuclear energy sources for only 3 % (BP PLC 2016). Africa's current energy needs are met through a mix of biomass and fossil fuels, with biomass accounting for approximately half of Africa's total primary energy supply. Coal and natural gas account for about 14 % each and oil approximately 22 %. Hydropower represents about 1 % of the total primary energy supply in Africa (IRENA 2015).

## 2.2   Trend 2020

*Global*  The global consumption of fossil and nuclear resources will increase significantly in the next 5 years. As a consequence, the nonrenewable primary energy sources will still dominate the markets in the near future. Highly populated emerging nations (especially China, India, and Brazil) will increase their industrial sector, resulting in increased energy demand. Furthermore, the living standards especially in—but not limited to—these countries will be a driver for growing energy consumption. These trends, however, will also be linked to economic factors, national/international conflicts, famines, and epidemics and may in turn impact resource availability as well as price fluctuations. In consideration of such conditions, fossil energy consumption may increase to 600 or even as much as 610 EJ. It can be assumed that the share of the different fossil sources will not change significantly as the overall distribution grows (BP PLC 2016).

*EU*  Assuming a development of primary energy consumption similar to the past 10 years, the estimated primary energy demand will decrease to 65 EJ in 2020 ($-1$ %/a). Despite the decrease in primary energy consumption, fossil as well as nuclear energy sources will still dominate the fuel market. Oil will continue to be the most frequently used fossil fuel, with a market share of 33 %. Natural gas will cover 19 % of the European primary energy demand and coal 15 %. The demand for nuclear energy is estimated to decrease by 2020, especially as a result of Germany's phasing out of nuclear energy. By 2020 this technology will decrease to 8 EJ/a (11 % of European primary consumption). All told, the market share of nonrenewable energy sources is estimated to be approximately 78 % of the total primary energy demand (BP PLC 2016).

*Asia* Owing to an economic boom and increasing living standards in many Asian countries, primary energy consumption increased dramatically—averaging 4 % per year—in the past 10 years. Assuming the same trend for the next 5 years, Asian primary energy consumption could reach a level of 275 EJ in 2020. The increase will be evenly distributed across the different fossil energy sources, with coal being the most important primary energy source, with a contribution of more than 50 % to satisfy Asian primary energy demand. Oil will be the second largest contributor, with one quarter of the total Asian energy consumption. Furthermore, natural gas will contribute a sizeable share, 12 %, while all other primary energy sources will continue to play only a marginal role in the energy sector (BP PLC 2016).

*North America* Primary energy consumption in North America declined by 0.1 % each year over the last 10 years. Assuming this trend will continue for the next 5 years, North American primary energy consumption will have declined by 1–116 EJ in 2020. The share of the different energy sources in the primary energy mix will not change dramatically. Oil will remain the most frequently used primary energy source (36 %), followed by natural gas (35 %). The increased consumption of natural gas will be a result of the strong efforts of the US government to push shale gas production. Following the average increase over the last 10 years, the use of coal will decrease, while nuclear energy use will grow. In 2020 coal will account for 13 % and nuclear for 8 % of primary energy production (BP PLC 2016).

*Rest of the world* Energy consumption in the rest of the world will continue to grow to 2020. Assuming a growth trajectory similar to that of the past 10 years (2 %/a), the primary energy consumption could increase to 147 EJ in 2020. Based on the average increase over the past 10 years, the consumption of oil will rise significantly. In 2020 it may account for up to 40 % of total primary energy consumption in these regions of the world. Natural gas, with a share of 35 % and coal with 12 %, will be the second and third most important energy sources. Owing to restrictions in many countries, the use of nuclear energy sources will decline to nearly 1 % (BP PLC 2016).

# 3 Utilization and Markets of Renewable Energies

Renewable energy can be harvested from various sources: solar irradiance (e.g., biomass and photovoltaic), energy stored in the Earth's layers (e.g., geothermal), climatic energy (e.g., wind), or planetary gravitation and motion (e.g., hydropower and tidal). Different technologies for converting renewable resources into energy are known and will be developed in the future. In what follows, their contribution to satisfying global electricity, heating, and fuel demand is discussed, and a detailed insight into important renewable energy markets is presented.

## 3.1 Renewable Electricity Production and Application

### 3.1.1 Status Quo

Water and wind were the first renewable energy resources used for generating electricity. Through the invention of the first steam engine in the seventeenth century, the use of biomass to produce electricity emerged as an additional option. In the last century a variety of technologies to generate electricity from renewable resources, such as wind (onshore or offshore), tidal energy, and wave and marine current power, as well as energy from geothermal and solar sources, were introduced. Table 2 gives an overview of actual global renewable electricity production.

*Global* At the end of 2015 the global installed capacity of renewable electricity was up to 1900 GW, and the potential electricity production was between 5960 and 6370 TWh. Water is still the most dominant RE source for electricity production, with 3950 to 4200 TWh being produced currently in hydropower plants. Hence hydropower has a share of 66 % in renewable electricity production. Wind contributes with 16–17 % and biomass with 11 % of global renewable generated electricity. Less important are solar and geothermal electricity production, with shares between 5–6 and 1–2 %, respectively.

*EU* At the end of 2015 approximately 515 GW of renewable power capacity was installed in the EU. These plants had a potential electricity generation of 1000 TWh annually. Most electricity was produced from hydropower (354 TWh; share of 35 %) and wind (also a share of 35 %), while electricity generation based on biomass (18 %) and solar (10 %) were less important. Electricity generated from RE sources contributed more than one quarter (25.4 %) of the EU's gross electricity consumption. A very high proportion is provided in Austria (68.1 %) and Sweden (61.8 %), where more than half of all the electricity consumed was generated from RE sources, largely as a result of hydropower and biomass (Eurostat 2015).

*Asia* The renewable electricity market in Asia is dominated by hydropower, with a share of 77 % (1 586 TWh). Wind is the second largest RE source with electricity output of 240 TWh (12 %). All other RE sources contribute only marginally to

**Table 2** Global renewable electricity production in 2015 and trend for 2020 [data partly estimated and based on Bloche-Daub et al. (2015)]

| Energy source | Capacity (GW) | | Electricity production (TWh/a) | |
|---|---|---|---|---|
| | Status quo 2015 | Trend 2020 | Status quo 2015 | Trend 2020 |
| Hydro | 1035–1080 | 1100–1200 | 3950–4200 | Approx. 5000 |
| Hydro energy (without tidal and wave energy) | 1035–1080 | 1100–1200 | 3950–4200 | Approx. 5000 |
| Tidal and wave energy | Approx. 0.6 | Max. 1 | 0.8–1 | Approx. 1 |
| Wind | 432 | 790 | 992–1004 | 1750–1800 |
| Onshore | 421 | 755 | ca. 950 | 1620–1646 |
| Offshore | 11 | 35 | 42–54 | 130–154 |
| Solar | 232 | 540 | 305–400 | 710–925 |
| Solar thermal energy | 5 | 10 | 10–13 | 20–25 |
| Photovoltaic | 227 | 530 | 295–386 | 690–900 |
| Geothermal | 13 | 16 | 76 | 92 |
| Biomass | 139–143 | 160–165 | 646–681 | 795–975 |
| Solid biofuels | 106 | 117–120 | 464 | 585–720 |
| Organic waste | 16 | 18–20 | 77–97 | 85–110 |
| Biogas | 17–21 | 25 | 105–120 | 125–145 |
| Sum, global | 1851–1900 | 2606–2711 | 5969–6361 | 8347–8792 |

renewable electricity production. Solar has a share of 4 % (86 TWh) and the other RE sources combined (e.g., geothermal, biomass) amount to 7 % (144 TWh/a) (Fig. 2) (BP PLC 2016). The most important markets for renewable electricity production in Asia are China, India, and Japan. With a renewable electricity consumption of more than 1400 TWh China is Asia's number one consumer of renewable electricity and leads the world. In particular, in 2015 electricity production from wind and solar surged in China to record levels (Coghlan 2016). Currently, China has 43.2 GW of solar capacity with a potential electricity generation of more than 40 TWh (BP PLC 2016; Martin 2016). Hydropower is the most important source of renewable electricity in China, with an annual generation of 1126 TWh and an installed capacity of 320 GW (International Hydropower Association 2016). Wind energy increased dramatically in China in the last year (74 % increase in installed capacity compared to 2015) and a cumulative capacity of almost 145 GW (Global Wind Energy Council 2016; National Bureau of Statistics of China. n.d.). So far electricity generated from geothermal and bioenergy applications plays a minor role in the Chinese RE system. In 2014 approximately 53 TWh of electrical energy were produced from these sources (BP PLC 2016). By the end of 2015 a total capacity of 100 MW from geothermal power plants was supposed to be installed in China (Nitkoski 2015). With 193 TWh annually India is the second largest Asian market for renewable electricity production. The highest

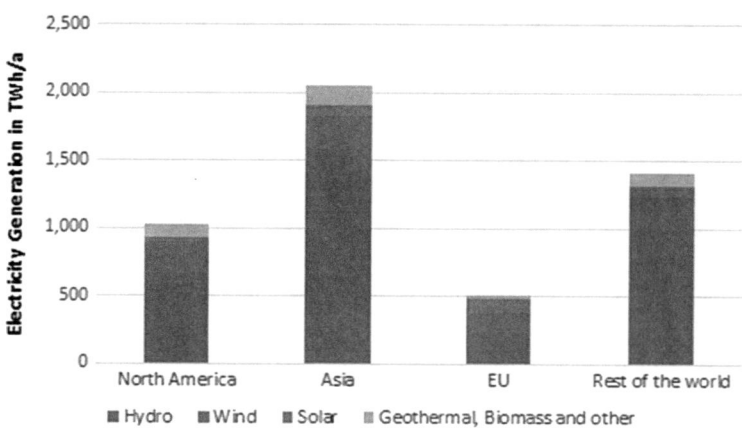

**Fig. 2** Electricity generated from different renewable resources in 2015 [illustration based on data from BP PLC (2016)]

share, at 124 TWh annually, is contributed by hydropower, followed by wind with 41 TWh, while geothermal and biomass together contribute to 21 TWh (BP PLC 2016). Currently 27 GW of wind energy are installed in India, and the country plans a further increase of wind energy capacity (Government of India 2016). Solar power plants have the second largest installed capacity, with 6.8 GW, followed by biomass power cogeneration plants (previously operated with bagasse from sugarcane processing factories) with 4.8 GW and small hydropower plants with an installed capacity of 4.3 GW. Furthermore, between 8 and 15 MW of biomass power capacity are installed in India (Government of India 2016).

*North America* In 2015 North America produced a total of 1032 TWh of renewable electricity. With 667 TWh annually, hydropower accounts for the largest part of renewable electricity, followed by wind (225 TWh) and solar (193 TWh). With the exception of hydropower production, the US market is the leading North American market for renewable electricity production, at 571 TWh. Only in terms of electricity production from hydropower is the USA surpassed, by Canada, which produced 383 TWh of hydroelectricity in 2015 compared to 254 TWh of hydroelectricity in the USA (BP PLC 2016). Thanks to US state and federal government incentives for RE production, the use of renewable energy sources (without hydropower) doubled between 2000 and 2014, bringing the share of RE sources for electricity production to 13 % in 2014 (EIA 2015). Most electricity is produced by hydropower (44 %) and wind (34 %; 193 TWh). Solar energy contributes only 7 % to US renewable electricity generation. Biomass, geothermal, and other renewable resources have a share of 15 % (biomass being the main contributor). With an installed capacity of around 3.6 GW, geothermal energy is of only marginal importance.

*Rest of the world* A total of 1410 TWh of renewable electricity was produced worldwide in 2015. The highest amount was provided by hydropower, with an annual production of 1231 TWh (87 %). All other renewable technologies play only a marginal role in electricity production. Biomass and geothermal contribute 7 %, wind 5 %, and solar only 1 % of the total renewable electricity production (BP PLC 2016). In the past few years, South American countries in particular have attracted investors for RE projects, and this trend is expected to continue (Pothecary 2016). So far, wind dominates in Brazil, while hydropower is the most often used RE source for electricity production in Sub-Saharan Africa and the Middle East (BP PLC 2016; IRENA 2015).

### 3.1.2   Trend 2020

Following the trend of the past 10 years, in 2020 renewable electricity could amount to 8347–8792 TWh (Table 2). With an estimated 5000 TWh of electricity generation by water, it can be assumed that hydropower will continue to be the most important means of renewable electricity production, followed by wind, with an estimated 1780–1805 TWh. Thus, wind may contribute nearly a quarter of renewable electricity generation in 2020. All other RE technologies will keep developing but will not be as relevant in the energy sector in the coming 5 years as wind or hydropower are on a global scale.

## 3.2   Renewable Heat Production and Application

### 3.2.1   Status Quo

Renewable heat generation is as old as humanity. Biomass use for heating purposes started with the discovery of fire. Even today the traditional form of biomass use in an open three-stone hearth for heating and cooking is employed in developing countries (Butt et al. 2013). Although biogenic fuels are used in some regions of the world in primitive and inefficient ways, the global trend is to use the Earth's renewable resources more efficiently and sustainably. Thus, the technologies to generate heat from biomass have developed significantly in the past few years. Besides biomass, solar thermal heat generation, often used in decentralized systems, has a long tradition. Similarly, near-surface geothermal heat generation has been used in some geologically rich regions in the form of heat pumps and other geothermal technologies in the past few centuries. Table 2 presents an overview of renewable heat production worldwide.

*Global* Heat production (usable heat) on a worldwide basis from RE technologies was between 26 and 28 EJ in 2015 (Table 3). With a share of 93 %, this figure is largely dominated by solid biofuels. Although all other options for renewable heat generation (near-surface and deep geothermal heat production, solar thermal heat production, and biogas) are of secondary importance, the development of these renewable heat technologies has increased significantly in the past few years.

**Table 3** Global renewable heat production in 2015 and trend for 2020 (data partly estimated)

| Energy source | Installed capacity (GW) | | Heat generation (PJ/a) | |
|---|---|---|---|---|
| | Status quo 2015 | Trend 2020 | Status quo 2015 | Trend 2020 |
| Solar thermal | 445 | 1000–1100 | 1250–1300 | 2200–2600 |
| Geothermal | 74.5 | 86 | 615 | 713 |
|    Near-surface (heat pumps) | 53 | 61 | 345 | 399 |
|    Deep | 21.5 | 25 | 270 | 314 |
| Biomass | 340 | 367–377 | 24,390–25,715 | 26,870–28,270 |
|    Solid biofuels | 315[c] | 340–350[c] | 23,980–25,165[a] | 26,420–27,670 |
|    Biogas | 25 | 27 | 410–550[b] | 450–600 |
| Sum, global | 860 | 1453–1563 | 26,255–27,630 | 29,783–31,583 |

[a]Including heat from combined heat and power plant (CHP) process of solid biomass and organic municipal waste: 567–762 PJ (2015); 623–847 PJ (2020)
[b]Including heat from CHP process biogas: 20–30 PJ (2015); 30–40 PJ (2020)
[c]Only modern bioenergy heating plants; data based on Bloche-Daub et al. (2015)

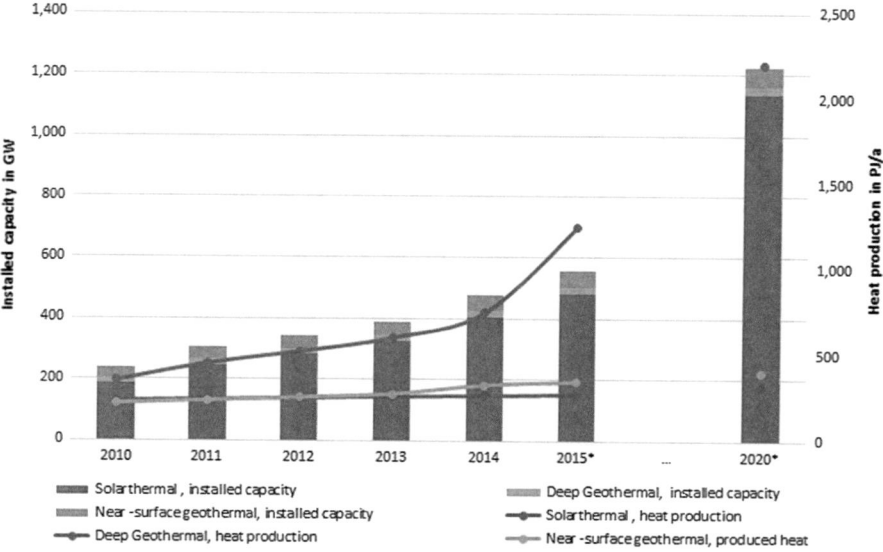

**Fig. 3** Global solar thermal and geothermal heat production and installed capacities [(*asterisk*) data estimated; illustration based on data from Angelino et al. (2014), Epp (2016), Lund and Boyd (2015), Observ'ER (2015), and Renewable Energy Policy Network for the twenty-first century (2016)]

Owing to the implementation of targets and incentive programs in different countries across the globe (e.g., China), the importance of solar thermal heating systems will keep increasing in the coming years (Fig. 3).

*EU* In 2015, between 4200 and 4350 PJ of renewable heat was produced and consumed in the EU. The highest amount, 95 %, was provided by solid bioenergy, with 4000 to 4100 PJ in 2015. All other renewable heat technologies contributed only marginal amounts of heat, for example, solar with 1.5 % or geothermal heat with 3.5 %. On a global scale, some European countries, such as Iceland, Norway, and Sweden, have some of the highest shares of renewable heat in their overall heat production (>50 %). Europe is also the leader in technology development, such as combined solar thermal systems, integration of solar thermal heating into district heating networks, or the use of renewable heat for industrial processes. Furthermore, the development of small-scale renewable heating applications based on geothermal energy or bioenergy are a main driver of the European renewable heating market (Renewable Energy Policy Network for the 21st Century 2016).

*Asia* Asian countries are the global leaders in the consumption of modern renewable heat. This is especially due to bio-heat used in the industrial sector in India and other Asian countries. Furthermore, China is the frontrunner in the direct use of geothermal and biogas for heat purposes. On top of that, China is the most important market for solar-based water-heating systems, with 70 % of the world-wide installed capacity in China (Renewable Energy Policy Network for the 21st Century 2016).

*North America* Owing to a decrease in biomass consumption in the industrial sector, renewable heat has been declining in North America since 2007. The most important market for renewable electricity and for renewable heat technologies is currently the USA. This is where the largest number of installed solar thermal water collectors can be found, which makes them the frontrunner in this technology not only in Northern America but globally, with an installed capacity of 4.5 % (Renewable Energy Policy Network for the 21st Century 2016).

*Rest of the world* The currently used renewable heat technologies differ strongly in the remaining countries of the world. The South American countries and Africa use mainly bio-heat, while the Middle East has increased the use of solar thermal water heaters in the past few years significantly. In particular, the use of traditional biofuels, such as forestry, agricultural residues, and animal excrement, continues to be the favored heating fuel for the lower class in, for example, Africa and South America. Though this traditional form of renewable heating production is rather inefficient, it still accounts for a significant share of renewable heating (Renewable Energy Policy Network for the 21st Century 2016) Besides biomass for heating purposes, the use of solar thermal heating systems has also increased in the last few years in Brazil, for example, which had an installed surface of 1.8 Mil m$^2$ in 2015 (Dawson 2015).

**Table 4**  Renewable fuel production global 2015 and trend 2020

| Energy source | Energy content (PJ/a) | |
|---|---|---|
| | Status quo 2015 | Trend 2020[a] |
| Bioethanol | 2218 | 2100–2300 |
| Biodiesel | 1050 | 1170 |
| Sum, global | 3268 | 3270–3470 |

[a]Date estimated, data based on Bloche-Daub et al. (2015)

### 3.2.2  Trend 2020

By 2020 the consumption of renewable heat could reach 28–31 EJ given the current trend (Table 3). Solid biofuels, such as woodchips and pellets, and traditional biofuels (e.g., forestry and agricultural residues and animal excrements) will remain the most important renewable sources for heat generation. These fuels could provide up to 87 % of total renewable heat. Solar thermal heat production (ca. 8 %) and geothermal energy (ca. 3 %) will continue to be of marginal significance to the energy sector.

## 3.3  Renewable Fuel Production and Application

### 3.3.1  Status Quo

Renewable fuel production is the most recent of the renewable energy technologies after heat and electricity. So far, only biogenic resources are being used for renewable fuel production and are mainly applied in the transport sector. As with fossil fuels, biofuels are used predominantly in liquid and gaseous form. Owing to the technical and economic challenges of alternative concepts, only individual projects exist for nonbiogenic renewable fuel production or application, such as solar-powered airplanes or hydrogen-storage technologies. Table 4 summarizes the current status of the two main biofuel production pathways and provides an outlook for use in 2020.

*Global*  Worldwide, in 2015, approximately 97 billion L of bioethanol (2218 PJ) and around 32 billion L of biodiesel (1050 PJ) were produced (Table 4); this amounts to a biofuel production of 3268 PJ in that year (Fig. 4).

*EU*  In Europe approximately 460 PJ was provided as biofuels in 2015. With a share of 78 % biodiesel makes up the largest part of it and is mainly produced in Germany and France. Rapeseed is the main energy crop used for biodiesel production in Europe. Because EU policy allows for the double counting of cooking oil and tallow so that targets may be met in the transportation sector, the use of these residues for biodiesel production will increase in the future.

*Asia*  In the last few years, biofuel production has increased rapidly in Asia. Most biofuels are produced in China, Indonesia, and Thailand. Palm oil is used as crop

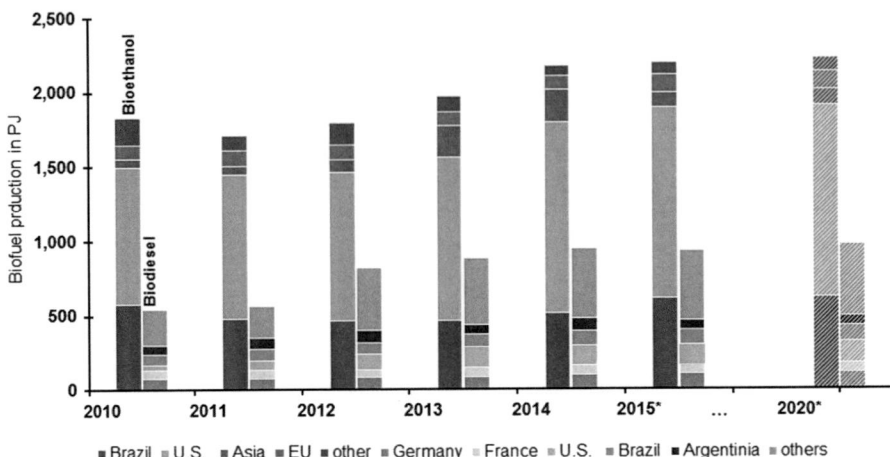

**Fig. 4** Biofuel production in selected countries/regions [(*asterisk*) data estimated, data based on Dawson (2015), OECD/Food and Agriculture Organization of the United Nations (2015)]

for biofuel production mainly in Indonesia, whereas China often cultivates sweet sorghum, cassava, and other nongrain crops. Currently, however, the Asian biofuel market is in a crisis. Restriction on biofuel imports to the EU, debates about direct and indirect land-use change, and problems with feedstock supply are among the main reasons for the crisis. The current shortage of feedstock supply in China, for example, has the biodiesel industry operating at 20–25 % of its capacity (Renewable Energy Policy Network for the 21st Century 2016).

*North America*  The USA is the leader in bioethanol production in North America; its production is mainly based on corn. In addition to bioethanol, the USA produces large amounts of biodiesel. Other than in Europe, where biodiesel is produced from rapeseed, US biodiesel is generated from soybeans.

*Rest of the world*  The South American countries Brazil and Argentina in particular are main producers of biofuel products. Here sugar crops are often used as feedstock for bioethanol production. All other countries play only a marginal role in global biofuel production, mainly owing to high production costs and a lack of land and water available for energy crop production.

### 3.3.2  Trend 2020

If no changes are made to the legal framework and policy support for biofuels by 2020, the produced capacity of bioethanol and biodiesel will stabilize at current levels. Hence, total biofuel production will reach 3.3–3.5 EJ/annually by 2020, with the share of biodiesel and bioethanol approximately on the same level as in 2015.

## 3.4 Relevance of Renewable Energy Sector

### 3.4.1 Status Quo

From the current total primary energy consumption of 551 EJ approximately 13 % is covered by RE resources (Fig. 5). As a result of the high amount of biomass used in large parts of the world for heat production by traditional means, solid biomass is the most important RE source. Biomass meet 6 % of total primary energy

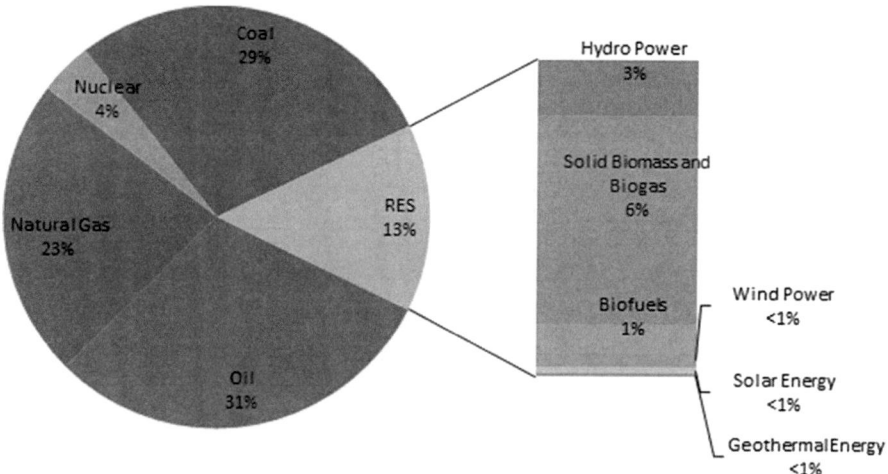

**Fig. 5** Contribution of different resources to global primary energy consumption, Status Quo 2015

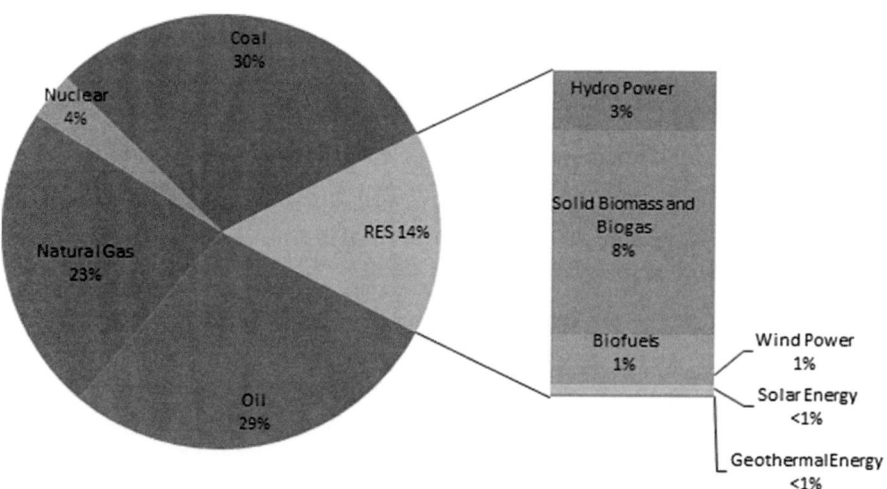

**Fig. 6** Contribution of different resources to global primary energy consumption, Trend 2020

consumption, while hydropower contributes 3 %. All other RE sources will be of only marginal significance to the global energy sector.

### 3.4.2  Trend 2020

Linked to the estimated increase in fossil-based energy demand, RE consumption will increase too. By 2020 regenerative energy sources could meet 14 % of the total world energy demand (Fig. 6). No significant changes will take place in the market share of the different RE sources compared to current levels. Thus, traditional and modern biomass, especially solid biofuels, will cover the highest demand, with 8 %. The contribution of hydropower to primary energy supplies will remain constant at the current 3 % level. All other renewable options, for example, wind and solar (respectively 4.6 and 3.6 EJ, or 1 % of world primary energy consumption) will not be of importance to the global energy sector.

## References

Aireg, e. V. (n.d.). Aireg (Aviation Initiative for Renewable Energy in Germany e.V.). http://www. aireg.de

Angelino, L., Dumas, P., & Latham, A. (2014). *The geothermal market in Europe—EGEC market report 2013/2014*. Brussels: European Geothermal Energy Council (EGEC).

Bloche-Daub, K., Witt, J., Janczik, S., & Kaltschmitt, M. (2015). Erneuerbare Energien—Globaler Stand 2014. *BWK, 67*, 6–7.

BP p.l.c. (2016). *BP statistical review of world energy*.

Butt, S., Hartmann, I., & Lenz, V. (2013). Bioenergy potential and consumption in Pakistan. *Biomass and Bioenergy, 58*, 1–408. doi:10.1016/j.biombioe.2013.08.009.

Coghlan, A. (2016, March 1). China set to surpass its climate targets as renewables soar. *New Scientist*. https://www.newscientist.com/article/2079179-china-set-to-surpass-its-climate-targets-as-renewables-soar/

Dawson, K. (2015, November). *Latest trends in the world traditional & renewable heating market*. https://www.bsria.co.uk/.../market-dynamics-in-world-heating-pr.

EIA. (2015, March). Renewable energy explained. *U.S. Energy Information Administration*. Accessed June 27, 2016, from http://www.eia.gov/energyexplained/index.cfm?page=renewable_home.

Epp, B. (2016, April 26). Big ups and downs on global market. *Global Solar Therman Energy Council*. Accessed June 13, 2016, from http://www.solarthermalworld.org/content/big-ups-and-downs-global-market.

European Commission. (2016, May 7). The EU Emission Trading System (EU ETS). *European Commission—Climate Action*. Accessed July 7, 2016, from http://ec.europa.eu/clima/policies/ets/index_en.htm.

Eurostat. (2015, May). Renewable energy statistics. *Eurostat—Statistic explained*. Accessed July 7, 2016, from http://ec.europa.eu/eurostat/statistics-explained/index.php/Renewable_energy_statistics#Electricity.

Global Wind Energy Council. (2016, April). Global statistics. *Global Wind Energy Council*. Accessed June 26, 2016, from http://www.gwec.net/global-figures/graphs/.

Government of India. (2016). *Executive summary power sector*. New Delhi: Ministry of Power, Central Electricity Authority New Dehli.

International Hydropower Association. (2016). *2016 Hydropower Status Report*. London: International Hydropower Association. http://www.hydropower.org/country-profiles/china.

IRENA. (2015). *Africa 2030: Roadmap for a renewable energy future*. Abu Dhabi: International Renewable Energy Agency. Accessed July 7, 2016, from www.irena.org/publications.

Lund, J. W., & Boyd, T. L. (2015). Direct utilization of geothermal energy 2015 worldwide. In *Proceeding World Geothermal Congress*. Presented at the World Geothermal Congress.

Martin, R. (2016, March). China is on an epic solar power binge. *MIT Technology Review*. https://www.technologyreview.com/s/601093/china-is-on-an-epic-solar-power-binge/

National Bureau of Statistics of China. (n.d.). Statistical Communiqué of the People's Republic of China on the 2015 National Economic and Social Development. *National Bureau of Statistics of China*. 29th February 2016. Accessed June 22, 2016, from http://www.stats.gov.cn/english/PressRelease/201602/t20160229_1324019.html.

Nitkoski, M. (2015, July). China's long-dormant geothermal energy sector may soon wake up. - *China Economic Review*. http://www.chinaeconomicreview.com/chinas-long-dormant-geothermal-energy-sector-may-soon-wake.

Observ'ER. (2015). *CSP und Solarthermie Barometer*. Brüssel: EurObserv'ER.

OECD/Food and Agriculture Organization of the United Nations. (2015). *OECD-FAO Agricultural Outlook 2015*. Paris: OECD Publishing.

Pothecary, S. (2016, May 16). Latin America becomes more attractive for renewable energy investment, as Europe suffers the reverse trend. *PV Magazine*. Accessed July 7, 2016, from http://www.pv-magazine.com/news/details/beitrag/latin-america-becomes-more-attractive-for-renewable-energy-investment--as-europe-suffers-the-reverse-trend_100024503/.

Renewable Energy Policy Network for the 21st century. (2016). *REN 21—Renewables 2015 Global Status Report*. Paris: Renewable Energy Policy Network for the 21st Century. Accessed April 14, 2016.

World Nuclear Association. (2016, January). Asia's Nuclear Growth. *World Nuclear Association*. http://www.world-nuclear.org/information-library/country-profiles/others/asias-nuclear-energy-growth.aspx.

**Karina Bloche-Daub** studied Engineering with a focus on public utilities (Baden-Wuerttemberg Cooperative State University Mannheim) and obtained her Master degree in Sustainable Energy Competence in 2011 (University of Applied Sciences Stuttgart). Between 2009 and 2011 she worked for two years as a project engineer for energy services and planning in Hamburg. Since finishing her Master degree Karina Bloche-Daub is working as a research assistant at the Deutsche Biomasseforschungszentrum gGmbH (DBFZ). Her research activity focuses on regional impacts of bioenergy use.

**Janet Witt** has been working at the Deutsche Biomasseforschungszentrum gGmbH (DBFZ) since 2002. During her studies of Building Service Engineering at the University of Applied Science in Erfurt she has specialized in engineering of heating, air conditioning & sanitary installations and got practical experience as a consultant during her work in an engineering office. Following, Janet Witt graduated in Project Management of Energy and Environment at the University of Northumbria in Newcastle (UK). At the DBFZ she is team leader of the working group "Biomass markets and Use" and coordinates national and international projects in the field of market development, market implementation and standardization of solid biofuels (ISO, DIN) as well as bioenergy process assessment. In her thesis at the Technical University in Hamburg she researched the optimization of the wood pellet durability during the production process and supply chain. Janet Witt is member of the steering committee of the European Technology Platform on Renewable Heating & Cooling and gives lectures in the field of renewable energy utilization options at the University of Applied Science in Erfurt.

**Volker Lenz** studied Aerospace and Aeronautic Engineering (University of the Armed Forces Munich) and Energy Economics (University of Applied Science Darmstadt). In 2010 he did his PhD in engineering at the Technical University Hamburg. From 2000 to 2005 he was Project Manager for renewable energies at the hessenENERGIE GmbH. From 2005 to 2008 he worked at the Institute for Energy and Environment as Project Manager in the department "Bioenergy Systems". Since 2008 he is the Head of the Department Thermo-chemical Conversion of the Deutsches Biomasseforschungszentrum gemeinnützige GmbH (DBFZ) in Leipzig. His research activity focuses on: fundamental and applied aspects of solid biomass fuel creation, development and improvement of small scale furnaces and boilers as well as on micro Combined-Heat-and-Power-Systems with an emphasis on efficiency and emission reduction. Since some years he has been focusing his work on the development, improvement and market integration of SmartBiomassHeat concepts using small scale biomass conversion systems to integrate the renewable energies in a future energy system and to stabilize the future energy supply with high amounts of fluctuating wind and photovoltaic energy by demand-driven energy converters for solid biofuels. Volker Lenz is member of national norming activities.

**Michael Nelles**  is an environmental engineer and studied Technical Environmental Protection (Technical University of Berlin). From 1994 to 1999 he was the Vice Director of the Department Waste Management of the Montanuniversität Leoben in Austria. From 2000 to 2006 he was full professor of Environmental Engineering of the University of Applied Science in Göttingen (Germany). Since 2006 he is full professor of Waste Management and Material Flow of the Faculty of Agricultural and Environmental Sciences of the University of Rostock, Germany. Since 2012 Prof. Nelles is also the Scientific Director of the German Biomass Research Center (DBFZ) in Leipzig. His research activity focuses on: fundamental and applied aspects of waste management with an emphasis on technological, environmental and economic aspects to mechanical, biological and thermal treatment systems of waste and biomass in different recycling and recovery routes. He is a member of national and international advisory boards of organisations in the field of waste management and biomass utilisation and also a board member of different national and international conferences and journals. He has authored over 400 journal articles and book chapters since 1994. His international activities focus on the Asian Region and in particular on China (Guest Professor in Beijing, Hefei, Shanghai & Shenyang; National Friendship Award 2011).

# Consumer Preferences for Renewable Energy

Roland Menges and Gregor Beyer

**Abstract**

In the face of the societal meta-topic of climate change, renewable energies promise solutions to the manifold challenges of mostly unsustainable lifestyles. This chapter is concerned with the concept of consumer preferences for renewable energy (RE) and provides an overview of the empirical literature on the matter. The chapter begins with a general discussion of the concept of preferences. It shows what assumption and preconditions must be accepted for individual preferences to unfold a normative character for energy politics and energy marketing that is in line with consumer sovereignty. The existing empirical literature on RE shows that there is a high social acceptance of RE. Beyond a general approval of RE, however, there is little consensus in the literature. This is in part a result of the complexity of the subject of investigation and the heterogeneous methods used in preference elicitation. Yet the core cause of diverging preferences for RE lies in the problem of public goods, which is shown here in its purest form. For the marketing of RE in competitive markets it is important that consumers make two conflicting demands. On the one hand, individuals derive benefits from the moral satisfaction of voluntary climate-friendly activities and RE development. On the other hand, they prefer political mechanisms that guarantee the development of RE by collective obligations that reduce or eliminate the possibility of free riding on other individuals' expenditures.

A previous version of this chapter has been published in Herbes, C.; Friege, Chr. (Hrsg): Marketing Erneuerbarer Energien. Grundlagen, Geschäftsmodelle, Fallbeispiele, 2015, Springer Gabler.

R. Menges (✉) • G. Beyer
Clausthal University of Technology, Department of Macroeconomics, Clausthal-Zellerfeld, Germany
e-mail: roland.menges@tu-clausthal.de

**Keywords**

Preference elicitation • Willingness-to-pay • Incentive compatibility • Stated preferences • Revealed preferences

## 1    Introduction

The increasing development of renewable energies (REs) is not just a central but also a deeply symbolic element of climate protection. In the electricity sector, the enlargement of RE capacities leads to structural effects and the crowding-out of conventional fossil power production facilities. This chapter examines how REs are perceived and evaluated by individual consumers. The concept of economic preferences is based on subjective reasoning. As such, this chapter discusses technical attributes of RE and their impact on larger economic scales only insofar as they effect individual utility.

- Following the perspective of environmental economics, the contributions of RE to environmental and climate protection are of no absolute value. Instead, the benefits and costs of RE are determined in relation to functional *substitutes* such as alternative climate protection policies (for instance, in the context of emissions trading).
- This includes matters original to the field of energy economics. For instance, the development of renewables potentially reduces energy security owing to production volatilities. Supply-side considerations of energy markets highlight the interplay of fossil fuels and RE more as a *complementary* interaction instead of a substitutional relationship. Hence, the integration of RE into energy markets raises fundamental questions about market design (such as the prioritization of the transmission of green electricity (GE) in transmission networks).

Most empirical studies show that the technical and abstract economic properties of RE development, such as those mentioned previously, are of only minor importance for consumers' assessments of RE. Contrary to ecological product attributes that are easy to communicate and grasp, these aspects are unnatural to the decision settings in which individuals reveal their preferences. Even if individuals are aware of the technical attributes of RE, private energy consumers are interested in the technical details of power supply only to a very limited degree.[1]

The majority of empirical studies cited in this chapter demonstrate that consumers perceive and support the development of RE as a step toward an environment- and climate-friendly energy supply. Individuals associate the promotion of RE with a sustainable economy and lifestyle, particularly when the

---

[1]This becomes evident in studies that quantify energy consumers' willingness to pay for energy security (Praktiknjo, 2014) or for the use of underground cables or overhead lines in transmission network development (Menges and Beyer 2014).

implications of RE development are discussed not only in the electricity sector but also in heat production and transportation (by means of so-called alternative fuels). Yet, there is little consensus on what conclusions may be drawn from these perceptions by politicians and utilities.

- In ecological economics, the normative question is raised of whether projects aiming to protect the environment and future generations' interests ought to be evaluated on behalf of present generations' individual preferences at all. Voiced concerns are that individual preferences may be instable or based on incomplete or false information. Constructs such as *consumer sovereignty* or *willingness to pay* are thus largely rejected in this discipline.
- In political economy, on the other hand, all policy goals are derived from individual preferences. Thus, the development of RE is a set target. However, it remains unclear which instruments should be used to endorse renewables, and individual preferences only rarely specify mechanisms to finance political programs. Relevant questions in this context include: Should renewables be subject to competition or should their advancement be forced "from the top" following the idea of the primacy of policy? Should consumers pay for RE development? If so, how may an acceptable or adequate financial burden be determined? Should the development be financed via tax revenues, public debts, or surcharges on energy prices? Are there reasons to exempt poorer households or businesses in international competition from financing RE policy?

The empirical literature on consumer preferences does not yield definite answers to these questions. This is because the procedures used and results obtained in various studies are not directly comparable. The *methods of data collection* employed vary strongly and imply different theoretical views on preference elicitation. Also, the concept of *renewable energy* is complex and interpreted heterogeneously, which hinders comparisons of different studies even if the same methodology is used. In view of the foregoing considerations, this chapter aims to provide a methodologically structured overview of the existing literature on consumer preferences for RE and the conclusions drawn from it. To do so, the next section specifies the concept of consumer preferences theoretically. The third section briefly discusses the methods of data collection commonly used in the empirical literature. The next section develops a framework for the elicitation of consumer preferences that incorporates different dimensions/understandings/ accesses/interpretations of the topic of renewable energies. This framework is then used to structure the existing literature on consumer preferences for RE and their results. The chapter ends with a conclusion and outlook in the sixth section.

## 2    On the Concept of Preferences

Individual preferences are neither directly nor objectively measurable. Preferences are much more complex than, for instance, opinions on or attitudes about desirable or not-so-desirable projects. Preferences form the core of the economic understanding of human behavior, yet the concept is used in many other areas of the social sciences.

Following methodological individualism, any value originates from the valuing individual. Any action, good, or thought (or the lack of these) contains value, when their realization contributes to the fulfillment of any given *need*. In this setting, *preferences* may be observed in defined, concrete decision settings as the ordering of alternatives based on their relative utility. Every decision setting is defined by an unlimited number of individual needs and a limited amount of resources available to fulfill these needs. This forces individuals to choose between the needs they wish to fulfill and the possible and available means of doing so. The results of these choices constitute what is commonly described as *behavior*.

The degree to which any given behavior—which in economics most commonly implies the purchase and consumption of goods—complements the fulfillment of subjective individual needs is called *utility*. Although the concept of utility is interpreted heterogeneously in different schools of economics (for instance, in terms of measurability: ordinal or metric), its existence always acknowledges individual preferences.

Preferences are assumed to be stable in the sense that they are exogenous and do not change over time. This assumption is less of an attempt to develop a realistic model of human behavior than a pragmatic methodological approach and may be interpreted as labor division in the social sciences. In psychology and sociology, the origins and developments of needs mark important research areas. In economics, however, the process of need formation is largely omitted. Instead, economists analyze observable behavior in the face of scarcity, though the idea that preferences may be adaptive and subject to change is not excluded in principle. In descriptive decision theory and in marketing, instable preferences are commonly remarked upon, and leaning or adaptive preferences are (under certain conditions) compatible with neoclassical theory (von Weizsäcker 2015).

The utility provided by a certain behavior is not objectively measureable, yet it is the only motive of any individual. Faced with a known set of restrictions, and thus a known set of possible choices, the observation of factual behavior allows the formation of preference orders (see Pindyck and Rubinfeld 2009 for the fundamentals of decision theory and the axioms used). Under the assumption that individuals are able to assess their possible choices and order them in sequence of decreasing utility, the observation of different choices under alternative restrictions and choice sets allows insights into the utility associated with any choice.

A central concept in the measurement of preferences and utility is the *willingness to pay* (WTP). Assuming that no individual will voluntarily lose utility, the utility an individual realizes with choice A in any given setting is equal to the maximum payment the individual is willing to make in order to realize that choice.[2]

---

[2]If in a given setting the price of a choice is lower than an individual's WTP for that choice, the individual will realize that choice and profit from consumer surplus. An alternative measure of the value or utility of a choice alternative is the willingness to accept (WTA). Here the utility of a

In other words, the utility of a choice or a good is determined by the willingness to abstain from alternative goods or choices.

At this point, one might ask in what way such fundamental theories are relevant for the evaluation of RE. The concept of consumers' WTP might be a correct indicator of the value of common consumer goods such as smart phones, and market prices are a socially accepted measure of the value provided by said products to society. However, there are good reasons to believe that consumers' WTP does not reflect the value of RE correctly:

- The value of RE may not be readily derived from purchase and investment behavior of consumers on conventional markets. These markets are usually regulated and to a large extent depend on political decisions exogenous to the respective markets.
- The provision and development of RE affects not just the interests and needs of individual consumers but also the needs and (possibly restrictions) of society as a whole.

Both of these aspects are discussed in detail with regard to the methods of preference elicitation in Sect. 3 The underlying idea that preference intensity for any given good is measured in opportunity costs is constitutive for all economic methods. In environmental economics, for instance, the concept is used to evaluate increases in environmental quality via environmental protective measures. Even though "environmental quality" is an immaterial good that is not traded on markets, and even though the potential benefits of improved environmental quality are shared by numerous individuals, preference intensity is evident in the amount of goods individuals are willing to give up in order to improve environmental quality. The idea is also used in marketing research and by energy utilities: In the development of new products such as eco-tariffs for electricity, consumers are regularly asked for their WTP or willingness to accept for adopting the still fictional product.

## 2.1    Consumer Preferences and Consumer Responsibility

In the research on sustainability, there is great emphasis on the normative question of the responsibility for sustainable consumption (Belz and Bilharz 2007). On one side are the market optimists. Ever since the formation of the so-called Lifestyles of Health and Sustainability (LOHAS), market optimists have believed that although consumers accept government regulation of consumption as necessary, true changes in consumption patterns rely on responsible individual action (Müller-Friemauth 2009). On the other side, market skeptics warn of a counterproductive "privatization of sustainability" (Grunwald 2010). In a similar manner, consumer

---

choice is equal to the minimum payment an individual must receive to abstain from that choice without realizing utility losses. See Weimann (2009) for an in-depth description of the so-called compensatory measures of welfare.

associations argue that political failures may not be compensated by ecologically correct consumption and place the responsibility for sustainability on politicians. A claim made here is that policy ought to "frame the conditions for sustainable economies and sustainable consumption by formulating rules and laws" (Lell 2012, p. 38). Considered a compromise between market optimists and skeptics, the *model of shared responsibilities* acknowledges the importance of both regulators and consumers for sustainability. Ultimately, the two areas are indivisible (Bilharz et al. 2011). Applying this concept to RE, the model suggests that RE should be promoted by government (for instance, via feed-in tariffs), while at the same time individuals ought to make concerted efforts to help promote RE (for instance, by changing electricity contracts).

From an economic perspective, this normative point of view is without consequence as long as it is not translated into economic incentives. Normative ideals are to be processed positively (Homann 1994) and only become morally justified when reflected in empirical conditions (Suchanek 2007).[3] Additionally, problems of environmental politics, such as climate change, are defined by the very observation that voluntary individual action leads to suboptimal allocations. The external effects of individual behavior constitute market failure and call for social and institutional regulation (Stübinger 2005, p. 132).

Economic theory discusses the relationship between individual behavior and sustainability from the perspective of altruism. Economic theories of altruism examine empirical *decisions* and behavior and their *social results* or consequences (see Ockenfels 1999 for an overview). On the individual level, these results depend on individual restrictions and preferences. On the social level, outcomes are a function of the interactions between individuals. Relating the economic understanding of altruism to the issue of responsibility for the social and ecological quality of decision outcomes, the question of whether individual and collective decisions are complementary or substitutionary to one another needs to be examined.

Climate protection activities are instruments to reduce the negative external effects of economic activities. As such, climate protection may be interpreted as a pure public good, the aggregate supply of which is equal to the sum of individual efforts of emission reductions and other climate protection activities.[4] On an individual level, altruism would motivate decision makers to comply with principles of sustainability. Since the *pure* altruist is indifferent to the source of public good provisions, the familiar social dilemma of free riding occurs in which market results are not socially optimal. Public provision of the public good—public

---

[3]The proposition of empirically determining opinions on the basis of responsibility for sustainability by means of surveys (Belz and Bilharz 2007, p. 40) appears to be of little value for the empirical assessment of the conditions of real behavior.

[4]The example of climate protection illustrates vividly that external effects and public goods are two sides of the same coin. Consumers of the common resource "environment" neglect the negative effects of their consumption on other consumers. This leads to overuse of the resource. If consumers reduce their consumption voluntarily, then all other consumers benefit from that reduction simultaneously. Also, no individual may be excluded from these benefits.

climate change activities—are perfect substitutes for individual provisions and are unable to resolve the dilemma. The resulting *crowding out* of individual efforts to protect the climate lead to constant aggregate provisions (neutrality theorem; see Bergstrom et al. 1986).

The coexistence of public and private climate change activities observed in the real RE market may be explained by the concept of *impure* altruists: Voluntary climate protection activities are of intrinsic value and provide a "warm glow of giving" (Andreoni 1990). If this is the case, voluntary contributions and third-party contributions are imperfect substitutes, so that the crowding out of individual activities is incomplete. [Crumpler and Grossman (2008) provide an overview of empirical findings on this matter; Croson (2007) even reports of crowding in, whereas Brooks (2000) and Menges et al. (2005) observe nonlinear relationships.]

An implicit assumption of the economic models of altruism is the reversibility of behavioral changes on the individual level caused by government activities. Under certain conditions, a policy featuring extrinsic incentives to promote a particular behavior may alter intrinsic motivation of individuals with lasting effect (Frey 1997). In such settings, there is a path dependency for collective and individual action that is explained by a subjective depreciation of moral activities (Akerlof and Dickens 1982) or a loss of reputation (Bénabou and Tirole 2006).[5]

## 2.2    Preferences, Attitudes and Behavior

In the social sciences, there are two rivaling approaches on the question of whether individual behavior may solve or reduce environmental problems.

- In economic theory, environmental problems are caused by the institutional conditions of individual behavior. According to the economic model of behavior, individuals react systematically to changing incentives or restrictions in accordance with their preferences. This means that changes made to relative prices (for instance via taxes) are possible solutions to environmental problems. The normative goal for policy is to create incentives that harmonize individual and collective rationale without offsetting consumer sovereignty.
- In sociology and psychology, however, the focus is placed on the concepts of *attitudes* and *morals*. If environmental problems originate in the conflict between short-term egoistic and long-term collective interests, in the majority of social sciences their solution requires adjustments in the individual attitudes

---

[5]Goeschl and Perino (2009) describe an experiment in which probands decided on individual payoffs usable for private consumption and real $CO_2$ certificates. They observe that the taxation of private consumption reduced the intrinsic motivation for certificate purchase. It was concluded that policymakers ought not rely on both individual WTP and extrinsic incentives simultaneously (Falk and Kosfeld 2003; Meier 2007). The finding of long-lasting or irreversible crowding out of intrinsic motivation is supported by many studies (Frey and Jegen 2001).

and morals. Many studies that identify a gap between the positive general attitude toward the environment and its protection on the one hand and factual behavior on the other stress the importance of information to strengthen the ecological knowledge of individuals.

A conclusive discussion on these two approaches may not be undertaken here (see for instance Menges et al. 2004a). However, there remains a question that is highly relevant for consumer research: What is the relationship between preferences and attitudes?

- In sociology and psychology, research on preferences for GE (e.g., Rowlands et al. 2003; Wortmann et al. 1996) is based on the hypothesis that individual environmental awareness determines the evaluation and choice of electricity contracts. This is why in these studies individual environmental attitudes are elicited in surveys and then used to explain factual behavior. To simplify: *awareness and attitudes are concepts to explain behavior*.[6]
- Economic approaches go in the opposite direction. In economic studies decision settings are artificially constructed to best simulate real restrictions and choice outcomes. *Behavior explains preferences*, that is, individual preferences are revealed by choices made, or, expressed in a more formal manner: preferences, which cannot be observed directly, can be reconstructed by analyzing observable choices.

In economic theory, preferences are thus often defined as "a decider's attitudes towards consequences or towards choice alternatives" (Eisenführ and Weber 2003, p. 31). This means that preferences are related to choice alternatives, whereas attitudes—such as environmental consciousness—are of a more general nature. They are regarded as predispositions of individual behavior and considered in the course of preference measurement. The operationalization of preferences in indi-

---

[6]Methodologically, these methods are doubtable when they resemble so-called participating observations. Participating observation is a method that was developed in the field of community work. It aims to motivate individuals to actively pursue their interest (e.g., Lüttringhaus and Richers 2003). In some research on the potentials of GE by Birzle-Harder and Götz (2010), group discussions were run that included members of the environmentalist group BUND that were invited as advisory experts (p. 21). The results of these group discussions suggest that consumers are willing to accept a price premium of 10% for GE. Furthermore, utilities and GE providers ought to cooperate with local climate protection groups when marketing their products (p. 35). In doing so, "engaged media" and "organs of engaged groups" ought to be considered at all times (p. 23). One might ask whether these results reflect true insights into individual consumer behavior and whether the method used fulfills standard criteria of reliability and validity. Such a blend of normative and positive questions also sheds a critical light on the tendency for transdisciplinary research, particularly in research on sustainability. This approach requires researchers to account for the interest of relevant stakeholders even in the theoretical phase of research conception.

vidual WTP provides insights into the structure of the underlying preferences for GE.

# 3    Methods of Preference Elicitation

The methods used to measure consumers' WTP may be categorized in three groups (Fig. 3.1). The first group of methods relies on market data. Field research such as market observations or field experiments yield more robust results than methods that rely on hypothetical decision settings and questions. These methods elicit individual behavior mostly free of biases (for instance, caused by interviewers) and are of high external validity (Skiera and Revenstorff 1999, p. 224). However, the collection of market data is cost-intensive (Hüttner et al. 1999, p. 50). Also, market data are usually highly aggregated, which makes it difficult to derive WTP on the individual level. Furthermore, it is problematic that preferences may only be determined for factual consumers. Information on nonconsumers is unavailable.

Second, if no "real" market data are available—for instance, in the case of preferences for nonmarketable goods—there are alternative methods, referred to as direct or indirect methods (Wricke and Herrmann 2002, p. 573; Gabor and Granger 1966, p. 45; Kalish and Nelson 1991, p. 328). A prominent example of direct methods is the *contingent valuation*. In this method, individuals are presented with two discrete choice alternatives that are identical except for the provision of the good that is to be evaluated. One of the choice alternatives is described as the status quo, and the test persons are then asked how much they would be willing to

| Market data | Experimental methods (e.g. test markets, field experiments) | | |
| | Non-experimental methods (e.g. market observations) | | |
| Preference data | Direct methods (e.g. contingent valuation) | | |
| | Indirect methods (e.g. travel cost analysis, conjoint-analysis) | | |
| Purchase offers | Auctions | Incentive compatible (e.g. Vickrey-auction) |
| | | Non-incentive compatible (e.g. highest-bid-auction) |
| | Lotteries | Incentive compatible (e.g. BDM mechanism) |
| | | Non incentive compatible |

**Fig. 3.1** Methods of WTP elicitation (Reproduced from Menges et al. (2004b), p. 249)

pay for realizing the alternative choice or scenario. WTP may also be measured using payment cards or referenda, either of which may be used once or multiple times. The characteristic feature of direct methods is that WTP is elicited for a variation of the good directly. In contrast, indirect methods determine WTP based on information about behavior and preferences for goods that are related but not identical to the good to be evaluated. Examples of indirect methods include hedonic regression, travel cost analysis, conjoint analysis, and their derivatives, such as the discrete-choice experiment. Hedonic regression uses market data on goods that differ in various attributes. Among these attributes is the one that is to be evaluated—for example, some environmental property—and in the comparison of the market prices of goods that are equal in everything except that attribute, evaluation becomes possible (Baumgartner 1997, p. 16). In travel cost analysis, the value of a good is derived from the costs individuals bear in order to consume a given good (e.g., a park). Conjoint analysis is a method in which individuals make multiple decisions between choice alternatives that differ in various attributes. Assuming that the utility provided by one alternative is defined by the sum of the utilities of its attributes, the conjoint method may be used to measure WTP for any given attribute (Weiber and Rosendahl 1997, p. 109).

In the third group of methods, individuals are presented with "real" purchase offers. Because of its incentive compatibility (Sect. 3.2), the Vickrey auction (second-price sealed-bid auction) (Vickrey 1961, p. 20) is the most commonly used method. The Becker–DeGroot–Marschak approach (BDM) (Becker et al. 1964) is another widely used method that employs a series of lotteries to determine the value of a given good.

## 3.1    The Special Case of Environmental Goods

When measuring individual WTP for environmentally friendly products, the fact that these products are public goods needs to be taken into consideration. This may be shown with the example of GE. The positive effects that a higher GE share in the electricity mix have on the environment are shared by everyone (nonexcludable), and the benefit any individual realizes is independent of other individuals' benefit (nonrivalrous). The positive theory of public goods postulates that in this setting, individuals behave strategically and disguise their true preferences by understating their WTP. However, this thesis has been confuted partially both empirically (Rondeau et al. 1999, p. 456) and theoretically (Andreoni 1989, p. 1448). The concepts of impure altruism, the warm glow of giving, and the purchase of moral satisfaction (Kahneman and Knetsch 1992, p. 64), discussed earlier, relate to the fact that individuals experience utility from moral behavior, which reduces the incentive to free-ride.

However, the assumption of the existence of nonuse values as well as altruistic or intrinsic motives contains strong implications for the methods of preference elicitation (Meyerhoff 2001, p. 393). Indirect methods such as travel cost analysis "only" account for prices and quantity and are unable to assess values that are independent from factual use (Degenhardt and Gronemann 1998, p. 1). For one, market data for the protection of environmental goods usually do not exist. Also, a price only reflects the WTP of the marginal consumer and thus does not mirror collective utility. This is why in the evaluation of environmental goods and public goods in general the method of contingent valuation is particularly useful. The controlled variation of provision levels of environmental quality (and environmental quality alone) enables the elicitation of environmental preferences in a controlled manner.

## 3.2    The Problem of Incentive Compatibility

WTP estimates are valid only if elicitation methods incentivize "true" statements or behavior. This means that data need to be free from noneconomic influences. Using the example of economic experiments, the so-called father of experimental economics Vernon Smith identifies various criteria that elicitation methods need to fulfill to be incentive compatible (Smith 1982). Test subjects ought to make choices autonomously (*privacy*), and the incentives provided need to outweigh any other potential incentives of decision making (*dominance*). Furthermore, the elicitation design needs to be *salient* in that incentive mechanisms should reward "better" choices. Finally, incentives need to be provided in a manner that incentives are always functional and not subject to satiation (*nonsatiation*). Elicitation methods that comply with all these conditions may be considered as making up an institutional framework in which individuals behave rationally in the economic sense.

Based on these criteria, methods of WTP elicitation that rely on hypothetical decisions are doubtable because they provoke a systematic overstatement of WTPs. "In choice experiments, customers can have a tendency to de-emphasize price, since they do not have to actually pay the price" (Goett et al. 2000, p. 27).[7] This is true of direct elicitation methods, such as contingent valuation (Cummings et al. 1986), and indirect ones, such as conjoint analysis (Roe et al. 2001, p. 917; Roe et al. 1996, p. 158).

Economic quasi-field experiments designed to determine the WTP for GE as pioneered by Menges et al. (2004a, 2004b, 2005) circumvent these issues by assigning real monetary value to hypothetical choices. Nonetheless, in compliance with a number of theoretical standards the validity of hypothetical methods may be increased significantly. In the case of the contingent valuation method, the *Report of the NOAA Panel on Contingent Valuation* (Arrow et al. 1993) contains a list of the possible causes of biases in stated preferences and names methodological criteria suited to counteract these causes. Since the issue of incentive compatibility may thus be nullified, methods of preference determination that rely on

---

[7]Hasanov (2010) acknowledges the hypothetical nature of the preferences for GE obtained from a telephone survey by using the term *payment readiness* instead of WTP.

hypothetical decision settings are widely spread, particularly because they are easily accessible and generally less expensive to conduct.

## 4     Consumer Preferences for Renewable Energy

Independently of the method used in preference elicitation, the need to specify the concept of RE in any elicitation format poses challenges. This section shows how spatial, temporal, and factual classifications of RE are used in preference and WTP measurement. In short, there is no objective assessment of RE in the sense of an all-encompassing model or concept. Instead, any preference measurement requires the specification of a decision setting that involves the characterization of RE by means of (more or less arbitrarily) chosen dimensions and aspects. This is important because preferences are revealed in and limited to specific sets of restrictions. Since these restrictions may be designed at will, there is a potentially infinite amount of possible approaches to the subject of RE.

This is in part due to the complexity and heterogeneity of the concept of RE. As an umbrella term, REs can be regarded from different angles. In an extensive meta-anaylsis on WTP estimates for GE, for instance, Sundt and Rehdanz (2015) identifiy eight different aspects of GE that are subject to evaluation. Individual preferences for RE may concern *energy sources* (e.g., wind power versus solar power) (Borchers et al. 2007). The term also comprises *technologies used in energy transformation* (solar heat versus photovoltaic) (Scarpy and Willis 2010) and in a broader frame encloses *political programs* and questions of *energy infrastructure* (Grieger and Cie Marktforschung 2013; Menges and Beyer 2014).

Even within a particular concept of RE, the aspects that are regarded in the decision settings offered in preference elicitation vary widely. This is quite evident in a comparison of studies measuring WTP for electricity from renewable sources. Henry et al. (2011), for instance, elicit WTP without specifically outlining the effects of changes to the generation system. Such effects are described in the study undertaken by Bigerna and Polinori (2014), who also mention the positive effects of GE on the environment. Roe et al. (2001) frame decision settings in an even more detailed fashion and list the greenhouse gas emissions of different power mixes. Other aspects of electricity generation from renewable sources that are considered in preference elicitation include effects on the labor market, energy security, national self-sufficiency, and landscape protection (Kaenzig et al. 2013). These examples demonstrate that the empirical literature on individual preferences is characterized by *framing effects* (Tversky and Kahneman 1986). Differences in the presentation of choice settings affect WTP measures, and the criterion of procedural invariance is not fulfilled.

Studies that determine WTP for RE also vary in the time span they consider. Time may be incorporated into elicitation techniques in various ways and increase the complexity of decision tasks. In most studies, time is excluded from decision settings, and decisions are made and come into effect immediately (e.g., Mozumder et al. 2011). In contrast, studies such as the one conducted by Guo et al. (2014) elicit

WTP for changes in the energy mix "in the next five years." In both cases the object of investigation is the WTP for renewable power supply, yet the individual studies assess different preference orders. In the case of incentive-compatible methods, another determinant of WTP may be the point in time at which opportunity costs come into effect. If WTP is measured via contingent valuation, the payment vehicle used in preference elicitation may imply immediate (Andor et al. 2014) or future spending (Abdullah and Jeanty 2011).

Uncertainty and risk are incorporated into decision settings in similarly heterogeneous ways. In the field of RE, uncertainty is a major determinant of individual decisions on short-term consumption and long-term investments (Soroudi and Amraee 2013). Whether and how uncertainties are addressed in preference elicitation has an important effect on WTP measurements.

Despite the manifold approaches to the subject of RE, the breadth and number of studies on consumer WTP may be reduced to two categories. In one category, consumers act on competitive markets (as consumers or investors), whereas in the other category individual consumers' preferences are elicited in a political context. This segmentation of the literature will be used in what follows to identify statements that are robust over varying elicitation methods and approaches to RE.

**Consumer Preferences in Competitive Markets**  Consumer decisions on RE in competitive markets in the different sectors may be subdivided into direct and indirect decisions. Households mainly consume energy in the forms of heat and electricity. The second largest consumption factor is transportation. Choices on *supply tariffs* thus constitute a meaningful choice for the direct demand for RE, whereas the choice of *transportation means* reveals preferences for RE indirectly.

*Consumer Preferences for Energy Supply*  A prominent and demonstrative example for consumer preferences regarding the direct supply of RE is the demand for GE. The subject has been intensively investigated[8] and approached from various angles. A large proportion of studies focus on the WTP for GE and the factors that may explain this WTP. A first baseline result is that even though GE is framed in different ways (Forsa 2011), there is a general positive WTP (e.g., Andor et al. 2014; Anselm 2012; Bigerna and Polinori 2014; Zorić and Hrovatin 2012; Henry et al. 2011; Grösche and Schröder 2011; Seung-Hoon and So-Yoon 2009; Menges et al. 2005). The magnitude of WTP varies broadly, however.

Regarding the factors that influence individual WTP for GE, a core result is that the preferences for eco-tariffs depend on the power source and electricity transformation technologies used in electricity generation. Different RE sources are valued differently. Electricity from solar power is preferred most, and electricity generated from wind is preferred to tariffs that feature hydropower. Furthermore,

---

[8]Sundt and Rehdanz (2015) identify 101 WTP studies for GE focusing on the English literature alone.

heterogeneous electricity mixes are met with a higher WTP than tariffs that rely on one power source only (Burkhalter et al. 2009). On a general note and in accordance with economic theory, WTP for electricity tariffs increases in accordance with the share of renewables they comprise (Menges et al. 2005; Grösche and Schröder 2011).

Owing to the homogeneous nature of electricity, consumers may not retrace the power source used in electricity generation easily. In light of the positive WTPs for GE and their shares in electricity mixes, it is surprising that consumers are hardly willing to pay for proofs of origin (Anselm 2012). Winther and Ericson (2012), Kaenzig et al. (2013), and Mattes (2012) argue that the existence of certificates, labels, and seals is mostly unknown and that the corresponding documents and concepts remain largely misunderstood and have aroused little interest. These findings lie in juxtaposition to the growing efforts of energy suppliers that consider certificates of power sources a successful tool of GE marketing (Reichmuth 2014).

Another segment of consumer research addresses the observation that in spite of a general positive WTP for GE, the number of consumers voluntarily and actively switching from conventional to green tariffs remains low. The barriers that are believed to hinder switching behavior in deregulated electricity markets range from high search and transaction costs over loyalties to former energy suppliers to insufficient financial incentives (e.g., Yang 2014; MacPherson and Lange 2013; Gamble et al. 2009; Sunderer 2006). Other barriers to switching are information gaps, misinformation, and misconceptions of potential consumers (see Sunderer 2006 on the *ipsative area of action*). Moreover, Boardman et al. (2006) show that consumers fear that a majority of GE tariffs are "rebranded conventional tariffs. This directly relates to the aforementioned findings of WTP for origin certificates and reveals a fundamental problem in GE marketing: On the one hand, consumers doubt the composition of green mixes; on the other hand labels and certificates apparently are not suitable instruments for reducing that skepticism.

Another strand of the literature on consumer preferences addresses instruments to overcome the barriers that prevent consumers from translating WTP into demand and formulating recommendations for the marketing of GE (e.g., Hübner et al. 2012). Wiser (1998), for instance, examines different organizational structures of GE suppliers and reports that GE marketing ought to be focusing on local target groups to build trust and customer relationships. This assessment is supported by Mattes (2012) and Bethke (2011), who show that WTP for GE increases with the proximity of generation facilities to consumers' places of residence. Other marketing recommendations concern the emphasis of subjective psychological aspects of the utility offered by GE. Marketing should appeal to consumers' sense of responsibility for the environment and underline the moral satisfaction that results from environmental protection (Herbes and Ramme 2014). Other studies refer to the manifold nonfinancial motives of potential GE consumers and highlight the relevance of market segmentation and target group focus in marketing (Rundle-Thiele et al. 2008).

*Consumer Preferences for Transportation*  The better part of energy consumption in the transportation sector stems from traffic in general and individual motor car traffic in particular (Dena 2012). Hence, consumer preferences for RE in the transportation sector are most commonly elicited in the field of alternative automotive fueling technologies. In contrast to the purchase of GE, preferences for such technologies are indirect measures for the preferences for RE. This conceptual difference shows up in the following robust result confirmed in a variety of studies: WTP for alternative drives is low (sometimes even negative) and is the polar opposite to the generally positive evaluation of WTP for GE (e.g. Axsen et al. 2013; Hackbarth and Madlener 2013; Jensen et al. 2013; Lo 2013; Ziegler 2012; Hidrue et al. 2011). Notably, preferences for alternative driving technologies are low regardless of whether consumers have gained real-world driving experience in test drives (Jensen et al. (2013).

Analogous to the heterogeneous preferences for different sources of GE, there are substantial differences in the preferences for alternative drives. A definitive order does not exist: Ziegler (2012) finds that consumer preferences are strongest for hybrid cars yet admits that this result is not shared in similar studies in the field. Hackbarth and Madlener (2013) observe that preferences for specific alternative drive technologies depend on the area of operation or the purpose of the vehicle. Natural gas vehicles and hybrid cars are less preferred in urban areas.

The low WTP values for alternative driving technologies may be explained by various additional factors not directly related to RE. In vehicle purchase, aside from purchase and running costs, the most important product attributes are engine power, range, and refueling possibilities/infrastructure. Since in general alternative drive cars meet these demands to a lesser extent than conventionally fueled cars, the negative verdict on RE in the transportation sector is conclusive. Interestingly, however, in almost all studies on individuals' WTP for alternative drive technologies, the level of $CO_2$ emissions is a significant explanatory factor (Hackbarth and Madlener 2013; Jensen et al. 2013). This indicates that the positive environmental effects of RE are a central motive for RE purchase and that, in principle, consumers do approve of alternative drive technologies. The fact that consumers still prefer conventional drive technologies is thus due to the fact that individuals weigh the individual costs of reduced range or power more strongly than the positive public effects of RE use. This conclusion is supported by WTP studies that contain self-disclosures of participants. Consumers who describe themselves as environmentally aware and who value engine power less express a higher WTP for alternative fuels (Hackbarth and Madlener 2013; Ziegler 2012).

**Consumer Preferences in Political Environments**  In making voting decisions, consumers exert an influence on the conditions that determine the development of RE markets, technologies, and policies. Consumers shape public opinion and form a collective will. Social acceptance is an essential condition for the success of political programs and realization of specific projects and may ultimately be reduced to individual preferences.

*Preferences and Political Programs*  Across the globe, governments actively promote the development of RE. Programs, measures, and concepts such as *energy transition* in Germany, *green home scheme* in England, or the *state-level renewable portfolio standard (RPS)* in the USA vary substantially, however. Programs differ in terms of specific goals, in terms of the instruments and measures to achieve these goals, and in the private and public resources committed to them.[9] Even though this implies difficulties in formulating general statements on consumer preferences for RE programs, the very existence of RE programs may be interpreted as an expression of consumer preferences for RE when consumers vote in fair and transparent democratic systems.

In Germany, preference elicitations find a strong preference for the active political support of the development of RE (Grieger and Cie Marktforschung 2013; Agentur für Erneuerbare Energien 2012; Christ and Bothe 2007). Yet the work presented by Grieger and Cie Marktforschung (2013) underlines the relevance of the aforementioned free-riding motive in RE support. There are significant differences in the WTP for an energy transition depending on whether the payment vehicle used in preference elicitation involves voluntary individual or compulsory collective efforts. If WTP is measured in a collective context that excludes the possibility of free riding, WTP is higher and more frequently positive.

The incentive-compatible experiments described by Menges et al. (2005) and Menges and Traub (2009) on the WTP for GE arrive at similar conclusions. Measuring not the absolute WTP but its reaction to external parameters, the researchers gained insights into the individual motives for GE purchase. Among the variations made to the experimental setup were the share of GE in the electricity mix and the conceptual decision modus; in one setting, individuals acted as individual energy consumers, whereas in the other setting individuals acted as voters who were making a collective choice (simulated using the median voter theorem) on the share of RE in the electricity mix. An important implication of these experiments is that individuals are concerned with the context in which the payments they made to promote RE come into effect. While a conjoint analysis conducted by Goett et al. (2000) suggests that consumers are only marginally interested in the social consequences of GE purchase and that WTP is mostly driven by the purchase of moral satisfaction, the aforementioned experiments imply the opposite. Consumers reveal distinctly different preferences for GE in individual and collective decision settings. If consumers were driven by individual warm-glow motives, then WTP would have had to be the same in collective and individual elicitation contexts. However, WTP was significantly and decisively higher in votes on collective, obligatory payments. In the face of the social prominence of the subject, the question of whether voluntary individual activities for climate protection may be considered substitutes for government activities must be answered in the negative.

---

[9]Refer to International Energy Agency (2014) for an overview.

Another important result of the aforementioned experiments is the nonlinear relationship between WTP and subjective opinions on the share of the electricity price that is reserved for RE promotion. Below a certain critical estimate on that share, higher electricity prices are met with a higher WTP for GE (crowding in). Individuals who perceive that the share of electricity price used for RE development is higher than that critical value react to increases in the electricity price with a reduced WTP (crowding out).[10]

The nonneutrality of government activities on individual behavior may be explained by the concept of impure altruism. Private and public activities are not complete substitutes. However, the nonlinearity of the rates of substitution has attracted significant research interest in the social sciences, including economics (e.g., Matiaske and Weller 2006).

The finding of free-riding motives and the relevance of the perceived effectiveness of payments for GE development suggest that preferences for GE are determined by social factors. Also, individual WTP seems to be driven by skepticism toward markets as allocation mechanisms. Together with the liberalization of electricity markets, the ongoing public debate on climate change has made environmental friendliness a relevant criterion in the choice of electricity suppliers. Yet the commercial success of GE has not met expectations. The results described earlier propose that this may be traced back to individuals' preferences for collectively obligatory regulation over market-based offers. This interpretation conforms to the political economy of environmental protection, which assumes that when in doubt, individuals prefer government regulation over incentivized voluntary action (Kirchgässner and Schneider 2003). In political economy, however, this preference for regulation is explained by individual irrationality in the form of *cost illusion*: utility-maximizing individuals would incorrectly assume that regulations do not alter individual mean incomes and vote accordingly. In the case of GE development, the experiments described here falsify that theory. In collective choices, individuals not only voted for higher shares of renewables in the electricity mix but also revealed a considerably higher WTP for GE then in decisions on individual contracts.

The free-riding motive is thus of great importance in the development of political RE programs. The findings reported earlier suggest that instruments establishing regulative constraint and thereby eliminating the possibility to behave strategically may lead to a faster development of RE than incentive-based instruments. Research on consumer preferences for specific policies supports this claim. While economic cost–benefit analyses of feed-in tariffs arrive at heterogeneous conclusions on program effectiveness and efficiency (Jenner et al. 2013; Butler and Neuhoff 2008), there is a clear consumer preference for such tariffs (Agentur für Erneuerbare Energien 2012).

---

[10]Similar results are available on individual donations to welfare organizations (Brooks 2000).

*Consumers and Projects of Public Interest* The political programs for RE promotion imply extensive changes to the energy economy. The reorientation of generation systems and the resulting need to adjust the energy infrastructure entail immense investment needs. As shown earlier, consumers are generally willing to accept cost increases on energy markets for the promotion of RE. However, consumers are far more critical to concrete investment projects aiming to secure energy security by means of power generation and distribution in a RE economy (Althaus 2012).

Resistance to the construction of RE supply capacities is most frequently voiced by residents in the vicinity of projected facilities. This observation is (not uncontroversially) (Wolsink 2012) conflated in the term *nimby* (not in my back yard), which expresses the idea that objections to construction are motivated by the mostly local negative effects of project realization.[11]

The fact that protests to the development of RE arise mainly locally is undisputed. More controversial is the question of the possible causes of nimby protests. Originally, local protest was interpreted as egoistic behavior (Esaiasson 2014). Relevant motives were the avoidance of visual and acoustic impairments by local facilities (Kontogianni et al. 2014) or the depreciation of real estate (Dear 1992). More recently, however, it has been shown that protests are also sparked by altruism and moral-ethical considerations. Bidwell (2013), for instance, examines protests to wind power stations. He demonstrates that resistance to construction is not due to individual and particular interests but rather collective interests. Study participants expressed concerns that windmill construction might pose financial risks for the local government. Another cause of protest lies in fundamental concepts individuals have of justice and inequality aversion. The siting of generation facilities is subject to natural and technical restrictions (e.g., solar exposure for photovoltaic or transmission loss). Thus, residents of locations that are favorable to facility operators are more likely to be impaired by potential negative—and more likely to benefit from potential positive—effects of RE projects. If the potential benefits and costs are high, this inequality may cause local protests (Pol et al. 2006).

Since local protests are a common and costly obstacle to RE development, significant efforts have been made to discover efficient instruments to their abatement. One such instrument is residential participation (Jami and Walsh 2014). The financial involvement in RE projects, for instance by means of cooperatives and other corporative forms, combats inequality aversion and calms local protests. Similarly, the inclusion of residents in planning and organizational processes is considered a proven means to increase the local acceptance of RE projects (Jones and Eiser 2010).[12]

---

[11]The concept of *nimby* is not limited to the context of RE. Other areas nimby is relevant to is the siting of landfills, cell towers, or even accommodation for the homeless.

[12]Other studies suggest that participation might have contrary effects on the social acceptance of RE projects. Menges and Beyer (2014) show for the case of transmission network development, that households who call for participation programs are significantly more likely to object to local grid construction plans than households that favor an supra-regional coordination of grid

In a discussion on nimby effects, it needs to be stressed that local protests to RE projects are mostly focused on plans and propositions. It has been shown that objections to RE projects decline significantly once projected plans are realized and residents familiarize themselves with the outcomes (van der Horst 2007). A similar effect may be observed in connection with the distance of a project site to individuals' homes. The *habituation effect* describes the counterintuitive observation that objections to RE projects increase with the distance of individuals' homes from project locations. Conversely, acceptance of RE projects increases with proximity of individual residences to project sites (Menges and Beyer 2014). A possible explanation for the habituation effect is the overestimation of impairments caused by the planned facilities that dissipates once individuals gain experience, which happens more quickly in closer proximity. This interpretation is the basis for another instrument used to increase the social acceptance of RE projects: road shows and discussions on network construction frequently organized by transmission network operators, which show vividly that communication and information represent effective means of avoiding local protests.

---

## 5    Conclusion

In consumers' eyes, RE is deeply symbolic. RE promises answers to many challenges of modern society. From an economic perspective, RE is not consumer goods in the classical sense but rather serves as an input in the production of energy services in the housing and transport sectors. As such, RE may be substituted by and compete with conventional energy sources.

Analyzing empirical estimates of WTP, there is a broad consensus that RE enjoys high social acceptance and that consumers are willing to pay for its advancement. Beyond this general approval of RE, there is little consistency in the results on selected topics in the field. This is in part caused by heterogeneous methods of data collection and interpretations of the concept of RE. However, the core of the problem lies in the nature of public goods. Compared to equally symbolic private goods such as organic fruits, the purchase of RE provides fewer private benefits. Why should individuals purchase products whose benefits—whether emission reductions or the protection of scarce resources—are shared globally? Economists answer this question with the concept of impure altruism. Individuals experience utility from altruistic behavior and moral consumption that complement the public effects of RE promotion.

From an economic perspective, the results of consumer research may be condensed to the following statements:

---

development. This finding suggests that participation mechanisms are subject to sample-selection-bias, as participation mechanisms may include an over proportional share of protesting households relative to the total population.

1. Consumers indicate and reveal a positive WTP for RE if the purchase of RE provides moral satisfaction. In areas where this moral surplus is low compared to the utility derived from other product features, the general acceptance of RE does not translate into market demand. This is the case in the area of alternative fuels in the transportation sector, for instance.
2. The social acceptance and a general WTP for RE may be considered unstable; free riding is an essential motive of impure altruists. Consumers purchasing RE for environmental protection reduce the incentive for other people to follow suit. Furthermore, it remains unclear how far voluntary consumer promotion of RE will go when energy prices increase as a result of structural changes of the energy market caused by increasing shares of RE in the energy mix. This issue is particularly important for low-income consumers who are more affected by increases in energy prices.
3. It is thus logical that the consent given to political support of RE development is especially stable in studies that suggest a binding collective effort. As soon as a credible scheme for obligatory RE support and its financing is provided, the free-riding motive is eliminated.

The analysis of consumer preferences for RE suggests some important conclusions for the marketing of RE in competitive markets. Strategies aiming to convert consumers' latent WTP for environmental protectoin into purchase behavior need to account for two conflicting aspects. On the one hand, marketing activities need to address and invoke nonmaterial individual warm-glow motives that result from voluntary environmental protection activities. At the same time, collective and binding aspects of RE development (for instance with regard to financing) need to be communicated. This is to reassure consumers that individual efforts made to promote RE may not be taken advantage of by other consumers, competitors, or even utilities.

## References

Abdullah, S., & Jeanty, P. W. (2011). Willingness to pay for renewable energy: Evidence from a contingent valuation survey in Kenya. *Renewable and Sustainable Energy Reviews, 15*(6), 2974–2983.

Agentur für Erneuerbare Energien e.V. (2012). Akzeptanz und Bürgerbeteiligung für Erneuerbare Energien. *Renews Spezial, 60.*

Althaus, M. (2012). Schnelle Energiewende—bedroht durch Wutbürger und Umweltverbände? Protest, Beteiligung und politisches Risikopotenzial für Großprojekte im Kraftwerk- und Netzausbau. In L. L. Ungvári (Ed.), *Wissenschaftliche Beiträge 2012* (pp. 103–114). Wildau: TH Wildau.

Akerlof, G. A., & Dickens, T. D. (1982). The economic consequences of cognitive dissonance. *American Economic Review, 72*, 307–319.

Andreoni, J. (1989). Giving with impure altruism: Applications to charity and Ricardian equivalence. *Journal of Political Economy, 97*(6), 1447–1458.

Andreoni, J. (1990). Impure altruism and donations to public goods: A theory of warm-glow giving. *The Economic Journal, 100*(6), 464–477.

Andor, M. A., Frondel, M., Vance, C., et al. (2014). Diskussionspapier: Zahlungsbereitschaft für grünen Strom—Die Kluft zwischen Wunsch und Wirklichkeit. *RWI Materialien, 79*(13).

Anselm, M. (2012). Grüner strom: Verbraucher sind bereit, für Investitionen in erneuerbare Energien zu zahlen. *DIW Wochenbericht, 79*(7).

Arrow, K., Solow, R., Portney, P. R., Leamer, E. E., Radner, R., Schuman, H., et al. (1993). Report of the NOAA panel on contingent valuation. *Federal Register, 58*(10), 4601–4614.

Axsen, J., Orlebar, C., Skippon, S., et al. (2013). Social influence and consumer preference formation for pro-environmental technology: The case of a U.K. workplace electric-vehicle study. *Ecological Economics, 95*, 96–107.

Baumgartner, B. (1997). Monetäre Bewertung von Produkteigenschaften auf dem deutschen Automobilmarkt mit Hilfe hedonischer Modelle. *Marketing ZFP, 19*(1), 15–25.

Becker, G. M., DeGroot, M. H., Marschak, J., et al. (1964). Measuring utility by a single-response sequential method. *Behavioral Science, 9*(3), 226–232.

Belz, F.-M., & Bilharz, M. (2007). Nachhaltiger Konsum, geteilte Verantwortung und Verbraucherpolitik: Grundlagen. In F.-M. Belz, G. Karg, D. Witt, et al. (Eds.), *Nachhaltiger Konsum und Verbraucherpolitik im 21. Jahrhundert* (pp. 21–52). Marburg: Metropolis.

Bénabou, R., & Tirole, J. (2006). Incentives and prosocial behavior. *American Economic Review, 96*, 1652–1678.

Bergstrom, T., Blume, L., Varian, H., et al. (1986). On the private provision of public goods. *Journal of Public Economics, 29*, 25–49.

Bethke, N. (2011). *Additiver Umweltnutzen als individuelles Entscheidungskriterium für die Wahl von Ökostrom.* Frankfurt am Main: Peter Lang.

Bidwell, D. (2013). The role of values in public beliefs and attitudes towards commercial wind energy. *Energy Policy, 58*, 189–199.

Bigerna, S., & Polinori, P. (2014). Italian households' willingness to pay for green electricity. *Renewable and Sustainable Energy Reviews, 34*, 110–121.

Bilharz, M., Fricke, V., Schrader, U., et al. (2011). Wider die Bagatellisierung der Konsumentenverantwortung. *GAIA, 20*, 9–13.

Birzle-Harder, B., & Götz, K. (2001). *Grüner Strom—eine sozialwissenschaftliche Marktanalyse.* Frankfurt a. M.: Institut für sozial-ökologische Forschung.

Boardman, B., Jardine, C., Lipp, J., et al. (2006). *Green electricity code of practice: A scoping study.* Oxford: Environmental Change Institute, University of Oxford.

Borchers, A. M., Duke, J. M., Parsons, G. R., et al. (2007). Does willingness to pay for green energy differ by source. *Energy Policy, 35*, 3327–3334.

Brooks, A. C. (2000). Public subsidies and charitable giving: Crowding-out or crowding-in, or both? *Journal of Policy Analysis and Management, 19*, 451–464.

Burkhalter, A., Kaenzig, J., Wüstenhagen, R., et al. (2009). Kundenpräferenzen für leistungsrelevante Attribute von Stromprodukten. *Zeitschrift für Energiewirtschaft, 33*(2), 161–172.

Butler, L., & Neuhoff, K. (2008). Comparison of feed-in tariff, quota and auction mechanisms to support wind power development. *Renewable Energy, 33*(8), 1854–1867.

Christ, S., & Bothe, D. (2007). *Bestimmung der Zahlungsbereitschaft für erneuerbare Energien mit Hilfe der Kontingenten Bewertungsmethode.* (EWI Working Paper Nr. 07/1). Cologne: EWI.

Croson, R. T. A. (2007). Theories of commitment, altruism and reciprocity: Evidence from linear public goods games. *Economic Inquiries, 45*, 199–216.

Crumpler, H., & Grossman, P. J. (2008). An experimental test of warm glow giving. *Journal of Public Economics, 92*, 1011–1021.

Cummings, R. G., Brookshire, D. S., Schulze, W. D., et al. (1986). *Valuing environmental goods: An assessment of the contingent valuation method.* Totowa: Rowman & Allanheld.

Dear, M. (1992). Understanding and overcoming the nimby-syndrome. *Journal of the American Planning Association, 58*(3), 288–300.

Degenhardt, S., & Gronemann, S. (1998). *Die Zahlungsbereitschaft von Urlaubsgästen für den Naturschutz: Theorie und Empirie des Embedding-Effektes.* Frankfurt a. M.: Peter Lang.

Deutsche Energie-Agentur GmbH (dena). (2012). Verkehr. Energie. Klima. Berlin: trigger. medien.gmbh.

Eisenführ, F., & Weber, M. (2003). *Rationales Entscheiden* (4th ed.). Berlin: Springer.

Esaiasson, P. (2014). NIMBYism—A re-examination of the phenomenon. *Social Science Research, 48*, 185–195.

Falk, A., & Kosfeld, M. (2003). The hidden cost of control. *American Economic Review, 96*, 1611–1630.

Forsa Gesellschaft für Sozialforschung und statistische Analysen mBH (forsa). (2011). *Erwartungen der Verbraucher an Ökostrom und Konsequenzen für Ökostrom-Labelkriterien.* Berlin: Forsa.

Frey, B. S. (1997). *Not just for the money: An economic theory of personal motivation.* Cheltenham-Brookfield: Edward Elgar Publishing.

Frey, B. S., & Jegen, R. (2001). Motivation crowding theory. *Journal of Economic Surveys, 15*(5), 589–611.

Gabor, A., & Granger, C. W. J. (1966). Prices as an indicator of quality: Report on an enquiry. *Economica, 33*(129), 43–70.

Gamble, A., Juliusson, E. A., Gärling, T., et al. (2009). Consumer attitudes towards switching supplier in three deregulated markets. *Journal of Socio-Economics, 38*(5), 814–819.

Goeschl, T., & Perino, G. (2009). *Combining Taxes and Moral Suasion for Resolving the Energy-Climate Nexus: Experimental Evidence of a Conflict* (Working Paper). University of Heidelberg.

Goett, A. E., Hudson, K., Train, K. E., et al. (2000). Customers' choice among retail energy suppliers: The willingness-to-pay for service attributes. *The Energy Journal, 21*(4), 1–28.

Grieger & Cie Marktforschung. (2013). Energieversorgung in Deutschland nach Fukushima. Accessed June 27, 2016, from https://www.grieger-cie.de/Marktforschung/Umfrage-Kernenergieausstieg-2011.pdf.

Grösche, P., & Schröder, C. (2011). Eliciting public support for greening the electricity mix using random parameter techniques. *Energy Economics, 33*(2), 363–370.

Grunwald, M. (2010). Wider die Privatisierung der Nachhaltigkeit—Warum ökologisch korrekter Konsum die Umwelt nicht retten kann. *GAIA, 19*(3), 178–182.

Guo, X., Liu, H., Mao, X., Jin, J., Chen, D., Cheng, S., et al. (2014). Willingness to pay for renewable electricity: A contingent valuation study in Beijing, China. *Energy Policy, 68*, 340–347.

Hackbarth, A., & Madlener, R. (2013). Consumer preferences for alternative fuel vehicles: A discrete choice analysis. *Transportation Research Part D: Transport and Environment, 25*, 5–17.

Hasanov, I. (2010). *Konsumentenverhalten bei Ökostromangeboten. Empirische Untersuchungen privater Stromkunden in Deutschland.* Essen: Duisburg.

Henry, O., Volschenk, J., Smit, E., et al. (2011). Residential consumers in the Cape Peninsula's willingness to pay for premium priced green electricity. *Energy Policy, 39*, 544–550.

Herbes, C., & Ramme, I. (2014). Online marketing of green electricity in Germany—A content analysis of providers' websites. *Energy Policy, 66*, 257–266.

Hidrue, M. K., Parsons, G. R., Kempton, W., Gardner, M. P., et al. (2011). Willingness to pay for electric vehicles and their attributes. *Resource and Energy Economics, 33*(3), 686–705.

Homann, K. (1994). Wirtschaftsethik in der Moderne: Zur ökonomischen Theorie der Moral. *Ethik und Sozialwissenschaften, 5*, 3–12.

Hübner, G., Müller, M., Röhr, U., et al. (2012). Erneuerbare Energien und Ökostrom—Zielgruppenspezifische Kommunikationsstrategien. Abschlussbericht zum BMU-Verbundprojekt (FKZ: 0325107/8).

Hüttner, M., von Ahsen, A., Schwarting, U., et al. (1999). *Marketing-management* (2nd ed.). Munich: Oldenbourg Wissenschaftsverlag.

International Energy Agency. (2014). *IEA Policy & Measures Database.* Accessed June 26, 2016, from http://www.iea.org/policiesandmeasures.

Jami, A. A. N., & Walsh, P. R. (2014). The role of public participation in identifying stakeholder synergies in wind power project development: The case study of Ontario, Canada. *Renewable Energy, 68*, 194–202.

Jenner, S., Groba, F., Indvik, J., et al. (2013). Assessing the strength and effectiveness of renewable electricity feed-in tariffs in European Union countries. *Energy Policy, 52*, 385–401.

Jensen, A. F., Cherchi, E., Mabit, S. L., et al. (2013). On the stability of preferences and attitudes before and after experiencing an electric vehicle. *Transportation Research Part D: Transport and Environment, 25*, 24–31.

Jones, C. R., & Eiser, J. R. (2010). Understanding 'local' opposition to wind development in the UK: How big is a backyard? *Energy Policy, 38*, 3106–3117.

Kaenzig, J., Heinzle, S. L., Wüstenhagen, R., et al. (2013). Whatever the customer wants, the customer gets? Exploring the gap between consumer preferences and default electricity products in Germany. *Energy Policy, 53*, 311–322.

Kalish, S., & Nelson, P. (1991). A comparison of ranking, rating and reservation price measurement in conjoint analysis. *Marketing Letters, 2*(4), 327–335.

Kahneman, D., & Knetsch, J. L. (1992). Valuing public goods: The purchase of moral satisfaction. *Journal of Environmental Economics and Management, 22*(1), 57–70.

Kirchgässner, F., & Schneider, F. (2003). On the political economy of environmental policy. *Public Choice, 115*, 369–396.

Kontogianni, A., Tourkolias, C., Skourtos, M., Damigos, D., et al. (2014). Planning gobally, protesting locally: Patterns in community perceptions towards the installation of wind farms. *Renewable Energy, 66*, 170–177.

Lell, O. (2012). Klimaschutz aus Verbrauchersicht. *Wirtschaftsdienst, 92*, 37–41.

Lo, K. (2013). Interested but unsure: Public attitudes toward electric vehicles in China. *Electronic Green Journal, 1*(36), 1–13.

Lüttringhaus, M., & Richers, H. (2003). *Handbuch Aktivierende Befragung—Konzepte, Erfahrungen, Tipps für die Praxis* (2nd ed.). Bonn: Stiftung Mitarbeit.

MacPherson, R., & Lange, I. (2013). Determinants of green electricity tariff uptake in the UK. *Energy Policy, 62*, 920–933.

Matiaske, W., & Weller, I. (2006). Kann weniger mehr sein? Theoretische Überlegungen und empirische Befunde zur These der Verdrängung intrinsischer Motivation durch extrinsische Anreize. In R. Rapp, P. Sedlmeier, G. Zunker-Rapp, et al. (Eds.), *Perspectives on cognition* (pp. 113–132). Lengerich: Pabst Science Publishers.

Mattes, A. (2012). *Potentiale für Ökostrom in Deutschland. Verbraucherpräferenzen und Investitionsverhalten der EVU*. Berlin: DIW Econ.

Meier, S. (2007). Do subsidies increase charitable giving in the long-run? Matching donations in field experiment. *Journal of the European Economic Association, 5*(6), 1203–1222.

Menges, R., & Beyer, G. (2014). Underground cables versus overhead lines: Do cables increase social acceptance of grid development? Results of a contingent valuation survey in Germany. *International Journal of Sustainable Energy Planning and Management, 3*, 33–48.

Menges, R., & Traub, S. (2009). Who should pay the bill for promoting green electricity? An experimental study on consumer preferences. *International Journal of Environment and Pollution, 39*, 44–60.

Menges, R., Schröder, C., Traub, S., et al. (2004a). Umweltbewusstes Konsumentenverhalten aus ökonomischer Sicht: Eine experimentelle Untersuchung der Zahlungsbereitschaft für Ökostrom. *Umweltpsychologie, 8*(1), 84–106.

Menges, R., Schröder, C., Traub, S., et al. (2004b). Erhebung von Zahlungsbereitschaften für Ökostrom—Methodische Aspekte und Ergebnisse einer experimentellen Untersuchung. *Marketing ZFP, 26*(3), 247–258.

Menges, R., Schröder, C., Traub, S., et al. (2005). Altruism, warm glow and the willingness-to-donate for green electricity: An artefactual field experiment. *Environmental and Resource Economics, 31*(4), 431–458.

Meyerhoff, J. (2001). Nicht-nutzungsabhängige Wertschätzungen und ihre Aufnahme in die Kosten-Nutzen-Analyse. *Zeitschrift für Umweltpolitik, 24*(3), 393–416.

Mozumder, P., Vásquez, W. F., Marathe, A., et al. (2011). Consumers' preference for renewable energy in the southwest USA. *Energy Economics, 33*(6), 1119–1126.

Müller-Friemauth, F. (2009). Setzen Konsumenten von grünem Strom nur auf Nachhaltigkeit oder auch auf andere Marktverhältnisse? *Elektrizitätswirtschaft, 108*, 34–35.

Ockenfels, A. (1999). *Fairneß, Reziprozität und Eigennutz—Ökonomische Theorie und experimentelle Evidenz.* Tübingen: Mohr Siebeck.

Pol, E., Di Masso, A., Castrechini, A., Bonet, M. R., Vidal, T., et al. (2006). Psychological parameters to understand and manage the NIMBY effect. *Revue européenne de psychologie appliquée, 56*, 43–51.

Pindyck, R., & Rubinfeld, D. (2009). *Mikroökonomie* (7th ed.). Munich: Addison-Wesley Verlag.

Praktiknjo, A. J. (2014). Stated preferences based estimation of power interruption costs in private households: An example from Germany. *Energy, 76*, 82–90. doi:10.1016/j.energy.2014.03. 089.

Reichmuth, M. (2014). *Marktanalyse Ökostrom.* Dessau-Roßlau: Umweltbundesamt.

Roe, B., Boyle, K. J., Teisl, M. F., et al. (1996). Using conjoint analysis to derive estimations of compensating variation. *Journal of Environmental Economics and Management, 31*(2), 145–159.

Roe, B., Teisl, M. F., Levy, A., Russell, M., et al. (2001). US consumers' willingness to pay for green electricity. *Energy Policy, 29*(11), 917–925.

Rondeau, D., Schulze, W. D., Poe, G. L., et al. (1999). Voluntary revelation of the demand for public goods using a provision point mechanism. *Journal of Public Economics, 72*(3), 455–470.

Rowlands, I., Scott, D., Parker, P., et al. (2003). Consumers and green electricity: Profiling potential purchasers. *Business Strategy and the Environment, 12*(1), 36–48.

Rundle-Thiele, S., Paladion, A., Apostol, S. A. G., et al. (2008). Lessons learned from renewable electricity marketing attempts: A case study. *Business Horizons, 51*, 181–190.

Scarpy, R., & Willis, K. (2010). Willingness-to-pay for renewable energy: Primary and discretionary choice of British households' for micro-generation technologies. *Energy Economics, 32*, 129–136.

Seung-Hoon, Y., & So-Yoon, K. (2009). Willingness to pay for green electricity in Korea: A contingent valuation study. *Energy Policy, 38*, 5408–5416.

Skiera, B., & Revenstorff, I. (1999). Auktionen als Instrument zur Erhebung von Zahlungsbereitschaften. *Zeitschrift für betriebswirtschaftliche Forschung, 51*(3), 224–242.

Smith, V. L. (1982). Microeconomic systems as an experimental science. *American Economic Review, 72*(5), 923–955.

Soroudi, A., & Amraee, T. (2013). Decision making under uncertainty in energy systems: State of the art. *Renewable and Sustainably Energy Reviews, 28*, 376–384.

Stübinger, E. (2005). *Ethik der Energienutzung—Zeitökologische und theologische Perspektiven.* Stuttgart: Kohlhammer.

Suchanek, A. (2007). *Ökonomische Ethik.* Stuttgart: UTB.

Sundt, S., & Rehdanz, K. (2015). Consumers' willingness to pay for green electricity: A meta-analysis of the literature. *Energy Economics, 51*, 1–8.

Sunderer, G. (2006). *Was hält Verbraucher vom Wechsel zu Ökostrom ab?* Trier: Universität Trier.

Tversky, A., & Kahneman, D. (1986). Rational choice and the framing of decisions. *The Journal of Business, 59*(4), 251–278.

Van der Horst, D. (2007). Exploring the relevance of location and the politics of voiced opinions in renewable energy siting controversies. *Energy Policy, 35*, 2705–2714.

Vickrey, W. (1961). Counter speculation, auctions and competitive sealed tenders. *Journal of Finance, 16*(1), 8–37.

Weiber, R., & Rosendahl, T. (1997). Anwendung der Conjoint-Analyse—Die Eignung conjointanalytischer Untersuchungsansätze zur Abbildung realer Entscheidungsprozesse. *Marketing ZFP, 19*(2), 107–118.

Weimann, J. (2009). *Wirtschaftspolitik—Allokation und kollektive Entscheidung* (5th ed.). Berlin: Springer.

von Weizsäcker, C. C. (2015). Adaptive Präferenzen und die Legitimierung dezentraler Entscheidungsstrukturen. In T. Apolte, M. Leschke, A. F. Michler, C. Müller, R. Schomaker,

& D. Wentzel (Eds.), *Behavioral Economics und Wirtschaftspolitik, Schriften zu Ordnungsfragen der Wirtschaft* (pp. 67–98). Stuttgart: Lucius & Lucius.

Wiser, R. H. (1998). Green power marketing: Increasing customer demand for renewable energy. *Utilities Policy, 7*(2), 107–119.

Winther, T., & Ericson, T. (2012). Matching policy and people? Household response to the promotion of renewable electricity. *Energy Efficiency, 6*(2), 369–385.

Wolsink, M. (2012). Undesired reinforcement of harmful 'self-evident truths' concerning the implementation of wind power. *Energy Policy, 48*, 83–87.

Wortmann, K., Klitzke, M., Lörx, S., Menges, R., et al. (1996). *Grüner Tarif. Klimaschutz durch freiwillige Beiträge zum Stromtarif*. Kiel: Energiestiftung Schleswig-Holstein.

Wricke, M., & Herrmann, A. (2002). Ansätze zur Erfassung der individuellen Zahlungsbereitschaft. *Wirtschaftswissenschaftliches Studium, 31*(10), 573–578.

Yang, Y. (2014). Understanding household switching behavior in the retail electricity market. *Energy Policy, 69*, 406–414.

Ziegler, A. (2012). Individual characteristics and stated preferences for alternative energy sources and propulsion technologies in vehicles: A discrete choice analysis for Germany. *Transportation Research Part A, 46*, 1372–1385.

Zorić, J., & Hrovatin, N. (2012). Household willingness to pay for green electricity in Slovenia. *Energy Policy, 47*, 180–187.

**Roland Menges** studied economics at the University of Kiel, where he also acquired his PhD in 1996 with a thesis in public finance, based on research in the field of experimental economics. Afterwards he worked for various research institutions in the field of energy and environmental policy. In 2006 he finished his post-doctoral lecturer qualification at the University of Flensburg. Since 2010 he is a full professor at the University of Technology Clausthal. His research fields are environmental and experimental economics.

**Gregor Beyer** holds a master's degree in business administration. He is a research assistant and PhD-student at the department of macroeconomics at Clausthal University of Technology, Clausthal-Zellerfeld, Germany. His research interest lies in consumer preferences for renewable energies and behavioural economics. In his PhD-thesis and in his recent research he evaluated energy efficiency policies and examines their behavioral effects.

# Direct Selling of Renewable Energy Products

Christian Friege

**Abstract**

For the distribution of renewable energy (RE) products (e.g., green power, heat from RE) direct selling is a very suitable strategy, especially as an element of multichannel distribution. Both product criteria (necessity of explanation, emotionality) and the business model (viability for sales commissions, win–win–win constellation) play a role in the choice of distribution model. Door-to-door selling, a very common classic sales technique, will be applied here—with good reason. The argument is supported by an example (green power) to illustrate how this can be implemented in practice.

**Keywords**

Direct selling • Direct sales • Multichannel distribution • Sales concept • Renewable energy

## 1 Problem

In a representative survey, TNS Infratest and the German Direct Selling Association (Bundesverband Direktvertrieb Deutschland, or BDD) recently identified that in the future consumers will be expected to buy more online but also in direct sales (BDD 2012). Such developments as anticipated by consumers suggest two implications for the commercialization of renewable energy (RE) products, which

A previous version of this chapter has been published in Herbes, C.; Friege, Chr. (Hrsg): Marketing Erneuerbarer Energien. Grundlagen, Geschäftsmodelle, Fallbeispiele, 2015, Springer Gabler.

C. Friege (✉)
Stuttgart, Germany
e-mail: cf@friege-consulting.de

usually are not offered in retail outlets. First, established direct sales of RE products will be further expanded. This is evident from a first look at the market in Germany, where for example the eco-energy provider Lichtblick labels its direct sales as a "key distribution channel" (Lichtblick 2009) and a number of other energy providers use direct selling for the distribution of green (and gray) power. Second, multichannel distribution strategies are becoming increasingly important, especially when online sales are combined with a form of personal selling. A good example of these multichannel strategies is the distribution of photovoltaic (PV) systems on a lease basis (Friege and Dharshing 2015).

Direct selling may have the potential to support marketing of RE products. Whether that is indeed the case, this paper will aim to answer the questions of what requirements must be met in detail and how direct sales of RE can support marketing. There will be an overview about basic concepts of direct selling and its incorporation into multichannel distribution strategies (Sect. 2), a detailed description, how RE products match the requirements for successful direct selling (Sect. 3), an example of RE marketing through direct sales (Sect. 4), and a perspective on future development (Sect. 5).

## 2      Basics of Direct Selling

## 2.1      Definition of Direct Selling

"Direct Selling is face-to-face selling away from a fixed retail location" (Peterson and Wotruba 1996, p. 2). Its key characteristics are therefore (1) direct and personal interaction between sales representative and customer (2) outside of a permanent sales outlet, for example, in a home, on the doorstep, or at a fair. Direct sales need to be distinguished from direct marketing. Direct marketing is "distance selling": direct, personal customer contact does not occur. Instead, direct marketing makes use of online stores, telemarketing, catalogs, direct response advertising, or e-mails, among others, to attract customers and to motivate them to make a purchase. Compared with direct marketing, the individual customer contact in direct sales is relatively expensive, despite the fact that mostly independent salespeople are involved.

It follows that direct sales tends to be a promising sales channel

– if the products to be distributed have a relatively high single value (e.g., Vorwerk[1] vacuum cleaners),

---

[1]The German company Vorwerk is a pioneer in direct selling. "Vorwerk's core business is the worldwide direct selling of high-quality products....When Vorwerk invented the Kobold hand-held vacuum cleaner in 1929, it was a technical sensation....Thanks to (direct) selling, the Kobold was soon a roaring success in Germany—so much so, in fact, that by 1936, Vorwerk was ready to set up its first international subsidiary: Vorwerk Folletto in Italy. Today, the turnover generated outside Germany was 66 percent, in direct sales it was even as much as 79 percent" (Vorwerk 2016).

- if there is a specific need for explanation (e.g., advantages and features of Tupperware), or
- if the product and its features can be particularly well tried out in a familiar environment (e.g., cosmetics products by Avon or Mary Kay).

RE products have two of these three characteristics. There is often a relatively high single value (e.g., PV systems). In addition, the products require specific explanations, both about the products themselves and about the extent to which the purchase is relevant for a reduction of $CO_2$ in the atmosphere. Clearly, during the first few years following the liberalization of the energy markets, this need for explanation triggered a direct sales approach for German green energy pioneer Lichtblick.[2] The lack of opportunity to try out RE products before purchasing can often be overcome through adequate marketing material.

In direct sales, three forms of distribution can be distinguished (Engelhardt and Jaeger 1998, p. 19ff.):

1. Doorstep selling, in which either single-stage (cold) doorstep selling is used without prior notification (e.g., Vorwerk vacuum cleaners) or two-stage doorstep selling, in which appointments are made prior to the visit at the door, mostly by telephone (e.g., wine sales). Sometimes, sales agents also use pop-up stands in malls or other public places or exhibit at consumer fairs;
2. Home service, in which goods are delivered out of a car with a prior reservation or spontaneously (like Eismann or Bofrost[3]); and
3. Home demonstration (often called a party plan), in which a host gives the sales representative the opportunity to present the goods to a group of friends/ acquaintances (e.g., Tupperware Party).

In all cases it must be ensured that a sufficient number of potential customers is geographically within reach in a reasonable time and at reasonable cost for most independent sales representatives.

For the distribution of RE products, home service obviously has no relevance. This does not necessarily apply to the party plan. However, while these have been tried from time to time with so-called electricity supplier change parties, the success of the method was generally so inadequate that a professional organization has never seriously approached this concept of direct selling in Germany. Important factors for successful home demonstrations, like entertainment value and fun factor, a presentation of different products that allows prospective buyers to try out products, and the likely request of at least one participant to do this "party" with a circle of his/her friends (sales lead), have so far for energy sales not been

---

[2]LichtBlick was a pioneer in the market in 1998 and today is Germany's largest independent green energy retailer (Lichtblick 2016).

[3]Bofrost and Eismann have been leading the German market for direct distribution of frozen food to the home for decades. Both are also active in other European countries.

established. For RE sales, there are few variations (e.g., green electricity, green gas, PV), no real "try-out," and lots of technical and factual information instead, with hardly any opportunity for games and entertainment. It remains to be seen whether there will be room for a party plan with new, innovative RE products in the future.

Classic doorstep selling clearly belongs in the repertoire of many energy providers. Not only Lichtblick but also other green electricity providers in Germany are selling their product using this distribution strategy: Naturstrom, Stadtwerke Iserlohn (Elementerra 2006), and Stadtwerke Stuttgart, among others, use the potential of direct sales exclusively for green electricity. Many other utilities also sell their green products in parallel with the fossil (gray) energy product, often using the same sales team.

Each of the three forms of direct selling focus simultaneously on selling the product, gathering leads to achieve more product sales, and attracting new salespeople. The result is a high growth potential through this sales channel, which can be implemented over several levels (multilevel[4]) relatively quickly and with relatively low investment and risk. This high growth potential with low risk is usually paired with above-average acquisition costs for each new customer, or cost per order (CPO). In addition, the sales team can be increased quickly, but at the same time it also tends to be rather unstable, as the sales representatives are independent and often only work part-time. It should be noted that the entire business model is not geared to the usual "win–win" constellation for supplier and customer but needs a "win–win–win" constellation, where in addition to supplier and customer the independent salespeople also need to show a profit. An additional challenge is the delicate balance that needs to be struck. On the one hand, the recruitment of additional partners through multiple levels of distribution as a driver of growth needs to be incentivized while on the other hand, the active distribution of the actual product needs to remain the primary objective in gaining sales. Creating an illegal so-called pyramid scheme should be avoided. The self-imposed standards of the German Direct Selling Association (BDD), which are revised regularly (BDD 2013) have proven very successful here as a guideline.

Finally, five tools essential to managing direct sales need to be introduced (Friege et al. 2013 p. 226f.; Friege 2016):

1. The *sales concept* fully describes what sales arguments and selling aids for what products will be used. Enforcing this concept is the precondition of successfully managing the sales organization, especially when new products or sales arguments are introduced.
2. The *remuneration system* is crucial to controlling the activities of the independent sales force. It needs to be designed so as to optimize product sales, partner recruitment, and building new leads. If, for example, the recruitment of new partners is financially more attractive than the product sale, the partners will focus primarily on the former. However, note that, especially for resellers who

---

[4]Often also referred to as multilevel marketing (MLM).

work part-time, social interaction with other members of a distribution unit is also of significant value for their motivation.

3. *Training* is crucial for the implementation of the sales strategy; additionally, the often complex remuneration system must be explained since both need to be effectively implemented.
4. *Controlling* both financial indicators and sales metrics is important for identifying opportunities to improve sales and sales team recruitment.
5. It is becoming increasingly important to integrate a *CRM system* into the sales process, if only because the client relationship should not exist solely with the sales partners and must be maintained even when the sales partner is inactive.

The BDD sees the 12 % growth of the industry in 2015 as a clear indicator of the viability of direct sales as a distribution channel (BDD 2016). The same applies to prospects worldwide, as the World Federation of Direct Selling Associations (WFDSA) reports in its 2016 annual report: in the period 2012–2015, the compound annual growth rate was +10.6 % in Asia Pacific, +4.8 % in the Americas, and +4.3 % in Europe—all well above inflation rates (WFDSA, 2016).

## 2.2    Direct Selling as an Element of Multichannel Strategies

Research conducted at the University of Mannheim (2014) shows that already 45 % of direct sales companies surveyed in Germany run some online business, and at least 22 % operate a flagship store. Hence, it can be assumed that multichannel distribution is important for direct sales companies. However, the specific challenges that this entails are not discussed in the literature. Risks and opportunities to pursue a multichannel distribution strategy in general (Fig. 1), however, are frequently a topic of discourse in business journals.

*Strategically*, a multichannel approach provides a company with the opportunity to achieve greater market coverage (Zhang et al. 2010; Schögel and Binder 2011). At the same time, there is the risk of underperforming in these additional channels compared to companies that are exclusively active there, the so-called pure plays (Schögel and Binder 2011, p. 184). For example, none of the online shops of bricks-and-mortar booksellers could ever really surpass the breadth of product (including used books) and delivery choices offered by Amazon. Whether a multichannel strategy in any case constitutes a competitive advantage is not clear, though there is potential for an "enduring competitive advantage" (Neslin and Shankar 2009, p. 73). This is particularly true if one considers the wide differentiation of purchasing behavior that is increasingly observed: information gathering, consultation, purchase decision, need for social interaction, communication with sellers—this is taking place with increasing frequency in different channels and thus gives companies operating several sales channels additional interactions and, thus, sales opportunities. Zhang et al. (2010, p. 169f.) in particular see the following three potentials in multichannel strategies: (1) access to new markets, which cannot be achieved with existing distribution channels, (2) increase in customer loyalty, and

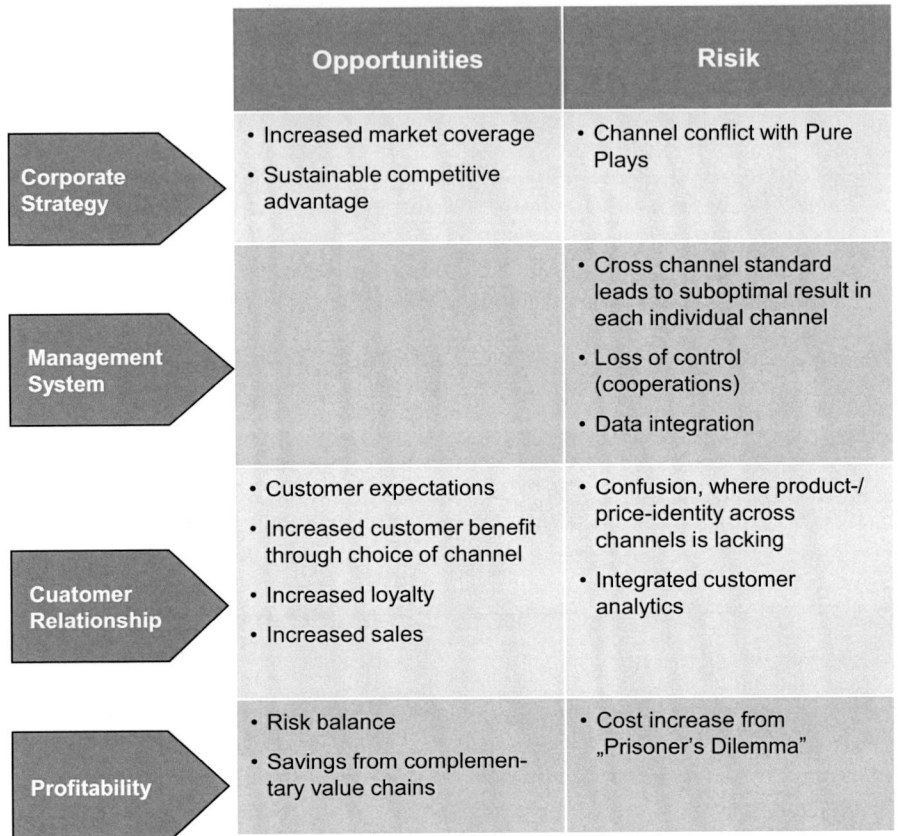

|  | Opportunities | Risik |
|---|---|---|
| **Corporate Strategy** | • Increased market coverage<br>• Sustainable competitive advantage | • Channel conflict with Pure Plays |
| **Management System** | | • Cross channel standard leads to suboptimal result in each individual channel<br>• Loss of control (cooperations)<br>• Data integration |
| **Cuatomer Relationship** | • Customer expectations<br>• Increased customer benefit through choice of channel<br>• Increased loyalty<br>• Increased sales | • Confusion, where product-/ price-identity across channels is lacking<br>• Integrated customer analytics |
| **Profitability** | • Risk balance<br>• Savings from complementary value chains | • Cost increase from „Prisoner's Dilemma" |

**Fig. 1** Opportunities and risks of a multichannel strategy [based on Neslin and Shankar (2009), Schögel and Binder (2011), Schögel et al. (2011), and Zhang et al. (2010)]

(3) strategic benefits from building and expanding the customer database and all related processes.

Multichannel strategies represent a challenge for *management systems*. First, because a uniform offer across all channels is a major requirement in multichannel distribution, this offer may end up being suboptimal in all channels (Schögel and Binder 2011, p. 184). Second, new channels are often operated together with partners, which can lead to a loss of control in these channels (Schögel and Binder, 2011). Finally, unified data management must be achieved across all channels, particularly with respect to customer data generated in different channels (Zhang et al. 2010, p. 172).

Regarding *customer relationships*, clearly a multichannel strategy—at least insofar as it relates to the inclusion of comprehensive online and social media coverage—is now expected by most customers. This is the case not only because the choice between different channels represents a benefit for customers (e.g.,

Schögel et al. 2011, p. 565f.). Significantly more important is that multichannel strategies lead to greater customer loyalty and higher revenues (Neslin and Shankar 2009, p. 72, with references therein). Such opportunities are reduced if a consistent experience across all channels cannot be offered (Schögel and Binder 2011), which in turn requires a seamless handover of customer data (Zhang et al. 2010).

All this should lead to higher *profitability* (e.g., Zhang et al. 2010). In addition to the previously discussed effects from increased loyalty and sales to the individual multichannel customer, benefits accrue from balancing risk across the different channels (Schögel and Binder 2011) as well as cost reduction possibilities when the value chains of the individual channels complement each other (Schögel and Binder 2011, p. 183). However, profitability is ultimately determined by the degree to which a company is forced by the competition's actions into additional channels and no sustainable competitive advantage could be established (Prisoner's Dilemma) (Neslin and Shankar 2009, p. 73).

In addition to the risks and benefits of a multichannel strategy in general as summarized in Fig. 1, some special considerations should be taken into account when using direct sales as a distribution channel. These apply both to established direct sales companies expanding into other channels and to companies actively running other distribution channels adding direct sales (Fig. 2). Establishing direct sales as an element of a multichannel strategy is very possible, but it requires skill and experience. It is especially important to apply the tools described in Sect. 2.1 (Definition of Direct Selling) in an optimal way. However, this also means that there will be limited competition—a unique position seems more attainable. For example, only a few of the more than 1100 energy distributors in Germany have imitated the doorstep sales model, which Lichtblick was so successful with. At the same time, virtually all sales organizations have an online presence.

**Fig. 2** Opportunities and risks of direct sales in a multichannel strategy

In addition, direct sales represents an active sales channel, where the amount of business can be controlled almost entirely. There is no dependency on advertising response, brand recognition (though that helps), or distribution partners; direct sales is a so-called push channel, which actively seeks out potential customers, and not a pull-channel, where the customer must be activated (e.g., through a Website). It can be expanded by adding additional sales representatives or an increased margin for the existing organization and will respond immediately. Therefore, direct sales is likely to compensate for a loss of control in other channels in the context of a multichannel strategy.

Nonetheless, direct selling is suitable only for certain products, especially those that need to be explained in a special way, have a certain monetary value (either as a single purchase or with recurring purchases as in electricity sales), and ideally can be tried out at home.

But above all, "perceived cannibalization" (Sharma and Gassenheimer 2009, p. 1076) is a risk encountered when establishing multichannel strategies with direct sales. Since the income of the sales force in direct sales depends entirely or largely on personal sales results, the salespeople regard it as a significant problem if the product can be purchased in a different channel and not exclusively from them as a result of the relationship they established with the customer. This concern is compounded by the traditional culture of direct sales as "single-channel distribution" (Friege et al. 2013, p. 227). Not only does the company then lack disciplinary means for independent distribution partners, but the sales representatives can walk away from the company on the spot. Hence, perceived cannibalization is a challenge in the deployment of a multichannel strategy encompassing direct sales.

Finally, exclusivity ("I can offer this product to you exclusively here and now and it is not available elsewhere") is an important argument for the conclusion of a sale in direct selling, which contradicts the principle of offer and price identity between channels. However, this risk can be countered through variations of the product (e.g., pack size, additional features, bundling, daily rate) in favor of the customer without significant impact on CPO.

In summary, a competitive multichannel offer including distribution through direct sales is an option that is likely to generate competitive advantages.

## 3 Direct Sales of Renewable Energy Products

### 3.1 Special Features of Direct Sales of RE Products

Why do customers choose RE products? The answer to this question contains information from a customer perspective on important aspects for the direct selling of RE products. Central to the motives of many customers, given economic viability, is a desire to contribute to a sustainable energy policy, to avoid $CO_2$, and to maintain a more sustainable lifestyle in general [e.g., overview in Herbes and Ramme (2014), p. 258ff., or regarding a willingness to pay for green energy, Herbes et al. (2015)]. Therefore, other than with conventional direct selling of products, in addition to product features, explanation of the environmental benefits for the customer plays a crucial role in the buying decision: The RE product is doubly in need of explanation and thus is particularly suitable for direct sales.

These characteristics of RE products need to be considered for direct sales (see also Friege and Herbes 2017):

1. The appropriate RE product requires a *double explanation*. In addition to the environmental value (see previous discussion), its basic properties must be explained and opportunities for questions must be allowed (for example, about a connection to a district heating network or a PV rooftop system), or the product can even be discussed (advice to save energy provided by the salesperson of a local green electricity supplier): There needs to be an added value from the selling process.
2. In addition, especially with RE products, often *conspicuous consumption* can be observed (e.g., Herbes and Ramme, 2014, p. 260), which can easily be supported in direct sales with appropriate information, such as by comparing $CO_2$ savings of the RE product with the equivalent car emissions.
3. The *lack of a live demonstration* of RE products can be compensated by appropriate sales aids (for example, looking at live data of an existing PV system through an app that is included with the offered PV solution).
4. Energy products are *low-involvement products*. Usually, consumers are hardly involved with supplier selection and the products are generally perceived as completely divorced from emotions. This is very different for typical direct selling products: cosmetics, wine, and lingerie, for example, are highly emotional and, hence, high-involvement products, and even the selection of a vacuum cleaner or a set of cooking pots will allow for more emotional involvement than electricity and heat or their production. But this lack of involvement can also be seen as an opportunity for direct sales. As a consequence, it is even more difficult to approach potential customers through direct marketing or advertising and other channels, and more often than not the results of such advertising campaigns are well below breakeven.
5. To a large extent, RE products are *credence goods*. Customers cannot try out green electricity before contracting, nor are they able to even notice after their purchase, using some sort of indicator, whether green energy is actually being delivered. The same applies to some extent to connections to district heating networks for the operation of "combined heat and power" plants or PV systems. Although you can check the heat in the house, it cannot be proven whether in fact a renewable source of biomethane is driving the cogeneration plant, whether only sustainably generated heat is being fed into the network, or whether the PV modules will work for 20 years. All this can be found out by the customer only with great difficulty, and some may find out very late or not at all. Against this background, the direct sales of RE products must—unlike with usual direct sales, where the product can be demonstrated and tested—develop indicators that generate and strengthen trust and credibility.

6. RE products will always *require some form of multichannel distribution* to facilitate the necessary explanations of technical details, ecological advantages, and determination of customer needs (including technical data) in detail. All this requires a comprehensive Internet presence in addition to personal explanations by the salesperson. It will also be advisable to not only rely on the consumer to collect the necessary data to calculate a detailed offer, but also check some of the data supplied in person as part of the sales process. Finally, this multichannel presence also makes it possible to address different customer groups in different stages of the purchasing process, for which—as a result of buying habits or geographic coverage—a single sales channel alone is insufficient.

## 3.2 Checklist for the Successful Direct Sales of RE Products

Ultimately and as a kind of summary of the argument made so far, the fundamental suitability of an RE product for direct sales can be determined on the basis of ten criteria (Table 1).

The way an existing RE product (green power) can be distributed using a multichannel strategy, including direct sales, is presented in the following section.

## 4 Green Electricity in Direct Sales

When applying the checklist shown in Table 1 to a common green electricity product that is to be sold in direct sales, a product solution might look like that outlined in Table 2.

While electricity per se does not have to be explained, it may be a "door opener" to offer—consistent with the sustainable positioning of the product—individual advice to save electricity (criterion 1). This conveys competence, provides the good feeling of profiting regardless of the purchasing decision, and gives potential customers free and useful information without making them feel pressured to buy. Greatly in need of explanation are the environmental benefits of the offer (criterion 2). Many companies in Germany use the image of the "power lake" here, where both green and fossil/gray electricity are injected. The more people order green energy, the more green energy will be fed into the lake and the greener it will become overall. To explain this and other features, some charts and diagrams should be used as sales aids, especially since green energy is not a product for which the unique selling proposition can be demonstrated or that the prospective customer can try out during a presentation, as is the case with many other products sold in direct sales (criterion 4). The complexity of the product is both the challenge for the sales person and the reason why it can be sold successfully in direct sales. Thus, it becomes crucial to emotionalize the product (criterion 7), which can be achieved by, for example, connecting the environmental benefits to the prospective buyer ("What planet do you want to leave to the next generation?") or by telling stories (true) and (recognizable as such) fairytales (too good to be true). Finally, the

**Table 1** Checklist for successful distribution of RE products

|     | Criteria | √ |
|-----|----------|---|
| 1.  | RE product in itself requires explanation | |
| 2.  | Ecological benefits of RE product requires explanation | |
| 3.  | Pricing of RE product (once or as an ongoing obligation) allows for appropriate commission payment | |
| 4.  | Lack of live demonstration can be compensated by sales aids | |
| 5.  | Distribution can be organized as a traditional (doorstep) sale through sales representatives | |
| 6.  | Geographic distribution of potential customers allows for people-based distribution | |
| 7.  | RE product can evoke strong emotional reactions | |
| 8.  | Perceived risk of credence good can be minimized | |
| 9.  | Business model supports win–win–win constellation | |
| 10. | Direct sales of RE products can be integrated into multichannel distribution | |

**Table 2** Checklist for green electricity in direct sales

|     | Criteria | Green electricity product | |
|-----|----------|---------------------------|---|
| 1   | RE product in itself requires explanation | Electricity does not need to be explained—yet; saving energy is a good door opener | √ |
| 2   | Ecological benefits of RE product requires explanation | Using simple imagery like the "lake" helps explain the physical basics of green electricity | √ |
| 3   | Pricing of RE product (once or as an ongoing obligation) allows for appropriate commission payment | Customer value allows for commission | √ |
| 4   | Lack of live demonstration can be compensated by sales aids | Charts and diagrams to support points 1 and 2 | √ |
| 5   | Distribution can be organized as a traditional (doorstep) sale through sales representative | Doorstep selling (single-stage/two-stage) | √ |
| 6   | Geographic distribution of potential customers allows for people-based distribution | Yes | √ |
| 7   | RE product can evoke strong emotional reactions | Not so much the product itself, but very much so the environmental value | √ |
| 8   | Perceived risk of credence good can be minimized | Certificates, guarantees, brand, and reputation of seller | √ |
| 9   | Business model supports win–win–win constellation | Yes | √ |
| 10  | Direct sales of RE products can be integrated into multichannel distribution | Yes, direct selling, Internet, and other distribution channels complement each other to differentiate the commodity "electricity" | √ |

risk associated with the purchase of credence goods must be minimized (criterion 8), which for green electricity is possible through

1. certificates, awarded by independent bodies, to attest the origin of the electricity;
2. labels, developed by the seller and not by an independent agency, to signal the quality of the product (e.g., an emblem "Genuine Green Electricity" attached to the sales material or displayed on a company Website, for example);
3. guarantees issued by the provider or a third party;
4. the positioning and brand of a seller that enjoys such a superior reputation that further certificates or guarantees are unnecessary. This is the strongest possible endorsement for a credence good, and the four established green electricity providers in Germany (EWS Schönau, Greenpeace Energy, Naturstrom, and Lichtblick), which have by and large an excellent reputation in the market, seem to be moving in this direction.

Direct selling of green electricity allows for a business model that has proven viable for a number of suppliers. The customer value is higher than the CPO (criterion 3), and a win–win–win constellation can be established (criterion 9), assuming that a sales agent is active during the whole month in order to achieve an income of more than € 2,000. Direct selling of green energy is always organized as a classic doorstep sale. The density of potential customers, especially in urban areas, will likely even make it possible to organize distribution without providing cars for the sales organization (criteria 5 and 6).

Finally, suppliers of green electricity usually operate Websites, other means of online marketing, and often further distribution channels and thus run a multichannel distribution. This multichannel distribution also contributes to the differentiation of green electricity from other commodities offered on the market (Friege and Herbes, 2017). For product/price combinations that are offered exclusively in direct sales, the requirement of product identity and price identity between channels can be partially neglected. Since offers exclusive to direct sales are not transparent to other customer groups, selling, for example, an exclusive bundle (e.g., adding some sort of insurance policy) or offering additional benefits (e.g., no base fee for the first three months) will not harm the multichannel strategy and will at the same time support the direct sales channel.

## 5    Summary and Outlook

It turns out that direct sales is a viable distribution channel for RE products, particularly in the context of multichannel marketing strategies. For successful implementation, however, the product and the marketing must meet certain requirements, as discussed in detail in this chapter. Generally speaking, it can be assumed that the importance of direct sales for the marketing of RE products will continue to increase in the future:

1. Where the development of RE markets moves out of the niche of experts and early adopters—and that obviously happens with deeper penetration of these markets—further growth will require more explanations of the ecological added value. This is a core focus of direct sales.
2. Many energy suppliers are currently developing strategies to introduce more complex products on the market, converting consumers into prosumers, where the boundary between energy production and consumption becomes fluid. Here significant additional benefits can differentiate such products, and direct selling is one of several options to organize the distribution. Direct sales are particularly suited to drive complex products.
3. Finally, new companies will enter the energy market, also with pure RE products or those very close to RE. Already, product offerings from companies such as Alphabet Inc. (parent company of Google) are available on the market, with Nest Labs selling a smart home product that competes directly with proprietary products of utilities or the energy industry in general. For these innovative product categories, it remains an open question which distribution channels will be successful in the future, but at this point, direct sales is at least worth considering.

In summary, RE products can be sold through direct sales as a channel within a multichannel strategy. The checklist of ten criteria proposed in this paper will help to assess the suitability of RE products for direct sales. In the future, it can safely be assumed, that additional and more innovative RE products will find customers through direct sales.

## References

BDD. (2012). Multi-Channel-Strategien—Direktvertrieb gewinnt weiter an Bedeutung. Accessed August 30, 2014, from http://www.direktvertrieb.de/News-detail.241.0.html?&tx_ttnews%5Btt_news%5D=581&cHash=7225e4e6060c7c568e068860edaf4f49.

BDD. (2013). BDD_Mitgliedsunternehmen erhöhen Verbraucherschutz. Accessed August 30, 2014, from http://www.direktvertrieb.de/News-detail.241.0.html?&tx_ttnews%5Btt_news%5D=726&cHash=b382c5e255ddca048f5f0348bc1c88cb.

BDD. (2016). BDD Mitglieder steigern Umsatz um 12 Prozent. Accessed May 11, 2016, from http://www.direktvertrieb.de/News-detail.241.0.html?&tx_ttnews%5Btt_news%5D=977&cHash=f04b4d712f06e5cde0cd6c754dcfbeef.

Elementerra. (2006). Elementerra beweist erfolgreiche Vermarktung von Stadtwerke-Produkten durch Direktvertrieb. Accessed August 30, 2014, from http://www.finanznachrichten.de/nachrichten-2006-03/6127152-elementerra-beweist-erfolgreiche-vermarktung-von-stadtwerke-produkten-durch-direktvertrieb-007.htm.

Engelhardt, W. H., & Jaeger, A. (1998). *Der Direktvertrieb von konsumtiven Leistungen— Forschungsprojekt im Auftrag des Arbeitskreises.* Bochum: Gut beraten—zuhause gekauft.

Friege, C. (2016). *Der Direktvertrieb in Mehrkanalstrategien.* Wiesbaden: Springer Gabler.

Friege, C., & Dharshing, S. (2015). Vivint Solar: Expanding into a European Market Governed by Different Regulation. University of St. Gallen Case Study Series. The Case Centre Ref. No. 315-211-1 (Case) No. 315-211-8 (Teaching Note).

Friege, C., & Herbes, C. (2017). Marketing renewable energy—Fundamental concepts. In C. Herbes & C. Friege (Eds.), *Marketing renewable energy.* Heidelberg: Springer.

Friege, C., Kraus, F., & Sahin, E. (2013). Direktvertrieb. *Wirtschaftswissenschaftliches Studium (WiSt), 42*(5), 224–230.

Herbes, C., Friege, C., Baldo, D., & Mueller, K.-M. (2015). Willingness to pay lip service? Applying a neuroscience-based method to WTP for green electricity. *Energy Policy, 87,* 562–572.

Herbes, C., & Ramme, I. (2014). Online marketing of green electricity in Germany—A content analysis of providers' websites. *Energy Policy, 66,* 257–266.

LichtBlick. (2009). LichtBlick-Direktvertrieb weiter auf Wachstumskurs. Medien-Mitteilung vom 18. März 2009. Accessed August 29, 2014, from http://www.lichtblick.de/medien/news/?detail=140&type=press.

LichtBlick (2016): About Us. Accessed July 16, 2016, from https://www.lichtblick.de/en/schwarmdirigent/about.

Neslin, S. A., & Shankar, V. (2009). Key issues in multichannel customer management: Current knowledge and future directions. *Journal of Interactive Marketing, 23,* 70–81.

Peterson, R. A., & Wotruba, T. R. (1996). What is direct selling?—Definition, perspectives and research agenda. *Journal of Personal Selling & Sales Management, 16*(4), 1–16.

Schögel, M., & Binder, J. (2011). Profitables Channel Management. In C. Belz (Ed.), *Innovationen im Kundendialog* (pp. 177–195). Wiesbaden: Gabler.

Schögel, M., et al. (2011). Multi-channel management im CRM. In H. Hippner et al. (Eds.), *Grundlagen des CRM* (pp. 559–597). Wiesbaden: Gabler.

Sharma, D., & Gassenheimer, J. B. (2009). Internet channel and perceived cannibalization. *European Journal of Marketing, 43*(7/8), 1076–1091.

University of Mannheim. (2014). Situation der Direktvertriebsbranche in Deutschland 2013— Studie im Auftrag des Bundesverband Direktvertrieb Deutschland e.V.; Zusammenfassung der Ergebnisse. Accessed August 31, 2014, from http://www.direktvertrieb.de/index.php?eID=tx_nawsecuredl&u=0&file=fileadmin/user_upload/MAIN-dateien/Kurzfassung_BDD_Marktstudie_2014.pdf&t=1409582296&hash=b69c53804abdb6c0a7f5bc9ac109a26806e746e4.

Vorwerk. (2016). Vorwerk Direct Sales. Accessed July 16, 2016, from http://corporate.vorwerk.de/en/portrait/direct-sales/.

WFDSA. (2016). Annual Report 2016. Accessed July 16, 2016, from http://wfdsa.org/download/advocacy/annual_report/WFDSA-Annual-Report-2016.PDF.

Zhang, J., et al. (2010). Crafting integrated multichannel retailing strategies. *Journal of Interactive Marketing, 24,* 168–180.

**Christian Friege** holds a doctorate in business administration from Catholic University Eichstaett. He was CEO of Lichtblick AG, Germany's leading green energy provider until 2012 and consulted in the field thereafter. Before joining Lichtblick he was Chief Customer Officer on the board of debitel AG and Chief Executive of BCA (a Bertelsmann subsidiary) in the UK. Currently, he is Chief Sales Officer on the board of CEWE Stiftung & Co KGaA. Throughout his managerial career he has always maintained a strong research interest in the field of marketing and sales. His academic publications have appeared in Journal of Service Research, Marketing Review St. Gallen and Energy Policy. Most recently he co-edited books in German on Marketing Renewable Energy and Financing Green Energy Projects together with Prof. Dr. Carsten Herbes.

# Market Segmentation for Green Electricity Marketing Results of a Choice-Based Conjoint Analysis with German Electricity Consumers

Stefanie Lena Hille, Andrea Tabi, and Rolf Wüstenhagen

**Abstract**

Consumers have the power to contribute to creating a more sustainable future by subscribing to green electricity tariffs. To reach consumers "beyond the eco-niche," identifying the drivers that positively influence the adoption of green electricity is of fundamental importance. This chapter examines various factors that help to explain the extent to which green electricity subscribers differ from those who display strong preferences toward green electricity but have not yet "walked the talk." By making use of a latent class segmentation analysis based on choice-based conjoint data, this chapter identifies three groups of potential green electricity adopters with varying degrees of preference for renewable energy. Findings indicate that sociodemographic factors play a marginal role in explaining the differences between green electricity subscribers and potential adopters, with the exception that actual adopters tend to be better educated. Analysis of psychographic and behavioral features reveals that adopters tend to perceive consumer effectiveness to be higher, tend to estimate lower prices for green electricity tariffs, are willing to pay significantly more for other eco-friendly products, and are more likely to have recently changed their electricity contract than nonadopters.

A previous version of this chapter has been published in Herbes, C.; Friege, Chr. (Hrsg): Marketing Erneuerbarer Energien. Grundlagen, Geschäftsmodelle, Fallbeispiele, 2015, Springer Gabler.

S.L. Hille (✉) • A. Tabi • R. Wüstenhagen
Institute for Economy and the Environment (IWÖ-HSG), Good Energies Chair for Management of Renewable Energies, University of St. Gallen, Tigerbergstrasse 2, 9000 St. Gallen, Switzerland
e-mail: stefanie.hille@unisg.ch

© Springer International Publishing AG 2017
C. Herbes, C. Friege (eds.), *Marketing Renewable Energy*, Management for Professionals, DOI 10.1007/978-3-319-46427-5_5

**Keywords**

Market segmentation • Choice-based conjoint analysis • Green electricity •
Latent class analysis • Green marketing

# 1    Introduction

With the liberalization of the electricity market in 1998, German residential
customers were given the opportunity to freely choose their preferred electricity
provider and electricity product from among several competitors. However, despite
the fact that recent consumer research in Germany shows that many German
citizens have strong preferences for renewable energy sources (Gerpott and
Mahmudova 2010; Kaenzig et al. 2013), the share of green electricity consumers
is still in the single-digit percentage range (Litvine and Wüstenhagen 2011). In
other words, even when consumers demonstrate a strong preference for green
electricity, they are largely passive when it comes to purchasing decisions.

The residential sector accounts for almost 25 % of total final electricity consumption in Germany (BMU 2009). Germany generated about 16 % of this amount
from renewable sources in 2009, with the share increasing to 23 % in 2012 (BMWI
2013). Generating electricity from renewable energy sources is fundamentally
important in a sustainable and secure energy system (Madlener and Stagl 2005)
and helps reduce dependence on foreign energy sources and hedges the risk
associated with movements in oil and natural gas prices.

In this regard, consumers have the power to express their desire for a more
sustainable future by subscribing to a green electricity tariff (Diaz-Rainey and
Ashton 2011). Identifying the consumer segments that are most receptive to
purchasing green electricity, exploring what product features customers in those
segments value most, and analyzing what distinguishes them from consumers that
have already switched to green electricity are fundamentally important if targeted
messaging is to be developed by policymakers and marketers that will reach
consumers "beyond the eco-niche" (Villiger et al. 2000).

To contribute to the existing literature in this field, the approach of a choice-
based conjoint (CBC) analysis was chosen to indirectly elicit the preferences of a
representative sample of the German population for electricity product attributes.
Based on CBC data, a latent class approach to market segmentation is then followed
by capturing market heterogeneity in attribute preferences across a full set of
attributes in order to identify segments with similar preferences (Desarbo et al.
1995).

Marketers can use the insights of this study to more effectively satisfy
consumers' needs by developing more effective and focused marketing strategies.
Another goal of the research is to identify a series of sociodemographic, psychographic, and behavioral factors that allow a distinction to be made between
consumers who have already subscribed to a green electricity tariff and those

who display strong preferences for a green electricity product but have not yet "walked the talk" (Litvine and Wüstenhagen 2011).

## 2    Related Research

Many studies in green marketing have attempted to define the characteristics of green consumers for segmentation purposes. In the marketing literature, these factors are often classified into the categories of geographic characteristics (e.g., geographic region), demographic and socioeconomic characteristics (e.g., age, sex, and household size), psychographic characteristics (e.g., values, lifestyle, and personality variables), and behavioral variables (e.g., purchase occasion) (Kotler and Keller 2006).

With regard to sociodemographic factors, several studies have focused on investigating the characteristics of potential consumers of green electricity who claim to be willing to pay a premium for green electricity. For example, several authors (Rowlands et al. 2003; Zarnikau 2003; Ek and Soderholm 2008; Diaz-Rainey and Ashton 2011) showed that earning a higher income tends to increase reported willingness to pay (WTP) a premium for green electricity. Gerpott and Mahmudova (2010) positively correlate household size with WTP for green electricity. In addition, several authors have claimed that WTP for green electricity tends to be positively correlated with a higher level of education (Rowlands et al. 2003; Zarnikau 2003; Wiser 2007; Ek and Soderholm 2008). Regarding other variables, Rowlands et al. (2003) showed that gender did not significantly explain a higher WTP for green electricity. Finally, several studies have concluded that younger consumers are also more willing to pay a higher premium for green electricity (Zarnikau 2003; Gerpott and Mahmudova 2010).

Although these briefly summarized findings are far from conclusive, they do indicate that sociodemographic variables have an influence on WTP for green electricity and offer an easy approach to segmenting the market. However, most authors agree that psychographic and behavioral characteristics are particularly important in explaining environmentally friendly behavior (Straughan and Roberts 1999).

In this connection, Diaz-Rainey and Ashton (2011) have pointed out that knowledge about green electricity has an important influence on preferences for green electricity. In addition, customers with a high price tolerance for green electricity have been characterized as having positive attitudes toward green electricity (Hansla et al. 2008), toward environmental protection (Gerpott and Mahmudova 2010; Rowlands et al. 2003), and greater concern for environmental problems (Rowlands et al. 2003).

In addition, perceived consumer effectiveness in an environmental context (i.e., the extent to which consumers think their own behavior might help to preserve the environment) has been shown to correlate with a stronger preference for green electricity (Rowlands et al. 2003).

In terms of behavioral variables, both Gerpott and Mahmudova (2010) and Wiser (2007) showed that consumers who are involved in pro-environmental activities have a higher willingness to pay for green electricity. Another relevant factor is whether a respondent has undergone a significant life event or change in status such as a divorce or relocation. For instance, Arnold (2011) determined that whether a respondent has relocated is (positively) correlated with their WTP a surcharge for sustainable products.

In contrast with research that has profiled potential adopters of green electricity, research into the profiles of subscribers to green electricity tariffs (Adopters) is relatively scarce; only a few studies have attempted to explore this topic so far (Rose et al. 2002; Clark et al. 2003; Arkesteijn and Oerlemans 2005; Kotchen and Moore 2007).

For example, Clark et al. (2003) found that participants in green electricity programs tend to have higher incomes and fewer members in their households than consumers that have not opted to purchase green electricity. In contrast, Kotchen and Moore (2007) found that demographic variables were not statistically significant in explaining adoption, although attitudinal factors such as environmental concern had a positive effect. In contrast, MacPherson and Lange (2013) found that respondents with incomes in the highest income quartile, those with higher levels of education, those who supported the Green Party, and those who exhibited strongly pro-environmental behavior were all more likely to have signed up to be supplied with green electricity. Moreover, Arkesteijn and Oerlemans (2005) found a negative correlation between the perceived difference in the price of gray and green power and the probability of adoption of green electricity. This finding is in line with a study by Clausen (2008) that found that green electricity buyers in Germany were overestimating the price of green electricity fourfold, whereas nonadopters were on average overestimating the real price by up to ten times. Previously, green electricity was typically sold at a higher price than conventional energy. Although the real price difference between green electricity and conventional power has significantly decreased over the last decade in Germany, consumers who have not opted for green electricity might still be implicitly assuming that electricity generated from renewable energy sources costs significantly more. This inaccurate perception of cost may be a component of the attitude–behavior gap, in that the perceived price difference results in an obstacle to the acceptance and uptake of the environmentally friendly product alternative.

# 3    Study Design

## 3.1    Investigating Stated Preference with Choice-Based Conjoint Analysis

This study makes use of a CBC analysis. This approach is very suitable for examining preferences for hypothetical products or attribute combinations when it is not possible to observe actual purchasing behavior or to measure preferences through revealed preference methods (Ewing and Sarigöllü 2000).

More precisely, this method simulates a real buying situation for respondents, where a choice must be made between several products. These products differ in their attributes, and respondents are required to choose one "package" from the choice set (Sammer and Wüstenhagen 2006). Consumer preferences for product attributes are implicitly derived from the stated choices through indirect questions.

CBC analysis has been used by several scholars to identify the attributes of an electricity product that are important to consumers (e.g., Kaenzig et al. 2013; Cai et al. 1998; Goett et al. 2000; Burkhalter et al. 2009). For example, a recent study by Kaenzig et al. (2013), on which this chapter builds, investigated the relative importance of different product attributes in the purchasing choices of German households. They found that price and electricity mix were the two most important attributes for the average customer, followed by the location of electricity generation, a price guarantee, certification with an eco-label, type of power provider (e.g., municipal utility or major national provider), and the terms of cancellation of the contract (notice period). Research by Burkhalter et al. (2009) revealed similar findings for the Swiss market. Swiss consumers also considered the electricity mix to be the most important attribute, followed by monthly electricity costs and the location of electricity generation. Other attributes, such as the electricity supplier, pricing model, eco-certification, and contract duration, only played a subordinate role. Rowlands et al. (2003) showed that price, reliability of power supply, and environmental features were the most important factors influencing the choice of a power supplier. Goett et al. (2000) also found that customers were vitally concerned about the provision of renewable energy. A recent study into preferences for electricity attributes in Germany highlighted that the most important product attributes for German customers, besides price and price guarantee, were that the energy provider should invest in renewable energy sources and that the power should be regionally generated (Mattes 2012).

## 3.2    Market Segmentation

Consumers often have diverse preferences. To effectively target the market, heterogeneity of buyer preferences should be considered in order to identify promising market segments. More than three decades ago two main approaches to market segmentation were identified (Wind 1978; Green 1977). With a priori segmentation respondents are classified into groups on the basis of demographic or socioeconomic variables; using post hoc segmentation respondents are clustered according to a set of interrelated variables (e.g., preferences associated with a product). With conjoint analysis such a post hoc segmentation approach to market segmentation can be followed by capturing market heterogeneity in attribute preferences across a full set of attributes in order to obtain segments with similar preferences (Desarbo et al. 1995). By knowing which sociodemographic, psychographic, and behavioral variables predominate in a segment, marketers are able to define marketing strategies that more closely match consumer needs.

## 3.3    Methodology

A preexisting data set of a representative sample of German electricity customers was used for the present study (Kaenzig et al. 2013). These data are a subsample of a larger representative consumer survey among 1257 German households that was carried out in June 2009 through the project seco@home. Respondents were surveyed using the computer-assisted personal interview (CAPI) method at the respondents' homes. This research makes use of a data set that is based on 4968 choice observations by a total of 414 respondents.

The survey design, the data collection process, and the sample are described in detail in Kaenzig et al. (2013). The choice experiment was set up in such a way that respondents received a series of 12 choice tasks involving comparisons of different electricity products with varying levels of attributes. By making use of a full-profile design, each choice task required respondents to choose from three different electricity products defined according to seven (randomly allocated) attributes. Data collected about the results of the choices were then used as input for a hierarchical Bayesian analysis, which allowed the estimation of part-worth utilities at the individual level. Kaenzig et al. (2013) show that the source of the energy and monthly electricity costs are the two most important decision-making attributes for the average electricity consumer. In addition, the authors reveal an implicit average WTP a premium of about 16 % for electricity from renewable sources. Instead of reporting on only one model of preferences that was built for all the respondents, the focus of the present study is to go a step further in the analysis of the same data set by identifying, via a latent class approach, the preferences and characteristics of different market segments for potential green electricity consumers. For this purpose, the following different sociodemographic, psychographic, and behavioral variables were selected (Table 1).

**Table 1** Variables selected

| Variable | Description |
|---|---|
| Demographics | *Gender* (1 = male, 2 = female) |
| | *Age* (years) |
| | *Household monthly net income* (1 = under 1000 euros, 2 = 1000–1499 euros, 3 = 1500–1999 euros, 4 = 2000–2499 euros, 5 = 2500–3499 euros, 6 = more than 3500 euros) |
| | *Education* (1 = no formal education, 2 = primary school, 3 = secondary school, 4 = polytechnical secondary school, 5 = college, 6 = high school, 7 = university degree) |
| | *Household size* (number of people living in household) |
| Psychographic and behavioral characteristics | *Relocation* (relocation within last 5 years (yes/no) |
| | *Switching of electricity tariff during last 5 years* (yes/no) |
| | *The following variables were chosen to measure sensitivity to environmental issues:* |
| | *Climate concern*: aggregate level of dis/agreement with the following statements (1: agree, 2: neutral, 3: disagree): |
| | – Humans are solely responsible for any effects of climate change that occur |
| | – As a consequence of climate change the quality of life of the population will worsen |
| | – Climate change threatens the livelihoods of humankind |
| | – There are no serious consequences from climate change |
| | *Support for eco-taxes and regulatory tools*: level of dis/agreement with the following statement (1: agree, 2: neutral 3: disagree): |
| | – Environmental protection should be ensured through the introduction of mandatory eco-taxes and other legislation |
| | *Trust in science*: level of dis/agreement with the following statement (1: agree, 2: neutral 3: disagree): |
| | – Science and technology will solve many environmental problems without requiring changes in our way of living |
| | *Perceived consumer effectiveness*: level of dis/agreement with the following statement (1: agree, 2: neutral 3: disagree): |
| | – As citizens we are able to contribute significantly to protecting the environment through our purchasing behavior |
| | *Awareness of green electricity labels*: TÜV, Grüner Strom Label, ok power (1: no knowledge of any labels 2: knowledge of at least one label, 3: knowledge of at least two labels) |
| | *Estimation of cost of green electricity*: what is your estimate of the cost of green electricity compared to conventional electricity? [1: much more expensive (more than 10 %), 2: slightly more expensive (up to 10 %), 3: same price, 4: slightly cheaper (up to 10 %), 5: much cheaper (more than 10 %)] |
| | *Willingness to pay (WTP) for eco-friendly products*: willing to purchase environmentally friendly everyday products (dichotomous, yes/no) |

**Table 2** Mean utility values for five segments

| | | Potential adopters | | | |
|---|---|---|---|---|---|
| | Adopters | Truly Greens | Price-sensitive greens | Local patriots | Likely nonadopters |
| Segment size | $n = 29$ | $n = 117$ | $n = 78$ | $n = 108$ | $n = 82$ |
| *Electricity mix* | | | | | |
| Mix 1 (60 % coal, 25 % nuclear power, 15 % unknown origin) | −179.0 (51.7)[a] | −179.8 (38.4) | −135.4 (35.2) | −25.9 (67.3) | −26.7 (45.7) |
| Mix 2 (60 % coal, 25 % nuclear power, 5 % hydropower, 5 % wind, 5 % biomass) | −105.69 (44.3) | −114.7 (26.4) | −67.6 (29.9) | −3.5 (47.1) | −8.7 (33.7) |
| Mix 3 (60 % coal, 25 % natural gas, 5 % hydropower, 5 % wind, 5 % biomass) | −4.08 (35.2) | −7.3 (34.3) | 31.4 (36.9) | 16.4 (45.6) | 8.3 (30.6) |
| Mix 4 (50 % wind, 30 % hydropower, 15 % biomass, 5 % solar) | 147.79 (54.8) | 141.0 (37.9) | 87.0 (36.0) | 10.6 (54.5) | 10.5 (38.3) |
| Mix 5 (100 % wind) | 140.99 (70.8) | 160.8 (42.9) | 84.7 (33.8) | 2.5 (67.9) | 16.8 (49.0) |
| *Power provider* | | | | | |
| Big, national provider | −6.56 (10.3) | −7.2 (10.8) | −7.0 (12.3) | −1.2 (19.7) | −2.8 (12.6) |
| Medium-sized, regional provider | 3.16 (15.6) | 4.0 (15.1) | −0.2 (14.6) | 2.4 (25.1) | 0.4 (15.2) |
| Municipal | 5.27 (13.6) | 4.7 (14.7) | 7.8 (20.1) | 3.4 (25.6) | 0.7 (17.3) |
| Specialized provider | −1.87 (15.1) | −1.5 (15.6) | −0.6 (15.6) | −4.6 (23.4) | 1.6 (17.1) |
| *Location of electricity generation* | | | | | |
| Local region | 16.05 (17.5) | 21.0 (21.8) | 21.0 (22.9) | 54.3 (38.3) | 16.9 (18.2) |
| Germany | 19.98 (16.7) | 18.7 (19.9) | 21.8 (23.2) | 53.4 (40.8) | 14.8 (19.7) |
| Switzerland | −3.76 (17.3) | −5.0 (20.6) | −10.4 (24.0) | −37.9 (42.0) | −11.8 (20.8) |
| Eastern Europe | −32.28 (26.5) | −34.7 (28.2) | −32.5 (30.2) | −69.8 (44.5) | −19.9 (24.9) |
| Monthly electricity costs | −7.98 (5.6) | −6.7 (3.2) | −11.1 (2.7) | −7.9 (4.1) | −18.83 (3.9) |
| *Certification* | | | | | |
| Ok power | 4.18 (16.5) | 1.1 (11.5) | 1.4 (14.0) | 2.5 (21.2) | −4.1 (13.4) |
| TÜV | 0.91 (12.7) | 3.1 (13.1) | 7.1 (13.9) | 4.6 (22.3) | −0.5 (12.5) |
| Grüner strom label | 2.72 (12.9) | 1.7 (12.7) | 6.6 (11.8) | 9.8 (19.0) | 11.3 (12.4) |

(continued)

**Table 2** (continued)

| | | Potential adopters | | | |
|---|---|---|---|---|---|
| | Adopters | Truly Greens | Price-sensitive greens | Local patriots | Likely nonadopters |
| No certification | −7.81 (14.9) | −5.9 (16.8) | −15.1 (16.7) | −16.9 (23.9) | −6.7 (18.2) |
| *Price guarantee* | | | | | |
| None | −10.04 (16.9) | −12.0 (14.1) | −21.3 (18.0) | −32.0 (28.1) | −25.0 (18.4) |
| 6 months | −5.57 (13.3) | −1.8 (13.9) | −0.3 (14.9) | 3.6 (23.0) | 1.5 (14.3) |
| 12 months | 9.35 (13.9) | 6.4 (15.3) | 9.1 (15.9) | 5.7 (27.0) | 10.9 (13.9) |
| 24 months | 6.26 (13.8) | 7.4 (17.1) | 12.6 (15.8) | 22.6 (29.3) | 12.6 (19.6) |
| *Cancellation period* | | | | | |
| Monthly | 4.66 (13.1) | 2.5 (14.3) | 4.4 (17.8) | 9.1 (25.1) | 6.2 (17.1) |
| Quarterly | 4.54 (10.0) | 4.2 (13.3) | −4.1 (14.6) | 0.3 (23.9) | −3.5 (11.9) |
| Bi-annually | −3.86 (14.1) | −0.9 (15.1) | 3.6 (15.5) | −5.3 (23.4) | −0.3 (14.3) |
| Yearly | −5.34 (11.8) | −5.8 (13.4) | −4.0 (17.2) | −4.1 (26.5) | −2.4 (14.7) |
| None (would not buy) | 158.30 (119.6) | 115.8 (106.3) | 61.8 (121.2) | 126.0 (170.3) | 99.5 (151.0) |

[a]Standard deviations are shown in parentheses

# 4    Results

We used Sawtooth software and its latent class module to reveal respondent segments with similar preference structures in the choice data (Sawtooth 2004). In this section detailed results from the hierarchical Bayes (HB) estimation for the different segments identified via the latent class segmentation procedure are presented first. A presentation of the sociodemographic, psychographic, and behavioral variables of the resulting segments then follows. Finally, the characteristics of potential adopters that were significantly different from those of consumers that had already adopted green electricity are described.

## 4.1    Preferences for Different Product Attributes

Table 2 shows the part-worth utilities and the corresponding standard deviations of five different market segments. As described earlier, four main profiles were

**Table 3** Relative attribute importance scores for five segments

|  | Actual adopters | Potential adopters | | | |
|  |  | Truly Greens | Price-sensitive greens | Local patriots | Likely nonadopters |
|---|---|---|---|---|---|
| Segment size | $n=29$ | $n=117$ | $n=78$ | $n=108$ | $n=82$ |
| Electricity mix (%) | 48.6 | 50.8 | 34.2 | 20.9 | 14.9 |
| Power provider (%) | 4.8 | 4.9 | 5.3 | 7.8 | 5.1 |
| Location of electricity generation (%) | 9.0 | 10.0 | 10.5 | 21.1 | 8.4 |
| Monthly electricity costs (%) | 23.2 | 19.4 | 31.8 | 22.8 | 53.8 |
| Certification (%) | 4.6 | 4.6 | 5.5 | 8.0 | 5.5 |
| Price guarantee (%) | 5.4 | 5.6 | 7.0 | 11.2 | 7.4 |
| Cancellation period (%) | 4.4 | 4.8 | 5.5 | 8.1 | 4.9 |
| Total (%) | 100 | 100 | 100 | 100 | 100 |

identified in the latent class analysis. A group of respondents ($n=29$) who had already subscribed to a green power tariff was excluded from the sample and thereafter given the name Adopters.

Part-worth utilities were rescaled and expressed as zero-centered diffs to increase comparability between groups. Positive values represent an increase in utility relative to the average level of that particular attribute, while negative values represent decreasing utility. Generally, utility values are dependent on the selected range of attribute levels and should thus primarily be used to compare the part-worth utilities of different levels of a given attribute.

Three out of the profiles included in Table 2 can be described as Potential Adopters based on their clear preference for electricity products derived from renewable energy sources. The remaining segment is the Likely Nonadopter segment that is fairly price-sensitive and places significant emphasis on the cost of the monthly electricity product when making purchasing decisions. Members of this group are the least likely to choose to buy green electricity at the moment. However, the current segmentation of those consumers with high price sensitivity into Likely Nonadopters will not remain valid if the cost of electricity from green electricity sources becomes less than or equal to that derived from conventional energy sources.

For the next step of the analysis we then used the individual part-worth utilities from the HB analysis and computed attribute importance scores for each segment (Table 3). These scores describe how much influence each attribute has on the decision to purchase. Importance scores are standardized to sum to 100 % across all attributes (Orme 2010).

As shown in Table 3, the most important attribute for the three segments Adopters, Truly Greens, and Price-Sensitive Greens is the composition of the electricity mix. The second and third most important product attributes are also identical for these three clusters (namely, the monthly electricity cost and the location of electricity generation, respectively). In contrast, Local Patriots consider monthly electricity costs to be the most important product attribute, followed by the location of electricity generation and the electricity mix. The Likely Nonadopter segment also holds the monthly electricity costs of an electricity product to be the most important attribute (54 %). To detect significant differences in the selected variables described among the five clusters, Mann–Whitney U nonparametric tests were performed. Truly Greens have product attribute preferences similar to those of Adopters ($p > 0.05$ when comparing preferences at all attribute levels). Adopters significantly favor an electricity mix consisting of renewable energy resources (mixes 4 and 5) compared to the other three clusters of Potential Adopters. Adopters also significantly negatively prefer electricity mixes containing fossil and nuclear energy resources (compared to all other clusters at $p < 0.05$, except for Truly Greens). Adopters are significantly less price sensitive compared to Price-Sensitive Greens. No significant differences compared to Truly Greens and Local Patriots could be found in this respect. When comparing the differences in preferences between the three segments of Potential Adopters, it may be seen that Price-Sensitive Greens attach less importance to the electricity mix and more importance to monthly electricity costs compared to Local Patriots. Local Patriots show the strongest preferences for local electricity generation (within their region or within Germany) compared to all other clusters ($p < 0.001$, all cases). Interestingly, the attribute "certification" was valued most by Local Patriots in comparison to the other identified segments, although the finding was not statistically significant. Nonadopters can likely be distinguished from other clusters by their low level of interest in green electricity and their high sensitivity to monthly electricity costs.

## 4.2 Market Segments Analyzed by Sociodemographic, Psychographic, and Behavioral Characteristics

The last step was to analyze whether differences existed between subscribers to green electricity tariffs and the different segments of Potential Adopters in terms of the characteristics analyzed. Mean values are summarized in Table 4.[1]

Sociodemographic characteristics such as gender, age, household net income and household size were similarly distributed across the five identified clusters, with the exception of level of education. The analysis, however, shows that Adopters were on average better educated, a finding that corresponds with existing research (one third of all respondents in this group have a university degree). In contrast, the share of respondents with a university degree from the other four clusters ranged between 7 and 12 %. Interestingly, the Truly Greens segment had on average the minimum level of formal education of all the clusters (almost 80 % of these respondents had only completed secondary education) yet enjoyed the highest

---

[1]Tabi et al. (2014), p. 214, details the $p$-levels and the test statistics of selected pairwise comparisons.

**Table 4** Market segments by sociodemographic, psychographic, and behavioral characteristics

|  | Actual adopters | Potential adopters | | | Likely nonadopters |
|  |  | Truly Greens | Price-sensitive greens | Local patriots |  |
| --- | --- | --- | --- | --- | --- |
| *Sociodemographic data* | | | | | |
| Gender, females (%) | 41.40 | 51.30 | 52.60 | 45.40 | 42.70 |
| Age (years) | 47.4 (14.1) | 49.1 (12.1) | 49.95 (14.5) | 51.29 (14.8) | 50.93 (12.9) |
| Level of education | 4.5 (2.1) | 3.2 (1.5) | 3.6 (1.8) | 3.5 (1.7) | 3.3 (1.6) |
| Level of income | 3.5 (1.9) | 3.8 (1.5) | 3.5 (1.5) | 3.2 (1.6) | 2.9(1.4) |
| Household size | 1.86 (0.9) | 2.08 (1.1) | 2.05 (1.2) | 2.18 (1.2) | 2.04 (1.1) |
| *Psychographic and behavioral characteristics* | | | | | |
| Relocation: yes (%) | 45 | 27 | 33 | 25 | 25 |
| Switch electricity contract: yes (%) | 69 | 12 | 17 | 20 | 16 |
| Level of climate concern: high (%) | 93 | 89 | 84 | 75 | 64 |
| Perceived consumer effectiveness: agree (%) | 90 | 66 | 69 | 59 | 46 |
| Trust in science: disagree (%) | 72 | 45 | 35 | 35 | 40 |
| Support for eco-taxes: agree (%) | 72 | 54 | 60 | 46 | 38 |
| Awareness of green electricity labels: 2 or more (%) | 21 | 14 | 12 | 9 | 5 |
| Estimated price premium for green electricity[a] | 2.3 | 1.9 | 1.9 | 1.9 | 1.7 |
| Willingness to pay for eco-friendly products: yes (%) | 79 | 53 | 42 | 37 | 21 |

[a]Average value of response to the question: What is your estimate of the cost of green electricity compared to conventional electricity? [1: much more expensive (more than 10 %), 2: slightly more (up to 10 %), 3: same price, 4: slightly cheaper (up to 10 %); 5: much cheaper (more than 10 %)]. The higher the average value, the lower the estimated price for green electricity.
Standard deviations are shown in parentheses.

average household net income of all the clusters (although not significantly different than that of Adopters). Worth mentioning here is the fact that while income in all clusters had a more or less normal distribution, 30 % of Adopters were placed in the highest and 40 % in the first and second lowest income categories. This finding should be further explored in future research to generate a better understanding of the nature of green electricity subscribers.

However, the average income of the Truly Greens segment differed significantly from the average of the Local Patriots and the Likely Nonadopters. Our results are therefore in line with those of many other authors (Rowlands et al. 2003; Zarnikau 2003; Gossling et al. 2005; Wiser 2007; Ek and Soderholm 2008; Diaz-Rainey and

Ashton 2011; Sagebiel et al. 2014) and reinforce the evidence that says that preferences for green electricity significantly differ across income groups. Among Adopters, 66 % live alone (i.e., have a smaller household size on average compared to respondents from other clusters), but no statistically significant differences could be found compared to Potential Adopters.

With regard to psychographic and behavioral characteristics, segments with a high preference for electricity mixes sourced from renewable energy could be characterized by their higher degree of concern for climate-change-related issues. Decreases in concern for climate change are correlated to decreases in the strength of preference for green electricity. However, no significant difference could be found between Adopters and Truly Greens in this regard. Significant differences were, on the other hand, found with variables that were used to examine sensitivity to environmental issues. More precisely, Potential Adopters are more likely to agree that science and technology will solve many environmental problems without requiring changes in our ways of living than Adopters. In addition, support for eco-taxes is also significantly higher with Adopters than it is with Local Patriots, whereas no significant difference could be found among Truly Greens and Price-Sensitive Greens.

In line with previous research that found that perceived consumer effectiveness plays a major role in forming pro-environmental behavior, a significant difference was identified between Potential Adopters and Adopters.

The perceived price level of green electricity (in contrast to conventional electricity products) differed significantly between Adopters and all other clusters. Only about 10 % of Adopters but 25 % of the Truly Greens and 43 % of the Likely Nonadopters believed that the cost of green electricity was 10 % or more than conventional electricity products. Whereas at the beginning of the process of liberalization of the electricity market in Germany green electricity was typically sold at a significantly higher price than electricity produced from conventional energy sources, the price difference has significantly decreased over the last decade. At the time this research was conducted, the costs of green tariffs in Germany showed high variability depending on the provider, with some offering cheaper green electricity than conventional electricity. Our results are therefore in line with previous research that showed that erroneous perceptions about the price difference between gray power and green power act to decrease the probability of the adoption of green electricity (Arkesteijn and Oerlemans 2005). Consumers who have not yet opted for green electricity may still be implicitly assuming that electricity generated from renewable energy sources is significantly more costly, even though reality tells a different story.

Awareness of green electricity labels also differed significantly between Adopters and two segments of Potential Adopters, the Price-Sensitive Greens and the Local Patriots. In addition, a weak (but nonetheless significant, at 10 %, significance level) difference between Adopters and Truly Greens with regard to the share of respondents who had changed residence within the last 5 years was found. Targeting consumers in the course of changing residence thus represents an interesting starting point for green power marketing, but the effectiveness of such a marketing campaign depends on the targeted segment.

Finally, the general WTP for eco-friendly products also differed significantly between Adopters and the three segments of Potential Adopters. This finding demonstrates that the marketing of premium-priced products has certain limits.

## 5    Conclusions and Recommendations

Many customers exhibit positive attitudes toward renewable electricity mixes, but only a small percentage of them have already opted for green electricity tariffs. The research described in this chapter was designed to reveal the characteristics that distinguish subscribers of green electricity tariffs from potential green electricity adopters. Based on the 4968 experimental choices of a representative sample of 414 German consumers, different consumer segments were identified based on their preferences for different electricity product attributes. Results suggest that the majority of respondents (80 %) have a clear preference for electricity mixes derived from renewable energy sources, but only 7 % of them had already translated their preferences into purchases of green electricity at the time the study was conducted. Correspondingly, the main goal of the research was to highlight how Adopters differ from those who show interest in renewables but have not yet subscribed to a green electricity product (i.e., Potential Adopters).

Demographic variables were found to play a marginal role in explaining the difference between Adopters and Potential Adopters, which corresponds to earlier research findings (Kotchen and Moore 2007). Results of this study show, however, that Adopters can be characterized by their significantly higher average level of education.

Our results suggest that it is particularly psychographic and behavioral factors that have great explanatory power when it comes to understanding why consumers who evince strong preferences for electricity produced from renewable energy sources do not act according to their preferences by opting to purchase green power (Fig. 1). For instance, estimates of the price difference between green and standard electricity tariffs is lower among Adopters than among Potential Adopters. In addition, Adopters demonstrate a greater awareness of green electricity labels than other segments, except for Truly Greens. Adopters also change their place of residence significantly more often than two segments of Potential Adopters and have more often recently switched their electricity tariffs. Adopters can be further characterized by their higher level of perceived consumer effectiveness compared to all other segments of Potential Adopters. Regarding price-related variables, Adopters, in contrast to the other segments of Potential Adopters, tend to be willing to pay significantly more for eco-friendly products.

For marketers, these findings indicate a major opportunity. Although the number of Adopters of green electricity might still be low, reported customer preferences suggest that there is significant potential for the number of adopters to rise. We can underline the role of a multitude of factors that could be exploited to convince consumers to seal the green power deal. Education seems to play a highly influential role in purchasing decisions and may also make a strong contribution to higher perceived consumer effectiveness. This highlights the necessity of better

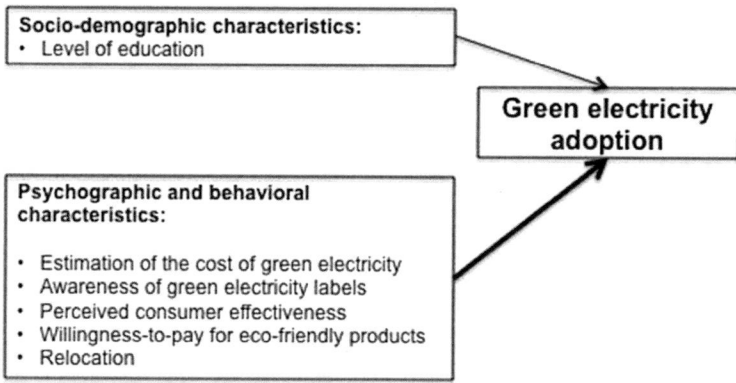

**Fig. 1** Determinants of green electricity adoption

communication about the actual impacts of opting for green power. Previous research shows that increasing perceived impact by providing information about social and private benefits can successfully modify purchasing behavior (Litvine and Wüstenhagen 2011). Our findings also draw attention to the existence of inaccurate perceptions about green electricity prices. Respondents were asked about the likely price premium between conventional and green tariffs. The majority of Potential Adopters estimated a premium of more than 10 %, even though the green tariffs at the time of conducting the survey did not always exceed prices for conventionally derived energy on the German market. This indicates that more accurate marketing communication about the actual price of green electricity could pay off in terms of increasing the uptake of green energy tariffs. Another interesting result for marketers is the strong preference of Potential Adopters for domestically produced electricity. This establishes the potential for the implementation of national or regionalized energy policies (such as introducing standards that require a declaration about the origin of the electricity source—although tradeoffs with the internal EU electricity market need to be considered). The Local Patriots segment identified in the research places almost the same emphasis on the location of power generation as it does on the cost of the electricity. Accordingly, advertising the regional origins of electricity might be particularly fruitful for this segment. The two segments Price-Sensitive Greens and Truly Greens do not differ with regard to most variables investigated, so they could be targeted with similar messages; however, the Price-Sensitive Greens are much more sensitive to increases in the cost of electricity. Power marketers could respond to these findings by targeting this segment with lower prices and a slightly lower share of green electricity in the mix.

For policymakers we can highlight that raising the level of perceived consumer effectiveness and increasing the feeling of being responsible for climate change can effectively constitute the core of environmental policies. For instance, Truly Greens might be targeted with awareness- raising campaigns that draw attention to the importance of taking individual action to safeguard the environment or the responsibility of humans for climate change. Findings show a low awareness of eco- labels among electricity consumers, which also represents an opportunity for policymakers who could elaborate on and further disseminate information about the different certification schemes that exist on the market.

**Acknowledgments**  This study is based on research funded by the German Ministry of Education and Research within the Socio-Ecological Research (SÖF) Project Social, Environmental and Economic Dimensions of Sustainable Energy Consumption in Residential Buildings (seco@home), contract 01UV0710. This work is also related to the Sciex Programme (project code: 12.163) and the Swiss Competence Center for Energy Research–Competence Center for Research in Energy, Society and Transition (SCCER–CREST). The authors would also like to thank the three anonymous reviewers of the journal *Ecological Economics* for their valuable comments and remarks on the original article.

# References

Arkesteijn, K., & Oerlemans, L. (2005). The early adoption of green power by Dutch households: An empirical exploration of factors influencing the early adoption of green electricity for domestic purposes. *Energy Policy, 33*(2), 183–196.

Arnold, D. (2011). *The influence of live events on sustainable consumption.* Bachelor Thesis, University of St. Gallen.

BMU (Federal ministry for the environment nature conservation and nuclear safety). (2009). *Electricity from renewable energy sources—What does it cost?* Accessed July 30, 2016, from http://www.bmu.de/fileadmin/bmu-import/files/pdfs/allgemein/application/pdf/brochure_electricity_costs_bf.pdf.

BMWI (Federal Ministry for Economic Affairs and Energy). (2013). Accessed July 30, 2016, from http://www.bmwi.de/DE/Themen/Energie/Energietraeger/erneuerbare-energien,did=20918.html.

Burkhalter, A., Kaenzig, J., & Wüstenhagen, R. (2009). Kundenpräferenzen für leistungsrelevante Attribute von Stromprodukten. *Zeitschrift für Energiewirtschaft, 33*(2), 161–172.

Cai, Y. X., Deilami, I., & Train, K. (1998). Customer retention in a competitive power market: Analysis of a double-bounded plus follow-ups questionnaire. *Energy Journal, 19*(2), 191–215.

Clark, C. F., Kotchen, M. J., & Moore, M. R. (2003). Internal and external influences on pro-environmental behavior: Participation in a green electricity program. *Journal of Environmental Psychology, 23*(3), 237–246.

Clausen, J. (2008). Betreiber von Solarwärmeanlagen und Ökostromkunden in der Klimaschutzregion Hannover. Forschungsprojekt Wenke 2. Accessed July 30, 2016, from http://www.borderstep.de/pdf/P-Clausen-Betreiber_von_Solarwaermeanlagen_und_Oekostromkunden_in_der_Klimaschutzregion_Hannover-2008.pdf.

DeSarbo, W. S., Ramaswamy, V., & Cohen, S. H. (1995). Market segmentation with choice-based conjoint analysis. *Marketing Letters, 6*(2), 137–147.

Diaz-Rainey, I., & Ashton, J. K. (2011). Profiling potential green electricity tariff adopters: Green consumerism as an environmental policy tool? *Business Strategy and the Environment, 20*(7), 456–470.

Ek, K., & Soderholm, P. (2008). Norms and economic motivation in the Swedish green electricity market. *Ecological Economics, 68*(1–2), 169–182.

Ewing, G., & Sarigöllü, E. (2000). Assessing consumer preferences for clean-fuel vehicles: A discrete choice experiment. *Journal of Public Policy & Marketing, 19*(1), 106–118.

Gerpott, T. J., & Mahmudova, I. (2010). Determinants of green electricity adoption among residential customers in Germany. *International Journal of Consumer Studies, 34*(4), 464–473.

Goett, A., Hudson, K., & Train, K. (2000). Customers' choice among retail energy suppliers: The willingness-to-pay for service attributes. *Energy Journal, 21*(4), 1–28.

Gossling, S., Kunkel, T., Schumacher, K., Heck, N., Birkemeyer, J., Froese, J., Naber, N., & Schliermann, E. (2005). A target group-specific approach to "green" power retailing: Students as consumers of renewable energy. *Renewable and Sustainable Energy Reviews, 9*(1), 69–83.

Green, P. E. (1977). A new approach to market segmentation. *Business Horizons, 20*, 61–73.

Hansla, A., Gamble, A., Juliusson, A., & Gärling, T. (2008). Psychological determinants of attitude towards and willingness to pay for green electricity. *Energy Policy, 36*(2), 768–774.

Kaenzig, J., Heinzle, S. L., & Wüstenhagen, R. (2013). Whatever the customer wants, the customer gets? Exploring the gap between consumer preferences and default electricity products in Germany. *Energy Policy, 53*, 311–322.

Kotchen, M. J., & Moore, M. R. (2007). Private provision of environmental public goods: Household participation in green-electricity programs. *Journal of Environmental Economics and Management, 53*(1), 1–16.

Kotler, P., & Keller, K. L. (2006). *Marketing management* (12th ed.). Upper Saddle River, NJ.: Prentice Hall.

Litvine, D., & Wüstenhagen, R. (2011). Helping "light green" consumers walk the talk: Results of a behavioral intervention survey in the Swiss electricity market. *Ecological Economics, 70*(3), 462–474.

MacPherson, R., & Lange, I. (2013). Determinants of green electricity tariff uptake in the UK. *Energy Policy, 62*, 920–933.

Madlener, R., & Stagl, S. (2005). Sustainability-guided promotion of renewable electricity generation. *Ecological Economics, 53*(2), 147–167.

Mattes, A. (2012). Potentiale für Ökostrom in Deutschland—Verbraucherpräferenzen und Investitionsverhalten der Energieversorger. Accessed July 30, 2016, from http://www.diw-econ.de/de/downloads/DIWecon_HSE_Oekostrom.pdf.

Orme, B. (2010). *Getting started with conjoint analysis: Strategies for product design and pricing research*. Madison: Sawtooth Software.

Rose, S. K., Clark, J., Poe, C. L., Rondeau, D., & Schulze, W. D. (2002). The private provision of public goods: Tests of a provision point mechanism for funding green power programs. *Resource and Energy Economics, 24*, 131–155.

Rowlands, I. H., Scott, D., & Parker, P. (2003). Consumers and green electricity: Profiling potential purchasers. *Business Strategy and the Environment, 12*(1), 36–48.

Sagebiel, J., Müller, J. R., & Rommel, J. (2014). Are consumers willing to pay more for electricity from cooperatives? Results from an online choice experiment in Germany. *Energy Research & Social Science, 2*, 90–101.

Sammer, K., & Wüstenhagen, R. (2006). The influence of eco-labelling on consumer behaviour—Results of a discrete choice analysis for washing machines. *Business Strategy and the Environment, 15*, 185–199.

Sawtooth Software. (2004). The CBC Latent Class: Version 3.0. Technical Paper. Sequim, WA: Sawtooth Software.

Straughan, R. D., & Roberts, J. A. (1999). Environmental segmentation alternatives: A look at green consumer behavior in the new millenium. *Journal of Consumer Marketing, 16*(6), 558–575.

Tabi, A., Hille, S. L., & Wüstenhagen, R. (2014). What makes people seal the green power deal?—A customer segmentation based on choice experiments in Germany. *Ecological Economics, 107*, 206–215.

Villiger, A., Wüstenhagen, R., & Meyer, A. (2000). *Jenseits der Öko-Nische*. Basel: Birkhäuser Verlag AG.

Wind, Y. (1978). Issue and advances in segmentation research. *Journal of Marketing Research, 15*, 317–37.

Wiser, R. H. (2007). Using contingent valuation to explore willingness to pay for renewable energy: A comparison of collective and voluntary payment vehicles. *Ecological Economics, 62* (3–4), 419–432.

Zarnikau, J. (2003). Consumer demand for 'green power' and energy efficiency. *Energy Policy, 31* (15), 1661–1672.

**Stefanie Lena Hille** is currently assistant professor at the Institute for Economy and the Environment at the University of St. Gallen in Switzerland. In her research she focuses on how to use concepts and theories from marketing, psychology and behavioral economics to better understand consumer behavior related to energy consumption and mobility. She acquired a PhD title in Marketing from the University of St. Gallen in Switzerland in 2012.

**Andrea Tabi** earned a PhD in Environmental economics and is currently doing a second PhD in ecology at the Department of Evolutionary Biology and Environmental Studies at the University of Zurich. Before, she was a PostDoc at the Institute of Economy and the Environment at the University of St. Gallen where she focused her research on choice modeling regarding the hydropower expansion in Switzerland. She investigated the economic, environmental, and social impacts of the expansion of hydropower production projected by the Swiss Energy Strategy 2050, as well as the economic trade-offs and preferences of Swiss population for the alternative hydropower expansion scenarios.

**Rolf Wüstenhagen** is the Good Energies Professor for Management of Renewable Energies at the University of St. Gallen. He completed his PhD in 2000 on marketing green electricity—from niche to mass market. His work on decision-making of energy consumers and investors has been published in leading academic journals. He has held visiting faculty positions at University of British Columbia, Copenhagen Business School, National University Singapore and Tel Aviv University. Since 2010, he is the Director of the University of St. Gallen's Executive Education Programme on Renewable Energy Management (REM-HSG).

# Introducing Green Electricity as the Default Option

Sylviane Chassot, Rolf Wüstenhagen, Nicole Fahr, and Peter Graf

**Abstract**

One of the key challenges in marketing (green) electricity is overcoming customer inertia. Recent insights from behavioral economics suggest that in the context of long-term decision making, this leads to a situation where consumers do not make the choices that are best for society or, in fact, their own long-term interest. Nudging consumers to more environmentally friendly decisions by introducing a green default may be an effective way out of this dilemma. This chapter reports on marketing research that was done with customers of a Swiss electric utility ahead of the introduction of a green default, combining eye tracking, choice tasks, and interviews. We also report on the successful implementation of the research results, which led to a significant increase in revenues available for investment in new renewable energy facilities, and discuss implications for communication, marketing, and organizational dynamics.

**Keywords**

Green default • Behavioral economics • Inertia • Eye tracking • Green power marketing

This chapter is a translated and revised version of an article first published in German: Chassot, S, Wüstenhagen, R, Fahr, N, Graf, P. 2013. Wenn das grüne Produkt zum Standard wird. Wie ein Energieversorger seinen Kunden die Verhaltensänderung einfach macht. Organisations Entwicklung, 3, 80–87.

S. Chassot • R. Wüstenhagen (✉) • N. Fahr
Institute for Economy and the Environment, Good Energies Chair for Management of Renewable Energies, University of St. Gallen, Tigerbergstr. 2, 9000 St. Gallen, Switzerland
e-mail: rolf.wuestenhagen@unisg.ch

P. Graf
St.Galler Stadtwerke, Vadianstrasse 8, 9001 St. Gallen, Switzerland

# 1     Introduction

The energy industry is undergoing fundamental change. The shift from the use of fossil and nuclear to renewable energy has been evolving over decades due to stricter climate laws and the political risks of dependence on oil and gas exporters. Since the nuclear accident in Japan in March 2011, the so-called energy transition has received enough attention from the public to proceed more decisively. Accordingly, the governments of Germany and Switzerland have decided to phase out nuclear power.[1]

For an electric utility company (EUC) this decision represents a major strategic challenge in the form of a call for fundamental change. So far, EUCs have provided their customers with an electricity mix consisting of the Swiss average of around 40 % nuclear energy. They are now required to replace this 40 % by investing in renewable energy and energy efficiency and, where necessary, using natural gas and electricity imports. Another challenge also presents itself at the other end of the value chain of an EUC: such renewable electricity must not only be produced and transported but also sold to customers in the form of appealing products. The case study described in this chapter highlights what must be considered a decisive step in the promotion of a more ecologically friendly product and describes how customers react to changes of the default product. The findings described herein also apply to the effects of "green defaults" in other industries.

According to a green electricity survey conducted in 2010, 507 of 730 Swiss EUCs offered electricity products generated from renewable sources, and around 15 % of customers used these products before the Fukushima accident. Sales of power products containing renewable energy amounted to 10 % of the total. The market share of renewable energy has been growing since its initiation in 1998 and with accelerating speed since 2011. In 2014, already 25 % of residential customers ordered or were nudged into a renewable energy tariff that was more expensive than the cheapest option. However, these eco-friendly products consist mainly of hydropower. Sales of new renewable energies, such as wind power or solar photovoltaics, remain below 2 %. If the nuclear phase-out is not to endanger climate protection targets, sales of new renewable energy must increase significantly. At this point, many utilities face a contradiction: On the one hand, decisions about the promotion of renewable energy have been made at the political level, and green energy is, according to surveys, desired by customers. On the other hand, customers' initiative to switch their electricity consumption to renewables remains rather low. What can companies do in this situation?

---

[1]Whereas Germany has decided to phase out all nuclear power plants by 2022, the precise date in Switzerland is still subject to debate.

## 1.1    If the Mountain Won't Come to Muhammad...

... then Muhammad must go to the mountain. In 2006, the Zurich municipal utility followed this principle by defining a green power product as the standard product for electricity customers who had not explicitly chosen a different option. Following its initiative, a dozen other Swiss utilities were inspired by the idea. Other utilities, however, shy away from taking this approach, often out of concern that they will come across to customers as patronizing and risk upsetting them.

The fact that preselection has a major impact on customer behavior is often illustrated using the example of organ donation. In Austria, the organ donation rate is 99 %; in Germany, it is 12 %. In Austria, the preselected option is pro donation, which means that the deceased are generally considered to have accepted that their organs will be donated in the event of their death, so relatives must actively disagree if they want to avoid organ donation (an opt-out system). Conversely, Germany follows an opt-in system. In the literature on consumer behavior, this is referred to as the *decision architecture* of a choice situation, where preselection is one of the most important elements, also known as the default option. American behavioral economists Richard Thaler and Cass Sunstein refer to this situation as a "nudge," which helps customers take wise decisions in complex choice situations (Thaler and Sunstein 2008).

## 1.2    Is the Green Default Legitimate?

According to a survey about renewable energies and green power (Hübner et al. 2012), customer laziness (inertia) is the key obstacle to switching to a different energy source. One study participant explained:

> *"That's called having a completion block. Writing to-do lists that one executes diligently, to then always transfer a few small items to the next list. And so some things are simply always pushed ahead. I hate paperwork. For me, it always grows, until I someday sit down and do it. And then there are a few things that I always put on top of the pile, but they disappear under the new submissions. So, I think this is actually quite normal. The outrage that I feel in-between, obviously, is not enough to make me sit down and do it all."* (media professional interviewed by Hübner et al. 2012)

The people this description applies to would benefit if their EUC switched their standard offering to a greener energy mix. Actively coming to a decision and selecting a new product requires more effort from customers than accepting a default. According to Sunstein and Reisch (2014), this is one reason why consumers often accept the default option. Moreover, two further obstacles are mentioned in the literature when it comes to switching:

- Customers perceive the "standard option" as being the one recommended by authorities, and they therefore accept it as the correct course of action;

- Psychologists have demonstrated the endowment effect. What one "owns" (the current option) is automatically of more (subjective) value than what one does not possess (a potential option).

Besides convenience and price, the free-rider problem is another obstacle to self-initiating customer changes when it comes to selecting an electricity tariff. Some customers are not willing to pay more for green energy as long as a majority consumes the cheaper mix; making the right choice in the wrong system is an option pursued only by a niche of ecologically concerned consumers (Karsten and Reisch 2008). Thus, making the sustainable product option the default is, according to Karsten and Reisch (2008), acceptable if the surcharge to customers is in a favorable ratio to social benefits of a more environmentally friendly product.

## 1.3    Examples from Other Industries

There are countless examples of companies steering the everyday decisions of their customers and employees by means of smart decision architecture—decisions by customers, as well as decisions by their own employees. An example of the latter relates to the paper consumption of an American university. When the default setting for university printers was changed from single- to double-sided, paper consumption dropped by 44 % (or 4650 trees) (Sunstein and Reisch 2014). This is an example of an environmentally friendly, cost-saving default change.

Another example comes from the American retirement scheme. Participation in the 401(k) pension plan is voluntary in the USA. Originally, employees paid into a company's own pension fund if they had made an explicit decision to do so. The result was low participation rates in many pension funds. In the late 1990s, an increasing number of pension funds adopted the principle of automatic participation in pension plans (opt out), leading to an immediate increase in participation rates by 35 % (Madrian and Shea 2001).

## 2    Case Study

## 2.1    Initial Situation of St. Galler Stadtwerke

In a referendum in November 2010, the population of the city of St. Gallen voted in favor of phasing out nuclear energy by 2050—a clear mandate for the St. Galler Stadtwerke (sgsw), which until then had been providing their customers with a default mix (*Basispower*) consisting of around 50 % nuclear power, 40 % hydropower, 8 % electricity from waste incineration, 1 % fossil fuels, and 1 % new renewables. In addition, all electricity customers were offered the opportunity to choose their preferred mix of hydropower, wind, and solar (Fig. 1).

| basispower<br>Unser Mixstrom | Annual consumption of customer | < 48 MWh | > 48 MWh<br>< 1,000 MWh | > 1,000 MWh |
|---|---|---|---|---|
| | High Tariff (7 am-8 pm)<br>Low Tariff (8 pm-7 am) | 10.7 Rp./kWh<br>7.0 Rp./kWh | 8.9 Rp./kWh<br>6.3 Rp./kWh | 8.3 Rp./kWh<br>5.8 Rp./kWh |

| | | Price Surcharge over basispower | Market Share | |
|---|---|---|---|---|
| | | | # customers | Sales [MWh] |
| aquapower<br>Strom aus Wasserkraft | Hydropower | 2.0 Rp./kWh | 7% | 7% |
| windpower<br>Strom aus Windkraft | Wind | 7.0 Rp./kWh | 0.2% | 0.06% |
| solarpower<br>Strom aus Sonnenkraft | Solar Photovoltaics | 75.0 Rp./kWh | 4.7% | 0.04% |

**Fig. 1** sgsw electricity product range in 2011 (before default change)

**Facts about St. Galler Stadtwerke**

- Municipal utility,100 % owned by the city of St. Gallen;
- Supplier and distribution grid operator for electricity, gas, heat, and water;
- Currently investing in fiber-optic telecommunications network;
- 260 employees, operating revenue of 204 million CHF, 503 GWh electricity sales, 58 million CHF investments (2015);
- Shareholder of SN Energie AG, together with six other regional energy supply companies operating hydropower plants, and holding shares in Swiss and French nuclear power plants;
- Public referendum in city of St. Gallen on 28 November 2010: 60 % majority in favor of nuclear phase-out by 2050 (assuming supply security guaranteed).

After the referendum, Peter Graf, head of Energy and Marketing at sgsw, faced the question of how to transform the will of the people into a new product range. One solution was to expand the marketing of existing renewable energy offerings. Various attempts have been made to do this since the launch of green power products in 2000, from postal mail to telephone sales, but without a satisfactory level of cost efficiency.

The other approach was to replace the existing standard product with a greener power product. But would this be demanding too much from customers, and would selling a more expensive product turn out to be a counterproductive strategy in an electricity market that was soon to be liberalized also for retail customers?

## 2.2     Referendum in 2010, Research Project in 2011

Shortly after the referendum in November 2010, sgsw was approached by the University of St. Gallen offering to collaborate in a research project to investigate the acceptance of so-called green defaults. The study was designed with the following objectives in mind:

- To find out what customers specifically look at when browsing flyers or the Internet for power products;
- to test whether a default would be accepted, even if it was not the cheapest product available.

### 2.2.1     Eye Tracking: Identifying Customer Requirements

To investigate what customers look for with electricity products (mix, price, label, endorsement by sgsw), an eye-tracking study was conducted. Using this research method, participants sit in front of a computer viewing an image—in this case a Website displaying power products—while a camera below the computer (the eye tracker) tracks the viewer's eye movements (see box for more information about the methodology). At the beginning of an eye-tracking session, participants were instructed as follows:

*Imagine you want to order a new electricity tariff. Below, one after another, you will be shown various advertisements, each with four power products. Take your time and decide which of the four products you prefer. Once you have made your choice, please use the left mouse button to click on one of the four selection boxes to confirm your choice.*

The subjects were shown nine websites with (hypothetical) sgsw power products on display. In Fig. 2 (first column), three of the nine websites are presented. The presentation of the price and preselection (default) varied in each case. The default was illustrated in three different ways:

- The preselected option was ticked in advance,
- The preselected option and the policy-endorsed power option matched,
- The policy endorsement supported a more ecological option than the preselected option.

Additionally, the effect of different ways to present energy costs was examined. The cost of electricity tariffs was

- Not mentioned,
- Expressed in Swiss Centimes (Rp.) per kilowatt hour,
- Expressed in Swiss Francs per month, based on the average paid by a St. Gallen household.

| Presentation in experiment | Customers' visual focus (Darker colour indicates longer period of observation) | Customer choices (N = 58) |
|---|---|---|
| **Preselection only** | | |
| | | |
| **Preselection and policy endorsement the same** | | |
| | | |
| **Policy endorsement higher than preselection** | | |

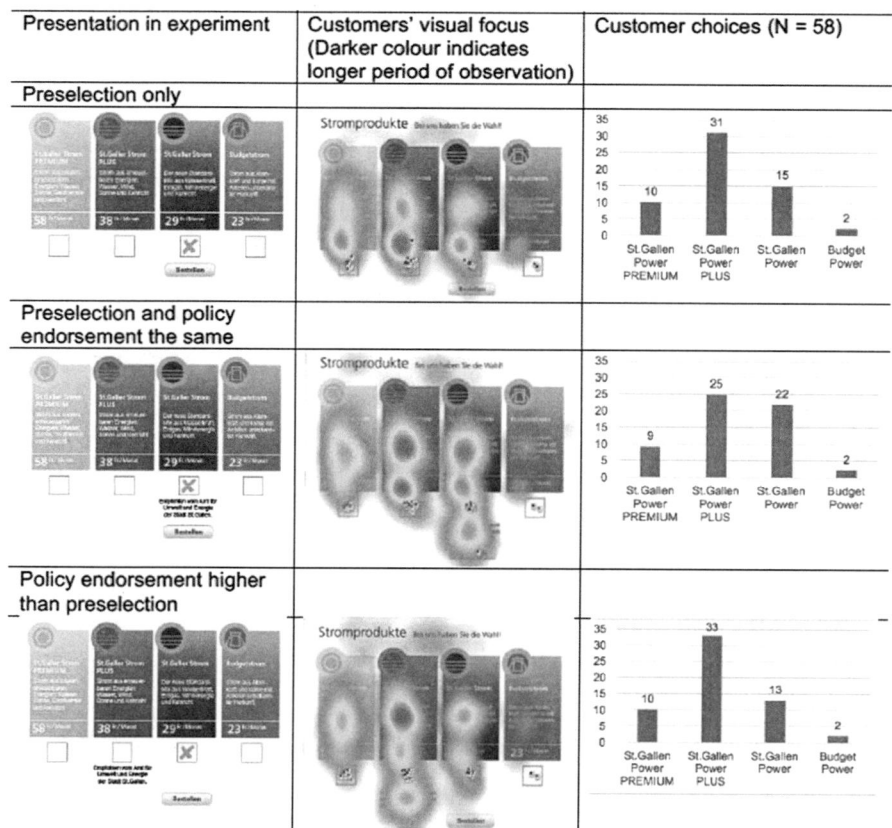

**Fig. 2** Tariff choice in experiment by type of default

The description of the energy mix of the four power products was kept the same for all nine images; each of the nine dummy webpages showed

- A "gray power" mix with nuclear energy, coal, and energy from unknown sources ("Budget Power");
- One with hydro, other renewables, and natural gas ("St. Gallen Power");
- A purely renewable product ("St. Gallen Power Plus"); and
- A premium renewable product with only hydropwoer, solar power, waste incineration, and geothermal energy ("St. Gallen Power Premium").

To validate the results of the eye-tracking study, participants also filled out a questionnaire about renewable energy. The study duration for each participant was 30–45 min. In May 2011, 66 sgsw customers participated in the study, 58 of whom were deemed valid for inclusion in the eye-tracking analysis. Selection of participants was carried out using quota sampling so that the sample was representative for the St. Gallen population in terms of age and gender.

**Eye Tracking as a Method of Market Research**
Eye tracking is a research method used to measure individuals' visual behavior. Using eye tracking (recording), the researcher can determine what elements of a visual stimulus, such as an advertisement, capture the attention of the viewer and what the consumer is reflecting on (the so-called eye-mind hypothesis). Thus, eye tracking makes it possible to assess which elements of a promotional stimulus are being observed (in our case, energy mix, price, default system, design), for how long, and what role they play in decision making (Djamasbi et al. 2008, p. 308f.; Poole and Ball, 2005, p. 3f).

## 2.2.2    Results of Research Project

Based on the recorded eye movements, the study was able to show what St. Gallen's electricity customers pay attention to when observing a description of electricity tariffs. All eye movements of each participant were recorded, and the individual-level information was condensed into a so-called heat map. The darker a spot on the heat map, the longer this spot was looked at on average (Fig. 2, column 2).

*Result 1* Electricity Mix and Price Are Considered the Longest
   In Fig. 2, column 2, it can be seen from the heat maps that

- The least ecological electricity tariff itself received almost no attention, even though it was cheaper than the default option;
- The description of the electricity mix and the price were looked at for the longest;
- The preselected option received less attention than the option with a policy endorsement.

*Result 2* Policy Endorsement Has a Greater Influence Than Ticking a Preselected Option
   Figure 2, column 3, illustrates the choices of the 58 participants. (Since the effect of the default is of interest here, differences due to price presentation are not discussed further.)
   In row 1 (preselection only), the majority opted for the product that was slightly more expensive than the default option. The product that was cheaper than the default option was virtually neglected. One participant who consistently chose the cheapest product explained his choice using the aforementioned free-rider attitude: he would not pay more for electricity from renewable sources as long as some other customers would not do so either.

In Fig. 2, row 2, a policy endorsement supporting the second cheapest item was added to the ticked preselection option, which was then selected more frequently than in the setting without the policy endorsement.

In the version with the preselection option for St. Gallen Power and a policy endorsement for the more expensive St. Gallen Power Plus (row 3), the majority of choices were guided by the policy endorsement, not the ticked preselection option.

In summary, use of a ticked preselection option seldom results in the selection of a less ecological product, but in many cases, the product selected is of higher ecological value than the option preselected with a tick. However, the policy endorsement had an even greater impact on tariff choice: some study participants switched their choice from St. Gallen Power to St. Gallen Power Plus if the policy endorsement supported the latter. However, the policy endorsement also acted in the opposite direction: some participants who had chosen St. Gallen Power Plus without reading a policy endorsement switched their choice to the less environmentally friendly St. Gallen Power when this was the policy endorsement.

*Result 3* Preselection, Not Paternalism

In the questionnaire following the eye-tracking experiment, participants were asked to reflect on their attitude toward electricity tariffs and choice behavior. Table 1 shows that only 4 out of 66 respondents would feel patronized if the standard mix was an eco-product. The majority of respondents rated the green default as a good or very good idea. No one feared having a supply problem following a switch to green electricity as the standard mix.

## 2.3 Implementation in 2012

In May 2011, Federal Councilor Leuthard communicated about the nuclear phase-out at the national level. It was clear at that time that the vote of November 2010 by the city of St. Gallen was not an outlier but an important proof of public support for sgsw to play a leading role in the energy transition. Simultaneously with the development of the new marketing concept, sgsw specified how to generate more renewable energy and feed it into the grid. The case illustrates that a true change of default involves not only the marketing department but also the supply side of an EUC.

Following presentation of the results of the eye-tracking study in June 2011, sgsw's concern that customers would feel patronized by a green default was empirically refuted and the joint research project of the University of St. Gallen and sgsw was completed.

In January 2012, the tariff change was implemented and four new products were created. All customers were informed that from now on they would receive the new St. Gallen Power Basic product unless they actively chose another product (former green electricity customers continued to receive a higher-quality product).

**Table 1** Acceptance of green defaults (frequency of answers)

| If the standard mix were affordable green electricity, not the current base power option, which of the following statements would most closely describe your opinion? | |
| --- | --- |
| I would like it very much | 13 |
| I would like it because many people would otherwise take no action about changing their energy mix, even if they were open to it | 42 |
| The change would be acceptable because there would still be the oportunity to opt for a different mix later | 7 |
| I feel patronized by the preselection option | 4 |
| There would be a supply problem if everyone chose to accept the preselection option | 0 |

To prepare sgsw customers for their new electricity product, an extensive communication campaign was implemented in the city of St. Gallen by the end of 2011. Details about the revision of the tariff and product were communicated via posters, brochures, a dedicated website, newspaper advertisements, personal mailings, and at the movies.

## 2.4    Customer Feedback and Financial Implications

At the St. Gallen Forum for Management of Renewable Energies in March 2012, Peter Graf reported on customer responses regarding the default change. While there were some negative reactions, which were partly political ("The energy transition must be brought about politically and globally!" or "Nuclear power phase-out, no thanks") or specifically directed at the presentation of the new products and details of the communications of sgsw ("exactly why the product is more expensive is not clear"), the majority of customers responded positively to the switch, as illustrated by comments like "Congratulations on this initiative, I am happy with it," as well as specific feedback about the related brochures and the forms of communication ("It's great that each and every person is able to decide which electricity source they will support!").

Quantitative customer response confirmed the results of the study carried out by the University of St. Gallen: The vast majority accepted the new default. Among residential customers, only 10 % switched back to the cheaper "nuclear power mix" (Fig. 3), whereas a larger share of the generally more price-sensitive corporate customers opted to switch back. Across all customer segments, the nuclear power mix accounted for 43 % of all electricity sold following the introduction of the new product range, while the new default product, St. Gallen Power Basic, accounts for 42 %, St. Gallen Power Eco 13 %, and St. Gallen Power Eco Plus 2 %.

**Fig. 3** Market share of new electricity products after default change (number of customers, residential, consuming less than 48 MWh p.a., March 2012)

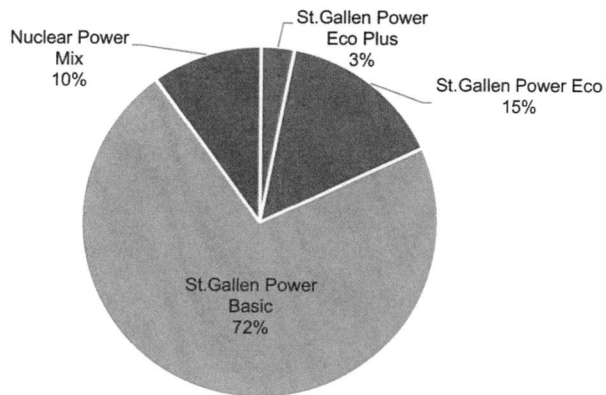

The price increase for the new default of roughly 1 Rp./kWh resulted in additional revenues of around 4 million CHF per year. sgsw uses this additional income to invest in renewable power generation, in particular in photovoltaics and wind power.

## 3  Conclusion

The example shows:

- When implementing a tariff revision, details matter. Careful planning of the product design in terms of product name, price display, colors, and mix is worthwhile;
- Widespread communication of easy-to-understand information about any proposed changes is crucial to avoid negative customer reactions—and increasing the enthusiasm of other customer segments about changes;
- After years of marketing efforts, sgsw was able to increase its share of green electricity customers from zero to more than 10 %. But by means of the default change, an additional 80 % of private customers were brought on board. Thus, a new default is much more effective at enhancing the diffusion of renewable electricity than incremental changes to the status quo.

Close cooperation between energy procurement, distribution networks, and marketing is also a prerequisite for success. Accordingly, launching a new product range is not just a marketing exercise but an issue of organizational development for EUCs.

sgsw's new default product still contains 30 % nuclear power, which will be replaced by renewable energy sources over the coming years—using sgsw's own production facilities as much as possible. The road to completely withdrawing from nuclear energy is long, but the end should be reached by 2035, according to the energy concept of the city of St. Gallen.

The same applies to the entire Swiss energy landscape. Some utilities are well advanced when it comes to implementing a new, forward-looking energy strategy, while others face a more difficult situation from the start or doubt the long-term viability of current political developments. The example of sgsw shows that it may well be worthwhile striving to overcome existing doubts by proactively participating in the fundamental changes in the Swiss energy industry and employing competences that have been built up over decades.

From a customer perspective, the introduction of the green default solves two problems at once: If every customer gets a green power product by default, EUCs have the answer to the free-rider argument, while customers, in return, have one less task to transfer from one to-do list to another. The St. Gallen case study shows how differences between changing customer preferences and a traditional product range can be aligned by redefining the standard offer. sgsw's example shows that implementing a green default can create positive organizational dynamics vis-à-vis both internal and external stakeholders.

# References

Djamasbi, S., Tullis, T., Siegel, M., Capozzo, D., Groezinger, R., & Ng, F. (2008). Generation Y & web design: Usability through eye tracking. *Proceedings of the Fourteenth Americas Conference on Information Systems (AMCIS)*, 1–11.

Hübner, G., Müller, M., Röhr, U., Vinz, D., Kösters, J., Simon, A., Wüstenhagen, R., Chassot, S., Roser, A., Gruber, E., Gebhardt, T., Frahm, B.-J. & Alber, G. (2012). Erneuerbare Energien und Ökostrom—zielgruppenspezifische Kommunikationsstrategien. In *Modul 1–Analyse der Konsumentenentscheidungen für Erneuerbare Energien und Ökostrom*. Abschlussbericht zum BMU-Verbundprojekt (FKZ: 0325107).

Karsten, J., & Reisch, L. A. (2008). Sustainability policy and the law. *German Policy Studies, 4*(1), 45–66.

Madrian, B. C., & Shea, D. F. (2001). The power of suggestion: Inertia in 401(k) participation and savings behavior. *Quarterly Journal of Economics, 116*(4), 1149–1187.

Poole, A., & Ball, L. J. (2005). Eye tracking in human-computer interaction and usability research. In C. Ghaoui (Ed.), *Encyclopedia of human computer interaction* (pp. 211–219). Pennsylvania: IGI Global.

Sunstein, C., & Reisch, L. A. (2014). Automatically green: Behavioral economics and environmental protection. *Harvard Environmental Law Review, 38*(1), 127–158.

Thaler, R., & Sunstein, C. (2008). *Nudge. The politics of libertarian paternalism*. New Haven: Yale University Press.

**Sylviane Chassot** holds a MSc in Economics from the University of Lausanne and a PhD in International Affairs and Political Economy from the University of St. Gallen. Her thesis adopted an economic psychology perspective to investigate renewable energy decision-making of consumers and investors. In 2012–2013, she was a Swiss National Science Foundation research fellow at the Department of Psychology, Norwegian University of Science and Technology, Trondheim. She now works as a journalist in Zurich.

**Rolf Wüstenhagen** is the Good Energies Professor for Management of Renewable Energies at the University of St. Gallen. He completed his PhD in 2000 on marketing green electricity—from niche to mass market. His work on decision-making of energy consumers and investors has been published in leading academic journals. He has held visiting faculty positions at University of British Columbia, Copenhagen Business School, National University Singapore and Tel Aviv University. Since 2010, he is the Director of the University of St. Gallen's Executive Education Programme on Renewable Energy Management (REM-HSG).

**Nicole Fahr** holds a Master's degree in Marketing, Services and Communication Management from the University of St. Gallen. Her award-winning M.A. thesis investigated the influence of green defaults on electricity consumer behavior. She is now working with one of Switzerland's leading retail companies.

**Peter Graf** is the Head of Energy and Marketing and a member of the board of directors of Sankt Galler Stadtwerke (sgsw), a mid-sized Swiss municipal utility. During his tenure with sgsw, starting in 2002, he has been responsible for several initiatives in product development and marketing of green electricity, including introduction of the company's green default in 2012.

# Certificates in Germany's Renewable Energy Market

Uwe Leprich, Patrick Hoffmann, and Martin Luxenburger

**Abstract**

- What green energy labels are of current significance in Germany's voluntary green power market and how do they differ from one another?
- What is the environmental co-benefit of a green power product, and what role does this play when marketing the product?
- What is the difference between a generation certification and product certification?

This chapter answers these and other questions relating to green power certification and provides an overview of the certificates currently available. The different approaches to certification are set out in a structured manner and their key features clearly summarized. The chapter explains the purpose of each certificate, describes the relevant issuing and certifying bodies, and details the requirements that the certificate holder or their products must meet. Particular focus is placed on assessing the marketability of the individual certificates in terms of the brand awareness of a specific label and its environmental requirements.

The chapter also identifies current challenges in the German green power market, introduces some of the new green power concepts, and assesses them in terms of feasibility. The range of mechanisms covered by these new concepts

This chapter is a partial translation of the article "Zertifikate im Markt der erneuerbaren Energien in Deutschland" from the original German. This version focuses, in abbreviated form, on the market for green power (eco-gas and biofuels are dropped in this translated version). In reaction to recent developments, Sect. 4, was revised and updated.

A previous version of this chapter has been published in Herbes, C.; Friege, Chr. (Hrsg): Marketing Erneuerbarer Energien. Grundlagen, Geschäftsmodelle, Fallbeispiele, 2015, Springer Gabler.

U. Leprich • P. Hoffmann (✉) • M. Luxenburger
IZES gGmbH, Altenkesseler Str. 17, 66115 Saarbrücken, Germany
e-mail: leprich@izes.de; hoffmann@izes.de; Luxenburger@izes.de

© Springer International Publishing AG 2017
C. Herbes, C. Friege (eds.), *Marketing Renewable Energy*, Management for Professionals, DOI 10.1007/978-3-319-46427-5_7

123

includes retiring carbon credits, the specific demands on system integration when dealing with fluctuating renewable energy sources, and the marketing of locally generated green power.

The chapter concludes with an outlook on the immediate future of green power certification schemes.

**Keywords**
Green power • Green electricity • Guarantee of origin • Certificate • Label

# 1    Introduction

The focus of this chapter is on certificates for retail products in the field of renewable energy, such as green power or green gas, as well as their generation. These certificates can, for example, take the form of quality labels, which are narrowly defined according to the RAL definition of word or design marks (RAL 2014), or of product standards, which are awarded, for example, by the TÜV companies in Germany and which use the logo of the certifying organization. Guarantees of origin (GOs) are also included in the considerations.

With the exception of the latter case, the certificates have in common that they define specific criteria in the form of requirements concerning quality and condition. Product providers must demonstrate that these requirements are met by the products if they want to obtain the relevant certificate. These requirements need not only apply to the concrete product but can also deal with its production or parts of the preceding value chain.

The certificates also share the commonality of being used as a quality seal and therefore fulfill two basic functions, which differ in relevance for the involved market participants (Manta 2012):

– *For prospective customers, certificates serve as a source of information concerning the quality of a specific product*: the requirements defined for the certificate assign specific attributes to the electricity and gas products, which are homogeneous products (commodities) in principle. This leads to differentiation and allows *prospective* customers to select product specifications based on their personal preferences. At the same time, the certificate provides credible proof of the specific attributes that are important to customers.
– *Producers and suppliers use them as a marketing tool in the context of their marketing strategies*: from the supplier perspective, gas and electricity products—which in their basic form are commodities and can only compete on the basis of price—obtain specific attributes through certification that are relevant for marketing.

The aim of this chapter is to provide a structured discussion of the various approaches to certification and to summarize the defining features. Each certificate considered will be analyzed with regard to the following aspects:

– Aim of certificate
– Requirements
– Provider and certification agent
– Target audience
– Assessment of marketability

The chapter initially addresses production and product certification. All certificates reviewed can be categorized under these two headings. Next, current challenges and possible solutions are discussed. The certificates discussed in this chapter are used in the German market, one of the leading markets for renewable energy.

## 1.1    Additional Ecological Benefit

The description of the certificates specifically focuses on the additional ecological benefit. As it turns out, this aspect is of particular importance concerning the differentiation and assessment of the renewable energy certificates. In what follows, the term *additional ecological benefit* captures all ecological contributions that are generated by the certified product and that exceed the requirement of purely renewable production. In general, this contribution is provided by the criterion of "additionality," which is understood as the provision of new facilities for the generation of renewable energy that are not compensated based on the relevant regime of government subsidies (Renewable Energy Act, *Erneuerbare-Energien-Gesetz,* EEG) (EnergieVision 2014a). Such an additional ecological benefit can also be demonstrated with other contributions, which will be discussed in more detail later in the chapter.

## 1.2    Marketability

When assessing the marketability of renewable energy certificates it must be kept in mind that the awareness level of end customers[1] is partially still very limited. In particular, those certificates on the market for green power that call for increased ecological demands on products and manufacturing are still not known to many electricity customers. According to a survey by DIW econ (2012) only every tenth respondent is familiar with quality labels for green power such as the *ok-power* label or the *green power* label. And even a large share of the consumers of green

---

[1]The business-to-business segment is not considered here.

power are not familiar with the labels for it: only every fourth respondent indicated knowing about the certificates. The green power certificates of the TÜV corporations, in contrast, are much better known, even though they differ substantially with regard to their criteria. A third of participant say they have already heard about TÜV seals for green power, and among those who consume green power, half of the survey participants are familiar with them (DIW econ 2012).

While this is certainly explained in part by the popularity of the institution TÜV,[2] it also shows that the demands of the labels, such as ecological criteria, type of GO, or green power models, are not the only influence on the degree of familiarity. Instead, labels are only one of several attributes of green power used by customers who must make a decision in favor or against a product. Among all the attributes of green power considered by DIW econ, the feature "presence of quality label" triggers the smallest willingness to pay among customers. More important for the participants were the regional base of the utility as well as the fact that it exclusively offers green power. Nonetheless, the study shows a statistically significant willingness among customers to pay an extra fee between 1 and 2 Eurocent per kilowatt-hour for certified green power products (DIW econ 2012).

The importance of the quality labels should therefore not be neglected from a marketing perspective. A look at comparable markets, where customers' ecological lifestyle serves as the primary parameter, underscores the importance of product certificates. Janssen and Hamm (2011), for example, show that customers' willingness to pay increases significantly in line with the demands of the labels used in the organic food sector once these labels have achieved a relevant degree of prominence. And the survey cited earlier also allows the conclusion that the relevance of ecological demands goes up for customers as their familiarity with the specific certificates increases (DIW econ 2012).

It can thus be concluded that both the ecological demands of the certificates and their familiarity are the key variables influencing their marketability. For that reason, the marketing value of the certificates described in this chapter was assessed along these two dimensions, as shown in Fig. 1.

As an example, the fictitious label A in the graph would be given a high marketability value since it combines a high degree of familiarity with strict ecological criteria.

Label B is also very familiar but only requires basic criteria with regard to the ecological demands. Its marketability value is thus only average, but it could be increased by tightening the ecological criteria.

Label C also only refers to fundamental ecological criteria in its basic version but offers different specifications, which can be selected by the customer. Since the label is not well known, its marketability, depending on the specification chosen, falls between low and medium.

---

[2]The "Technischer Überwachungsverein TÜV" (Technical Supervisory Association) is a leading technical service company in Germany, well known in Germany for providing (prescribed) periodical car inspection services.

**Fig. 1** Dimensions of marketability

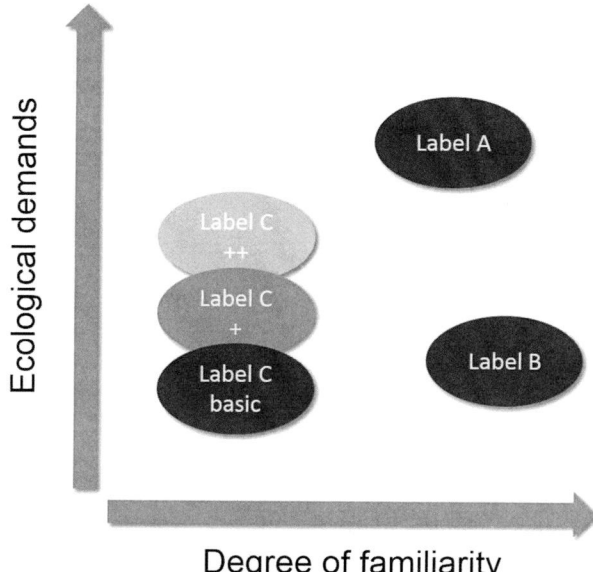

In this manner, it was possible to position all certificates considered concerning their marketability. This positioning reveals information about possible actions needed to positively influence the marketing potential.

If a certificate could not be assessed, this was noted in the text.

## 2  Certification of Green Power Generation

The certificates discussed in what follows only relate to the electricity generation process. They either fulfill the obligation with regard to the GO in line with EU Directive 2009/28/EG or serve as the basis for product certificates, which are discussed in the following chapter. Unlike the labels discussed in the following chapter, these certificates cannot be categorized as business-to-customer quality labels. Nonetheless, they can be used in a similar fashion when marketing products for the end user. This requires an appropriate positioning in the context of marketing.

From the perspective of the producer, two types of certificates can be distinguished in the green power market: GOs in the sense of Directive 2009/28/EC and freely developed generation certificates.

GOs attest to the producers of electricity from regenerative sources where and in what manner the electricity was generated. In Germany, GOs are only awarded for electricity that is not remunerated in the context of the EEG, but sold explicitly as renewable energy. In this process, the GO assures that the ecological attributes of the electrical power cannot be used more than once in the marketing process. They are invalidated for labeling purposes following a sale (UBA 2013c). GOs can be traded (a) jointly with or (b) separately from an actual electricity delivery. In case a, the dealer buys the corresponding volume of regenerative electricity from the plant operator in addition to the GO. In case b, he purchases the GO from a supplier and a corresponding volume of electricity from another source and sells the combination as green power to end customers.

Green power products on the voluntary green power market in Germany are based primarily on quantities of green power that are produced in other European countries and are traded in the form of GOs. For that reason, such products are highly marketable in the business-to-business segment. For end customers, GOs only play an indirect role, since those customers base their buying decisions not on the certificate but on the relevant attributes of the specific green power product being offered.

An assessment of the marketing perspectives just described is still of interest. Providers always have the option to explicitly mention GOs in the context of their product marketing, even though this may not be their original purpose.

GOs constitute the most basic criterion for green power: regenerative generation without a demand for any additional ecological benefit (Sect. 3.1). For that reason and due to the very low degree of familiarity, the marketing value from the perspective of the end customer must be considered as low. This assessment does not hold for the production certificates by TÜV Süd, since these are not original GOs in the sense of Directive 2009/28/EC. They were created by TÜV Süd as the foundation for the marketing of the corresponding product certificates developed by TÜV Süd and must therefore be considered separately (Sect. 2.1.2).

## 2.1    Guarantee of Origin"

### 2.1.1    Purpose and Requirement of Certificate

GOs are traded in the European Energy Certificate System (EECS). A GO states that 1 MW-h of regenerative energy was produced and fed into the grid. Since GOs on the basis of Directive 2009/28/EG only serve to label electrical power, they are not quality certifications in the proper meaning of the term (UBA 2013a). The information contained is limited to specific identification data for the production site, the amount of energy, possibly information concerning subsidies received as well as the issue date, issuing country, and identification number of the certificate (UBA 2013b).

In some European countries it is possible to supplement the GO with a note about the fulfillment of labeling criteria if a provider of a green power label is registered with the Association of Issuing Bodies (AIB) as an Independent Criteria Scheme (ICS).

The introduction of the EECS, the GO registry, and the corresponding trading with GOs is based on the implementation of EU Directives 2001/77/EG and 2009/28/EG. The EECS thus became mandatory and is a form of legal regulation (AIB 2014b). Since the start of the GO registry at the Federal Environmental Agency on 1 January 2013, every electricity company in Germany that sells green power to its customers is required to use GOs in the context of labeling electricity (European Energy Exchange 2013a).[3]

Since June 2013, GOs have been offered for trading at the European Energy Exchange (EEX). Traded are GOs from water power of the Scandinavian region (Norway, Sweden, Finland, and Denmark) and the alpine region (Switzerland, Austria, and Germany), as well as GOs for wind power from the North Sea region (Germany, Denmark, the Netherlands, and Belgium) (European Energy Exchange 2013b).

### 2.1.2 Provider/Certification Agent

GOs are managed and issued by the responsible bodies in the respective countries, the so-called issuing bodies. The European institutions that manage a registry of GOs are organized in the AIB. This nonprofit society domiciled in Belgium has also developed the rules for the EECS and provides the registries with the so-called hub as an electronic interface. This electronic interface makes it possible to transfer GOs across countries (AIB 2014c). Since 2013, Germany has also operated a national GO registry, housed by the Federal Environmental Agency, and has participated in the system as a user of the hub. However, it is not a member of the AIB (UBA 2014).

## 2.2 *Erzeugung EE* (TÜV Süd Industrie Service GmbH)

### 2.2.1 Aim of Certificate

With its own standard, CMS 83, TÜV Süd certifies the generation of electricity from renewable energy sources (*Erzeugung EE*). The certificates can be used as the basis for issuing country-specific GOs that conform to EU regulations, but they mainly serve as the basis for the proprietary product certificates of TÜV Süd (Sect. 3.1.3).

The certification always references concrete sources of energy generation and guarantees to the buyer that the electrical power is generated from renewable sources of energy. The electrical energy generated is usually purchased by intermediaries, but also directly by major consumers. The optional additional modules *Erzeugung EE+* and *Erzeugung EEneu* furthermore certify that additional requirements are met with regard to the simultaneity of consumption and the generation of the renewable energy as well as the age of the facility (TÜV Süd 2013a).

---

[3]The EECS has almost completely replaced the outdated Renewable Energy Certificate System (RECS). Across Europe only five countries continue to accept RECS certificates during a transition period. As the German registry of GOs is coming into effect, it is no longer possible to use RECS certificates in Germany.

### 2.2.2 Requirements

The standard *Erzeugung EE* is broken down into general, specific and optional requirements. The general requirements relate to the alignment of the corporate policy with the goal of climate protection, a correct communication of the certification, and the organization needed to supply all required information and documentation. Key elements of the specific requirements include the clear traceability of the regenerative energy source and the registration of the certified quantity of electricity based on the net principle. It is given by the net production provided to the grid minus the self-consumption, which was obtained externally, the pumping work of pumped-storage power plants, and all long-term delivery obligations, which explicitly call for delivery from or for the certified power plants (such as compensation in kind/restitution/servitude and deliveries from concessions) (TÜV Süd 2013a).

Optional requirements are defined for the assurance of promises of production and performance (module *Erzeugung EE+*) as well as the proof of a new installation (module *Erzeugung EEneu*). As a consequence, the recipient of the certificate must be in a position to always satisfy a predefined schedule with the certified pool of installations or to demonstrate that the production capacities are new installations in the sense of the standards of TÜV Süd. Both optional modules allow the delivery of the certified electrical energy in the form of tradable certificates (TÜV Süd 2013a).

### 2.2.3 Provider/Certification Agent

The certification agent *klima und energie* of TÜV Süd Industrie Service GmbH is in charge of the certification. The recipient of the certificate names an audit representative, who provides the necessary information.

### 2.2.4 Target Audience and Marketability

The certification of generation based on the TÜV Süd standard plays a major role, especially in connection with the optional additional modules for demonstrating the origin of the green power product certificate *EE01* and *EE02* of TÜV Süd (Sect. 3.1.3).

Assuming that all TÜV Süd product certificates are based on the certification of generation presented here, according to the Green Power Survey 2013 of the journal *Energie & Management*[4] that currently approximately 12 % of all providers of green power use *Erzeugung EE*.

---

[4]The results of the survey reflect only a portion of the entire market for green power. Of 824 suppliers contacted, 261 companies provided data. A total of around 470 different rates for green power were analyzed in the survey. Based on the assessment of the editors of *Energie & Management*, these data still represent the most important participants in the market for green power (Energie & Management 2013).

## 3 Certification of Green Power Products

While the previous chapter focused on the certification of energy generation, this chapter deals with certificates for green power products sold on the market.

All certificates described in what follows—also called green power labels—are based on the green power models established by the green power industry. They describe different ways to provide additional incentives for the expansion of capacities to generate renewable energy that go beyond the EEG. A description of the models that are currently available on the market can be found in the info box "Green power models."

The criteria covered by a green power label are not necessarily the only ones that the electricity product possesses. Several utilities offer green power products for which they define their own criteria in addition to the ecological requirements of the label used. These self-imposed standards are expected to enhance the ecological effect of products (e.g. Greenpeace Energy 2012).

In addition, all green power certificates—with the exception of the label *Grüner Strom*—allow the providers to select among different levels of rigor in line with their individual aspirations. In particular, the two TÜV corporations allow a large degree of choice. This freedom to choose, which is convenient to the provider, also implies that a single label can represent different ecological quality claims. Thus customers cannot simply assume that two green power offerings with the same green power label will necessarily fulfill the same requirements. As a consequence, their active involvement is required to arrive at an informed decision.

Table 1 provides a comprehensive overview of the green power labels analyzed.

**Table 1** Overview of product certificates in the field of green power

| Certificate | Certifying agent | Green power model | Degree of familiarity | Ecological requirements | Assessment of marketability |
|---|---|---|---|---|---|
| Grüner Strom label | Grüner Strom Label e.V. | Fund model with provider component | Low | High | Medium |
| ok-power label | EnergieVision e.V. + independent auditors | Depending on specification provider, fund, or initiation model | Low | High | Medium |
| Produkt EE01/ EE02 | TÜV Süd Industrie Service GmbH | Fund model with provider component/ provider model with fund component | High (TÜV) | Medium to high | Medium to high |
| Geprüfter Ökostrom | TÜV Nord Cert GmbH | Depending on specification provider model and fund model with provider component | High (TÜV) | Low to high | Medium to high |

**Green Power Models**

1. Provider model

    On the basis of the provider model, a provider of green power guarantees to its customers the provision of electricity from regenerative production. A widely used version of the provider model states that a specified portion of the electricity delivered must be from new plants. Normally this covers plants that are not older than 6 years. This is intended to provide an incentive to investors and operators of renewable energy plants to expand the production capacities for renewable energy (IZES 2014b). Since the physical delivery of a specific quantity of green power to customers via the public network is technically impossible, the utility needs to prove that it has sufficient ownership rights in green power attributes (Öko-Institut 2007). This proof relies on the rules for the labeling of electrical power based on GOs, which can be traded and exchanged independently of the physical delivery of electricity quantities (EU-Directive 2009/28/EG).

2. Fund model

    The fund model charges a specific premium on the price for the end user. The margins obtained in this manner are collected in a fund that is used for investments in new regenerative production facilities that cannot be operated profitably under the EEG (Hamburg Institut Consulting 2013).

    Electricity delivered on the basis of the fund model does not have to be based on physical or accounting-based (per certificate of origin) renewable energy production. If electricity from renewable sources is delivered, this is considered to be a combination of the fund model and the provider model (Öko-Institut 2007).

3. Initiation model

    In the case of electricity delivery on the basis of the initiation model, suppliers provide their customers with electricity that is—as in the case of the fund model—not necessarily based on regenerative production. The additional ecological benefit is supposed to be derived from the specific activity of the utility in initiating plants based on renewable generation. The use of existing means of refinancing such as the EEG is allowed. This is supposed to bridge the gap between government-supported renewable energy and the voluntary green power market (IZES 2014b). Additional requirements can be included with the help of green power labels. As an example, utilities that want to be certified according to the criteria of the *ok-power* label must demonstrate that 60 % of the quantities of electricity delivered are generated in a regenerative fashion from plants that were self-initiated and provided to the grid (Öko-Institut 2014).

## 3.1   *Grüner Strom Label* (Grüner Strom Label e.V.)

### 3.1.1   Aim of Certificate

The *Grüner Strom Label* e.V. (GSL e.V.) certifies green power products on the basis of the fund model with a provider component. Products that are to be certified according to the criteria of GSL e.V. are required to contain defined quantities of renewable energies. In addition, a specified amount is channeled toward the setup of regenerative facilities or measures to increase energy efficiency and infrastructure measures aimed at systems integration of renewable energy (Grüner Strom Label 2014a).

### 3.1.2   Requirements

At the moment one certificate is offered, the *Grüner Strom Label Gold (GSL Gold)*. In addition to different requirements for the supplying companies, the following fundamental demands must be met by the electricity product in question (Grüner Strom Label 2012a):

> Exclusive provision of regenerative electricity as well as the investment of a specified amount in projects in the field of renewable energy.

The certificates only allow the combined delivery of electricity. Not accepted are products where the source of the GO and the source of the physical delivery differ. To obtain a certification, the offering company needs to provide GSL e.V. with the required information, which is then validated by GSL e.V. If the criteria are met, the label is awarded for the remainder of the certification period (Grüner Strom Label 2014b). Upon completion of the first period and subsequently every other year, documents forming the basis of an evaluation by an independent scientific institute need to be submitted. In the next step, GSL e.V. makes a decision about the renewal of the certification (Grüner Strom Label 2014c).

### 3.1.3   Provider/Certification Agent

GSL e.V. is supported by seven nonprofit organizations, including the European Association for Renewable Energy (Eurosolar), three environmental organizations (Bund für Umwelt- und Naturschutz BUND, Naturschutzbund Germany NABU, Deutscher Naturschutzring DNR), a consumer advocacy group (Verbraucher Initiative), and the German chapter of International Physicians for the Prevention of Nuclear War/ Physicians for Social Responsibility and Responsibility for Peace and Sustainability, an initiative of scientists in the natural sciences.

### 3.1.4   Target Audience and Marketability

Among German customers of green power, the *Grüner Strom Label* has a high degree of familiarity and popularity. Based on the Green Power Survey 2013, approximately 29 % of all green power providers include the *Grüner Strom Label* in their offering (Energie & Management 2013). Outside of this group, the label is known only to a few electricity customers (DIW econ 2012). This, combined with the strict ecological requirements, leads to an assessment of medium marketability.

## 3.2 *Ok-power Label* (EnergieVision e.V.)

### 3.2.1 Aim of Certificate

According to the information provided by EnergieVision e.V., the *ok-power* label aims at transparency and consumer protection in the market for green power. The aim of the criteria is to give the assurance that all green power offerings that carry the label provide a guaranteed benefit to the environment. This is to be accomplished specifically via the contractually agreed provision of electricity from renewable sources and a contribution to the expansion of electricity generation from renewable energy (EnergieVision 2014a).

### 3.2.2 Requirements

In the opinion of EnergieVision e.V., the contractually agreed provision of green power without an expansion of electricity generation from renewable energies is not sufficient to provide a benefit to the environment. For that reason, additional criteria were defined for the *ok-power* label, and two of them are highlighted as the decisive elements. The first element is the demand to minimize the negative ecological consequences of the production facilities, for example by providing fish ladders for hydropower plants. The second element is the independent validation of the information provided by the electricity companies during the course of the certification as well as the correct information of the customers concerning their products.

The certification can be implemented for all three relevant types of green power models (Sect. 3.1). For each model, specific requirements are in place, for example concerning the type and age structure of the production facilities in the provider model (EnergieVision 2014a).

### 3.2.3 Provider/Certification Agent

The *ok-power* label is awarded by the association EnergieVision e.V. The association is supported jointly by Verbraucherzentrale Nordrhein-Westfalen (a consumer advocacy group) and the Öko-Institut (an institute for ecological research). The certification process mainly consists of five steps. Following an assessment of the suitability of the green power offering on the basis of the criteria of the label, a contract is concluded between the provider of the green power and EnergieVision e.V. In the next step, an independent expert is tasked with thoroughly assessing the green power offering and making sure it aligns with the criteria. The certificate of the assessor is submitted to EnergieVision e.V. and checked again. The information is reviewed again at the end of the calendar year (EnergieVision 2014b).

### 3.2.4 Target Audience and Marketability

The certification *ok-power* specifically targets all suppliers of green power that utilize one of the three business models—initiation model, provider model, and fund model. Just like the *Grüner Strom* label, the *ok-power* label is very well known in the green power arena, and 32 % of all suppliers of green power offer products that carry the label (Energie & Management 2013). As in the case of the label *Grüner Strom* this degree of popularity does not carry over to the majority of electricity customers. In light of the demanding ecological requirements, the marketability is thus again assessed as medium.

## 3.3 *Produkt EE01/EE02* (TÜV Süd Industrie Service GmbH)

### 3.3.1 Aim of the Certificate

TÜV Süd currently certifies green power products with its proprietary standard CMS 80 in two versions: *Produkt EE01* (provider model with optional fund model component) and *Produkt EE02* (fund model with a provider model component). Both product certifications aim at supporting the maintenance and expansion of renewable energy sources through the commitment of all certified producers to climate protection and an expansion of renewable energy. With the motive of unburdening transregional transport lines, both certifications allow the addition of the optional module "regional focus," which certifies minimum shares of regional and renewable energy generation.

### 3.3.2 Requirements

In two separate criteria lists (*EE01*, *EE02*) TÜV Süd outlines the requirements for the certification of the green power product. The criteria lists are structured according to general requirements (corporate policy, communication, and organization), specific requirements, and the optional additional module "regional focus." Both certificates (*EE01* and *EE02*) contain the requirement that the green power must be based fully on renewable energy sources and can be traced back to uniquely identifiable sources. With its introduction, the proof of the electricity source must be based on the GO register. A minimum of two thirds of the possible positive price difference of the green power product not justified by additional costs of including renewable energy must be directed at advancing climate protection. If the "regional focus" module is selected, a minimum share from regional electricity sources of 60 % of annual consumption must be satisfied. Further requirements are presented separately for the *EE01* and *EE02* products.

– *EE01:* For the energy balance a period of at most 12 months is in effect for renewable production. Concerning the age of the installations it must be assured that 30 % of production facilities are at most 36 months old at the time the certificates are initially granted. Overall, an installation can be kept in the portfolio for 120 months after it begins operation. As an option to satisfy the share of new installations, the certified company can elect to make a contribution per kilowatt hour of electricity sold into a support fund.

- *EE02*: The primary requirement follows from the simultaneous provision and utilization of green power. Depending on the customary time units of the national energy sector, the shortest possible unit must be selected.[5] An additional price premium to support new installations is optional.

In general, the criteria lists provide several options to introduce additional and possibly stricter criteria. In that sense, the catalogs can be seen as basic requirements, which can be made more challenging based on the wishes of the distributors of green electricity.

### 3.3.3 Provider/Certification Agent

The certification agent *"klima und energie"* of TÜV Süd Industrie Service GmbH certifies the adherence to the published list of criteria. Auditors are trained by employees of TÜV Süd and participate in an annual exchange of experiences.

### 3.3.4 Target Audience and Marketability

The proprietary standard CMS 80 is a product certification that generally targets electricity providers that want to sell green power products (TÜV Süd 2013c, d). Additional optional requirements can be used by the certified party to increase the value of its products. Consequently, the ecological requirements of this label are ranked as medium to high, depending on the individual structure chosen. The high degree of familiarity of TÜV Süd stems originally from the certification of automobiles or the testing of fairground rides; a close association with the electricity sector can be assumed (DIW econ 2012). Since the awareness level can thus also be ranked highly, the marketability of the green power products can be assessed as medium to high. At the moment, the two labels are represented in the portfolios of about 18 % of the relevant providers who participated in the survey (Energie & Management 2013).

## 3.4    *Geprüfter Ökostrom* (TÜV Nord Cert GmbH)

### 3.4.1 Aim of the Certificate

The directives for certification based on the proprietary standard A75-S026-1 describe the criteria for awarding the label *Geprüfter Ökostrom* (= certified green power) by TÜV Nord Cert GmbH. According to TÜV Nord, the certification is based on the wishes of consumers who demand a larger contribution of the providers of green power toward strengthening and supporting regenerative production facilities as well as an increasing electricity share from newer regenerative

---

[5]Quarter hours in Germany.

production facilities. The standard also allows the alternative certification of a provider or fund model.

### 3.4.2 Requirements

The list of criteria for the various standards distinguishes requirements concerning proof of production and origin, accounting treatment, and marketing of the certified electricity product, as well as customer communications. It is pointed out explicitly that the criteria that are listed and certified are mainly minimum requirements and can be augmented in line with customer demands.

The electricity that is used in the context of the certified electricity product must be generated fully from renewable energy sources (according to the definition of the national legislators). The proof of the electricity source must be provided via the GO register. An additional contribution to deepening the market for renewable energy generation is demonstrated either via a share of 33 % of the electricity provided from installations that are no older than 6 years or via an investment in the addition of new capacities for regenerative electricity generation. The balance between electricity consumption and delivery must be achieved after at most 12 months. All steps of the certified electricity between generation and consumer must be documented without exception; if certificates are used for this documentation, the route of the certificates is checked for transparency.

### 3.4.3 Provider/Certification Agent

Following a request by the customer, the certification process of TÜV-Nord Cert GmbH involves five steps. To assess whether certification is possible in principle, documents are checked initially. Next an audit is conducted on the premises of the provider. This activity is documented in a report that is handed over to the head of the certification unit for assessment and approval. In case of a positive decision, the certificate is issued. Two more monitoring audits take place during the 3-year contract term.

### 3.4.4 Target Audience and Marketability

TÜV Nord identifies companies that produce green electricity from renewable energy sources and market green power products to end customers or other utilities as the target audience for this voluntary certification (TÜV Süd 2013c). In line with the products of TÜV Süd, the recipient of the certificate can specifically increase the value of its own products via additional and optional requirements.

As with TÜV Süd, the degree of familiarity of TÜV Nord can also be assessed as high. Owing only to the low degree of ecological requirements in the minimum version, this valuation metric again falls into the range between low and high, depending on the individual specification chosen. In the overall assessment, a medium to high marketability can be assumed.

In the green power industry, the certified green power label *Geprüfter Ökostrom* of TÜV-Nord is about as widespread as the *Grüner Strom Label* (29 % of green power offerings). But it must also be pointed out that the label in some cases certifies products that additionally carry the *ok-power* label or *Grüner Strom Label* (Energie & Management 2013).

## 4  Current Challenges and Possible Solutions[6]

The German green power industry currently witnesses a crisis which can be explained primarily by two factors: stagnating numbers of customers (Kübler 2014) and increased skepticism of customers concerning the actual ecological benefit of the green power models offered (Hamburg Institut Consulting 2013).

The lack of customer growth can mainly be explained by the fact that more and more traditional utilities are switching parts of their customer portfolios towards green power without charging a premium for this offering. This is made possible by the purchase of very inexpensive GOs abroad, which allows the "greening" of the original gray electricity offering and its marketing as green power. Since these types of green power products normally do not fulfill any additional ecological requirements such as sourcing from new installations, the purchased GOs mostly come from old hydro power plants, which were built at the time without any concern for green power (IZES 2014b).

It can nonetheless be assumed that a part of those existing customers who have already contemplated the thought of switching to a provider of green power will reconsider their wish to switch and take their business elsewhere. The missing ecological benefit—as already described in Sect. 1—matters only for a small number of customers.

The traditional providers of green power that previously benefitted from the willingness of ecologically conscious electricity customers to switch offerings are losing some of their potential clients as a result of this strategy—about 20 % according to estimates (Köpke 2013).

The willingness of customers to switch is additionally reduced by the fact that even green power products with strict ecological criteria are increasingly being viewed skeptically (Kübler 2014). On the one hand, this is due to the fact that the dubious benefit of "greening" gray electricity has presented the entire industry with a credibility problem. On the other hand, the criticism is based on the fact that the initially expected effect on new renewable energy capacities based on the demand for green power massively lagged behind expectations. Especially in comparison to the success of the EEG, the impact remains very small (Hamburg Institut Consulting 2013).

---

[6]In reaction to current developments, this chapter was reworked and modified compared to the German version. Discussions of the ecological electricity market model and the customer market model were deleted since both proposed models have lost relevance. The treatment of the green power market model, which was developed from the previous two models, was updated based on current developments. Newly added was the model of regional green power labeling, which was recently (March 2016) introduced into by the German Federal Ministry for Economic Affairs and Energy (BMWi) in the context of a position paper (BMWi 2015a).

This was realized in the industry, and as a consequence, different modeling approaches to leaving the crisis behind are currently being discussed. Three of these approaches are comprehensively summarized in the following chapters. While the first model targets an added ecological benefit via the European emissions trading mechanism, the other models focus primarily on the systems integration of renewable energy or on regional linkages of electricity offerings as additional ecological benefits.

## 4.1    Climate Electricity Model

The climate electricity model is based on an approach presented in late 2013 by IZES GmbH. The idea is to directly use consumers' willingness to pay, which exceeds the price of an average conventional product from fossil resources for the purchase of certificates in the European emissions trading mechanism.

The model was developed against the background of a price increase for these emission certificates that have massively disappointed expectations. The steady reduction in the availability of certificates has therefore not materialized and no or only a few positive effects for climate protection in the form of emission reductions have been accomplished. Because of the numerous freely distributed certificates, surpluses from the economic crisis, and significant external credits for lowering emissions during the second trading phase (2008–2012), only moderate price increases can be expected over the long term. The climate electricity model is supposed to act as a corrective measure.

A provider would have to structure his products in such a way that the customer, similar to the fund model (Sect. 3.1), pays a defined premium per unit, which is used exclusively for the purchase of emission certificates from the EU emissions trading mechanism. The certificates purchased will be put aside and invalidated in the official register of the emissions trading authority. Thus these certificates would no longer correspond to $CO_2$ emissions, and the available quantity in the emissions trading mechanism would be reduced for all participants. With this product, consumers are given an opportunity to directly intervene in the market-based instrument of emissions trading and, thus, to accomplish a reduction in the emissions cap that is independent of political targets.

The difference to compensation products[7] lies in the fact that the resulting ecological effect does not target a compensation of energy use but instead makes a contribution to the successive improvement of the effects of emissions trading. The stronger these effects are, the greater is the ecological contribution that follows from an increase in the $CO_2$ price owing to the increased scarcity of the certificates.

---

[7]Compensation products are offered mainly in the market for ecological gas. They involve the purchase and cancellation of an amount of emissions rights that corresponds to the emissions volume generated during the use of the (natural gas) product.

Fundamentally, a coupling with emissions trading in line with the model is possible for every product. In the opinion of IZES, products such as electricity and gas, which are characterized by a particularly high level of $CO_2$ emissions during production or consumption and are consumed in large quantities, are particularly suitable for the implementation of the idea. This is justified both by the expected effects and by marketing considerations (IZES 2013).

## 4.2    Green Power Market Model

The model is a combination and refinement of two previous suggestions for green power models (ecological electricity market model and customer market model[8]), which were rendered obsolete with the publication of the green power market model (GMM). In 2014, Naturstrom AG, Greenpeace Energy, Elektrizitätswerke Schönau (EWS), Clean Energy Sourcing, and MVV Energie AG jointly spoke out in favor of GMM and jointly advertise it on a common homepage.[9]

The goal and aim of the GMM is to create the possibility of using green power that was generated on the basis of the EEG for the direct delivery to the end customer in a manner that is clearly traceable—without violating the prohibition of dual use (§80 Renewable Energy Act 2014). The GMM can be interpreted as a supplement to the market premium model of the EEG. The additional ecological benefit of the green power market model stems from the systems integration of renewable energy via electricity distribution.

This is supposed to become possible via the option for the distributors to leave the redistribution system of the EEG, as long as a defined minimum share of EEG electricity is purchased directly and without support from the redistribution scheme of the EEG from facility operators. The minimum shares of EEG electricity as well as electricity from wind and photovoltaic installations are specifically based on the current nationwide ratio of production and consumption by the end user subject to levies.[10] To maintain the cost neutrality of the electricity, which the utilities can take into account when satisfying the minimum shares, the average EEG compensation (the average cost of the total volume of the electricity subsidized via the EEG system) must be paid. Differences due to the inclusion of EEG facilities with higher

---

[8]The ecological electricity market model was presented jointly in early 2014 by Elektrizitätswerke Schönau (EWS), Greenpeace Energy and Naturstrom AG, three German pioneer providers of green power. It specifically targeted ambitious providers of ecologically generated electricity. Also early in 2014, Clean Energy Sourcing GmbH (CLENS) published a proposal for the "market integration of electricity from renewable energy through incorporation into the competition for customers" (in short, the customer market model). To assure topicality and comprehensiveness, the two models are not discussed in more detail at this point.

[9]www.gruenstrom-markt-modell.de

[10]In contrast to the complete provision of the minimum share via the fluctuating generation of renewable energy, as required in the ecological electricity market model, the requirement of the customer market model (shares of controlled and fluctuating production) was adopted.

or lower compensation payments are settled between the utility and the Renewable Energy Account or the responsible operator of the transmission grid. The fulfillment of the minimum shares must be assured in the annual accounting. In addition, a penalty payment (integration payment) was introduced in the amount of 2 Eurocent per kilowatt hour for those quantities of EEG electricity that could not be integrated on the basis of quarter-hourly values. Distributors now have an interest in avoiding these penalty payments, which leads to the search for forms of flexible compensation that are as cost effective as possible. The remaining surplus cover, which can be sold, for example, via the exchange or covered via balancing energy and penalties, can be credited for the fulfillment of the annual minimum shares. Since facility operators do not receive any payments from the EEG transfer system, they should receive GOs for that electricity and be allowed to sell it as electricity from renewable energy sources (CLENS 2014d, e).

In a letter from the Federal Minister of Economics and Energy to the German Bundestag, the GMM was officially rejected by the responsible ministry in October 2015 (BMWi 2015b). This was received largely with disappointment by the green power industry. The reason for the rejection was that the model was assessed to be "*presumably not cost-neutral, extremely complex and without relevant value added for the energy sector.*" Instead, the Ministry of Economics and Technology considered regional aspects to be more expedient for the acceptance of the further local advancement of renewable energy and announced the development of a model for the regional labeling of electricity that is supported by the EEG (BMWi). A corresponding position paper was published in March 2016. The proposal for regional green power labeling outlined in that paper is summarized in the following section.

## 4.3    Regional Green Power Labeling

A position paper by the German Federal Ministry for Economic Affairs and Energy (BMWi), published in March 2016, outlines a model for the labeling of green power that is supposed to enable participants to separately market electricity that was subsidized on the basis of the EEG as green power (BMWi 2015a). In contrast to the GMM, the focus is not on the systems integration of renewable energy, but rather on its regional distribution, where producers and consumers must come from the same area. In the opinion of the ministry, this new regional element of green power will have a favorable effect on the acceptance of the energy revolution at home.

According to the proposal, regional proximity exists if the customer is located in the same area where the production facility is situated. The paper discusses the advantages and disadvantages of fixed regions (such as administrative districts) and of moving regions, where a specific radius around the consumer is defined. In the latter version, it would also be possible to include in the system foreign installations that are situated close to the border; this assumes the inclusion of foreign installations in the EEG.

The labeling of power as green is supposed to be implemented via so-called regional certificates (which depend on the quantity of electricity provided). For installations that want to participate in the labeling, it is issued and managed by the German GO register. This GO register is also responsible for validating the regional linkage.

In the assessment of the ministry, this regional labeling of green power complements the electricity labeling that is currently in place. In addition to the current determination of the mix of energy sources and the corresponding labeling of electricity, a utility can now also label a certain share of the electricity as regional.

Trading of regional certificates is supposed to be possible along the contractual delivery chain, and the attributes of the electricity can only be acquired jointly with electricity delivery.

Installations that comply with the EEG might be able to earn additional revenue from marketing the aspect of regional production. These revenues are expected to benefit the customer, who is paying the EEG surcharge, and not the distributers: installations that offer tenders[11] can price the aspect of regional production in their bids and will thus be placed more favorably in the process. For installations that do not participate in the tender process, the position paper suggests a reduction of the market premium by 0.1 Eurocent per kilowatt hour. In the assessment of the ministry, both versions will provide relief to the surcharge account of the EEG.

A detailed draft of the regional certification of green energy as well as the evaluation of the conformity with European law are still outstanding at the time of writing (March 2016).

## 4.4   Closing Assessment and Outlook

As described in Sect. 1, the additional ecological benefit is an important aspect for the success, also with regard to marketing, of an electricity product. The previously presented modeling approaches offer new and, in some respects, quite different starting points for the generation of such an additional benefit. Especially as the design moves beyond incentives for capacity expansion that are hard to quantify, these products could make an important contribution as the industry struggles to overcome the current period of stagnation.

In the case of the climate electricity model, the effectiveness mainly depends on the agenda for the continuation of the European emissions trading scheme (ETS). Since the current trading period, which ends in 2020, is characterized by an expected excess of certificates of 2–2.8 billion, it is very questionable whether a product that is structured along these lines can already have a significant effect in the short run (Neuhoff and Schopp 2013).

---

[11]The determination of the market premium will most likely commence in 2017 and will be based exclusively on a tender procedure.

However, the approach looks promising as we approach the fourth trading period, which starts in 2021. Its concrete structure remains hard to assess. A key question relates to the continued use of certificates from emission reduction products in the ETS. In addition, it must be determined to what degree the planned market stability reserve in the ETS would work against the mechanism of the climate electricity model. Based on the current discussion and proposals, the aim of the reserve is to maintain stability in the quantity of certificates that are auctioned off in a range between 400 and 833 million (European Commission 2014). Should it be implemented as currently discussed, the anticipated effect of a climate electricity product would be reduced significantly.

Prior to the negative assessment by the German Federal Ministry for Economic Affairs and Energy, hopes that the GMM would be implemented were based on the power to issue ordinances of the German EEG from 2014. It makes it possible to issue an ordinance that allows for the implementation of the GMM and concretely addresses the implementation of "a system for the direct marketing of electricity from renewable energy sources," where the electricity can be labeled as "electricity from renewable energy sources" (Bundesregierung 2014). The clear rejection of the GMM by the Ministry of Economics and Technology has made it clear that the implementation of the model or the ordinance are a distant object at the moment.

The trend currently appears to be moving in the direction of regional marketing instead. With the model concept for the regional labeling of green power, the Federal Ministry for Economic Affairs and Energy (BMWi) has identified the regional aspect as the most important issue (BMWi 2015a) and is pursuing an agenda. The design of a regional green power label is also covered by the previously mentioned power to issue ordinances under the EEG.

Whether the currently discussed models will contribute to a further diversification or even a repositioning of the green power industry will thus become apparent in the near future. What is certain, however, is this: There will be completely new and, in some instances, complex requirements for the structuring of future certificates.

## References

AIB. (2014a). Certification FAQ—Are there certificates for all types of electricity? Accessed June 25, 2014, from http://www.aib-net.org/portal/page/portal/AIB_HOME/CERTIFICATION/C_FAQ/Types_of_certificate.

AIB. (2014b). Certificates supported. Accessed June 26, 2014, from http://www.aib-net.org/portal/page/portal/AIB_HOME/CERTIFICATION/Types_certificate.

AIB. (2014c). Accessed July 27, 2014, from http://www.aib-net.org/portal/page/portal/AIB_HOME.

AIB. (2014d). Accessed September 12, 2014, from http://www.aib-net.org/portal/page/portal/AIB_HOME/FACTS/Market%20Information/Accepted_certificates.

BMWi. (2015a). Regionale Grünstromkennzeichung. Eckpunktepapier. Accessed March 21, 2016, from http://www.bmwi.de/BMWi/Redaktion/PDF/P-R/regionale-gruenstromkennzeichnung-eckpunktepapier,property=pdf,bereich=bmwi,sprache=de,rwb=true.pdf.

BMWi. (2015b). Schreiben des deutschen Wirtschaftsministers Sigmar Gabriel an den deutschen Bundestag. Berlin vom 13.10.2015. Accessed March 22, 2016, from http://images.klimaretter.info/filestore/1/6/2/3/3_fab9aa3260714bd/16233_ff3794420982e3c.pdf?v=2015-10-30+17:15:48.

Bundesregierung. (2014). Beschlussempfehlung und Bericht des Ausschusses für Wirtschaft und Energie (9. Ausschuss). Drucksache 18/1891 vom 26.06.2014, Berlin.

CLENS. (2014a). Marktintegration von Strom aus Erneuerbaren Energien durch Einbeziehung in den Wettbewerb um Kunden. Accessed July 09, 2014, from http://www.clens.eu/fileadmin/Daten/Veroeffentlichungen/140211_Kundenmarktmodell_CLENS.pdf.

CLENS. (2014b). Echtzeitwälzung: Strom aus Erneuerbaren Energien in den Wettbewerb um Kunden integrieren. Presentation slides 3. IZES Energiekongress, 12.03.2014. Saarbrücken.

CLENS. (2014c). Versorgung von Stromkunden mit Strom aus Erneuerbaren Energien. Presentation slides 4. MCC-Kongress Erneuerbare Energien, 06.05.2014, Berlin.

CLENS. (2014d). Das Grünstrommarktmodell—Vorschlag für ein optionales und kostenneutrales Direktvermarktungsmodell zur Versorgung von Stromkunden. Accessed July 09, 2014, from http://www.clens.eu/fileadmin/Daten/Veroeffentlichungen/Gruenstrommarktmodell_CLENS.pdf.

CLENS. (2014e). Grünstrommarktmodell: EEG-Strom in den Wettbewerb um Stromkunden integrieren. Accessed July 10, 2014, from http://www.clens.eu/fileadmin/Daten/Veroeffentlichungen/Praes_Gruenstrommarktmodell_I.pdf.

DIW econ. (2012). Potentiale für Ökostrom in Deutschland. Accessed July 04, 2014, from http://diw-econ.de/en/wp-content/uploads/sites/2/2014/03/DIWecon_HSE_Oekostrom.pdf.

Energie & Management. (2013). Übersicht der Ökostromtarife in Deutschland. Auswertung der jährlichen Ökostromumfrage der Fachzeitschrift. Energie & Management. Accessed July 04, 2014.

Energie & Management. (2015). E&M Powernews: Gabriel kippt Grünstrom-Marktmodell. Accessed March 22, 2016, from http://www.clens.eu/fileadmin/Daten/Mediathek/Pressespiegel/151020_EM-Gruenstrom-Marktmodell.pdf.

EnergieVision. (2014a). Kriterien für das Gütesiegel "ok-power"für Ökostrom. Accessed July 09, 2014, from http://www.ok-power.de/fileadmin/download/Kriterienkataloge/ok-power-Kriterien_7-3_v2.pdf.

EnergieVision. (2014b). Für Energieversorger—Zertifizierungsablauf. Accessed July 09, 2014, from http://www.ok-power.de/energieversorger/infos.html.

European Energy Exchange. (2013) Rahmenbedingungen für Herkunftsnachweise in Deutschland. Accessed July 11, 2014, from http://cdn.eex.com/document/136091/Haendlerworkshop_GoOs_DE.pdf.

European Energy Exchange. (2013b). Handel von Herkunftsnachweisen ab 06. June 2013 an der EEX. Accessed June 25, 2014, from https://www.eex.com/de/about/newsroom/news-detail/eex--handel-mit-herkunftsnachweisen-startet-am-6--juni/61016.

European Commission. (2014). Questions and answers on the proposed market stability reserve for the EU emissions trading system. Accessed July 22, 2014, from http://europa.eu/rapid/press-release_MEMO-14-39_en.pdf.

Greenpeace Energy. (2012). Kriterien von Greenpeace für sauberen Strom. Accessed July 16, 2014, from http://www.greenpeace-energy.de/fileadmin/docs/zertifizierung/gp_kriterien.pdf.

Grüner Strom Label. (2012a). Criteria 2012. Accessed July 20, 2016, from http://www.gruenerstromlabel.de/fileadmin/dateien/PDF-Dokumente/GSL_Criteria_2012.pdf.

Grüner Strom Label. (2012b). Kriterienkatalog 2012. Accessed July 20, 2016, from http://www.gruenerstromlabel.de/fileadmin/dateien/PDF-Dokumente/GSL_Criteria_2012.pdf.

Grüner Strom Label e.V. (2013). Erste Biogasprodukte erhalten Grünes Gas Label. Accessed July 02, 2014, from http://www.gruenerstromlabel.de/aktuelles/newsansicht/?tx_ttnews[tt_news]= 154&cHash=d7779129b55daf3aa756392553c43f89.

Grüner Strom Label. (2014a). Häufig gestellte Fragen (FAQ). Accessed July 02, 2014, from http://www.gruenerstromlabel.de/faq/.

Grüner Strom Label. (2014b). Weg zur Zertifizierung. Accessed July 02, 2014, from http://www.gruenerstromlabel.de/gruener-strom-label/fuer-energieversorger/weg-zur-zertifizierung/.

Grüner Strom Label. (2014c). Ablauf der Zertifizierung. Accessed July 02, 2014, from http://www.gruenerstromlabel.de/gruener-strom-label/fuer-energieversorger/ablauf-der-zertifizierung/.

Hamburg Institut Consulting (2013) Weiterentwicklung des freiwilligen Ökostrommarktes. Accessed June 30, 2014, from http://www.ok-power.de/uploads/media/Projektbericht_ Zukunft_fOEM_final_v2.pdf.

IFEU. (2013). Ökologische Bewertung von Ökogas-Produkten—Hintergrundpapier für den EnergieVision e.V. Accessed July 04, 2014, from http://www.ok-power.de/uploads/media/ Hintergrundpapier_Oekogas-Produkte_01.pdf.

IZES. (2013). Verzahnung von Energievertrieb und Emissionshandel—Ein Vorschlag am Beispiel "Klimastrom". Accessed July 04, 2014, from http://www.izes.de/cms/upload/publikationen/ IZES_Diskussionspapier_Klimastrom.pdf.

IZES. (2014a). *Netzwerk Elektromobilität Rheinland-Pfalz—Modul 6a: Ökostrom für Elektromobilität*. Saarbrücken: Im Auftrag des Ministeriums für Wirtschaft, Klimaschutz, Energie und Landesplanung (MWKEL).

IZES. (2014b). *Ökostrom in Klimabilanzen*. Saarbrücken: Im Auftrag des EnergieVision e.V.

IZES. (2014c). *Netzwerk Elektromobilität Rheinland-Pfalz—Ökostrom für Elektromobilität*. Saarbrücken: Im Auftrag des Ministeriums für Wirtschaft, Klimaschutz, Energie und Landesplanung (MWKEL).

Janssen, M., & Hamm, U. (2011). Zahlungsbereitschaft und Verbraucherpräferenzen für Produkte mit unterschiedlichen Öko-Zertifizierungszeichen. In G. Leithold, K. Becker, C. Brock, S. Fischinger, A.-K. Spiegel, K. Spory, K.-P. Wilbois, & U. Williges (Eds.), *Es geht ums Ganze: Forschen im Dialog von Wissenschaft und Praxis Beiträge zur 11. Wissenschaftstagung Ökologischer Landbau, Justus-Liebig-Universität Gießen, 15.-18* (pp. 279–280). Berlin: Verlag Dr. Köster.

Köpke, R. (2013). Deutliche Abkühlung auf dem Ökostrommarkt. Energie & Management. Accessed July 15, 2013, 9.

Kübler, K. (2014). Leistet man durch den Kauf von "Ökostrom" einen Beitrag zur Energiewende in Deutschland? *Energiewirtschaftliche Tagesfragen, 3*(2014), 43–46.

LBD. (2014). Gutachten zur energiewirtschaftlichen Bewertung des Ökostrom-Markt-Modells. Accessed July 3, 2014, from http://www.lbd.de/cms/pdf-gutachten-und-studien/1403-LBD-Gutachten_Oekosstrom-Markt-Modell.pdf.

Manta, M. (2012). *Bedeutung von Gütesiegeln. Einfluss von Involvement auf die Bedeutung von Gütezeichen im Produktbeurteilungsprozess* (pp. 5–10). München: FGM-Verlag.

Neuhoff, K., & Schopp, A. (2013). Europäischer Emissionshandel: Durch Backloading Zeit für Strukturreform gewinnen. *DIW Wochenbericht Nr., 11*, 3–11.

Öko-Institut. (2007). Green power labelling—Final Report from the project "Clean Energy Network for Europe" (Clean-E). Accessed June 30, 2014, from http://www.oeko.de/oekodoc/ 1480/2007-230-en.pdf.

Öko-Institut. (2014). Zertifizierungsmodelle—Initiierungsmodell. Accessed June 30, 2014, from http://www.ok-power.de/energieversorger/zertifizierungsmodelle.html.

Ökostrom-Online. (2014). RECS-Zertifikat. Accessed June 05 2014, from http://www.oekostrom-online.com/recs-zertifikat.php.

RAL. (2014). Guidelines for quality marks. Accessed July 20, 2016, from http://www.ral-guetezeichen.de/fileadmin/lib/pdf/guete/englisch/RAL-Guidelines_for_quality_marks_0808.pdf.

TÜV Nord. (2014c). Certifying Criteria for "Certified Eco Power" according to the TUV NORD CERT Standard A75-S026-1. Accessed July 22, 2016.

TÜV Süd. (2013a). TÜV SÜD Standard CMS 83 (Version 08/2013) Zertifizierung der Erzeugung von Strom aus Erneuerbaren Energien. Accessed June 10, 2014, from http://www.tuev-sued. de/uploads/images/1332939911881816900083/generationee-e.pdf.

TÜV Süd. (2013b). Ökostrom-Zertifizierung von TÜV SÜD für EU-weiten Handel zugelassen. Accessed June 10, 2014, from http://www.tuev-sued.de/tuev-sued-konzern/presse/ pressearchiv/oekostrom-zertifizierung-von-tuev-sued-fuer-eu-weiten-handel-zugelassen.

TÜV Süd. (2013c). TÜV SÜD Standard CMS 80 (Version 07/2013) Zertifizierung von Stromprodukten aus Erneuerbaren Energien mit mindestens 30% Neuanlagenanteil (Produkt EE01). Accessed July 20, 2016, from http://www.tuev-sued.de/uploads/images/ 1338873780525196720762/ee01-e.pdf.

TÜV Süd. (2013d). TÜV SÜD Standard CMS 82 (Version 07/2013) Zertifizierung von Stromprodukten aus Erneuerbaren Energien mit zeitgleicher Lieferung (Produkt EE02). Accessed July 20, 2016, from http://www.tuev-sued.de/uploads/images/1338873773865501850299/ee02-e. pdf.

U.S. Environmental Protection Agency. (2008). EPA's Green Power Partnership—Renewable Energy Certificates. Accessed July 11, 2014, from http://www.epa.gov/greenpower/ documents/gpp_basics-recs.pdf.

UBA. (2013a). Was unterscheidet einen Herkunftsnachweis von einem Ökostromlabel? Accessed June 25, 2014, from https://www.umweltbundesamt.de/service/uba-fragen/was-unterscheidet- einen-herkunftsnachweis-von-einem.

UBA. (2013b). Welche Angaben enthält der Herkunftsnachweis? Accessed June 25, 2014, from http:// www.umweltbundesamt.de/service/uba-fragen/welche-angaben-enthaelt-der-herkunftsnachweis.

UBA. (2013c). Herkunftsnachweise sorgen für Durchblick im Ökostrommarkt. Accessed June 27, 2014, from http://www.umweltbundesamt.de/sites/default/files/medien/press/pd13-002_ herkunftsnachweise_ sorgen_fuer_durchblick_im_oekostrommarkt.pdf.

UBA. (2014). Personal inquiry by email and phone September 2014.

**Uwe Leprich** studied economics at the University of Bielefeld and the University of Athens, Georgia, USA. In 1993 he graduated on the subject of "Least-cost Planning" and from 1987 to 1995 was a research assistant in the energy field of the Öko-Institut, Freiburg, and furthermore responsible for energy policy in Hessian Ministry for Environment, Energy and Federal Affairs. Since 1995 Uwe Leprich is professor at the University of Technology and Economics of the Saarland (HTW), responsible for Economic Policies and Energy Systems in the department of Economics. In 1999 he was a co-founder of the IZES (Institute for Future Energy Systems); scientific director of IZES since 2008. From 2001 to 2002 he was expert member of the Enquête Commission of the 14th German parliament for "sustainable energy supply". Since 2010 Alternate Board Member of the Agency for the Cooperation of Energy Regulators (ACER) of the EU. Specialties: framework of the energy sectors with main focus on liberalisation and regulation of the electricity sector; Instruments of national and international energy and environmental policy; future actors and market roles in the regenerative economy.

**Patrick Hoffmann** studied Industrial Engineering at the Umweltcampus Birkenfeld of the University of Applied Sciences Trier, specialisation field Environmental Planning. From 2002 to 2008 he was marketing manager of SilverCreations Software AG. Since 2008 he is research assistant in the work field energy markets of the IZES (Institute for Future Energy Systems) with main focus on analysis of user groups in the field of energy efficiency services. Furthermore he is project and consortium leader of various energy-industry studies and research projects focusing on issues such as energy efficiency, green electricity, smart metering and electric mobility.

**Martin Luxenburger** studied industrial engineering at the University of Technology and Economics of the Saarland (HTW). He gained practical experience in the fields of power plant location study and emissions trading at the MVV Energie AG as well as accounting energy quantities in the VSE Group. Since 2010 he is research assistant in the work field energy markets of the IZES (Institute for Future Energy Systems) as well as project and consortium leader of energy related business and market processes, further development of the energy-economic regulations (with focus on renewable energies law), system and market integration of renewable energies, business case development, energy trade as well as system services.

# Part II

# Special Markets and New Business Models

# Marketing of Biomethane

Carsten Herbes

**Abstract**

Biomethane is playing an important role in the transition to a sustainable energy system. It is renewable and can be easily stored, and its availability can be matched to times of peak demand. It benefits from an existing infrastructure for transport, storage, and use, namely, the public gas grid, gas-based heating systems in millions of private households, and a powerful fleet of compressed natural gas (CNG) vehicles around the world. However, its marketing is complex since biomethane serves four distinct markets: to generate power in combined heat and power units, to heat private households and businesses, to fuel CNG vehicles, and to supply material for the chemical industry. Each of these markets has different competitors, legal frameworks, and customer requirements to which biomethane providers must adapt their marketing mix and strategies. To shed light on the factors involved in such marketing, we first look in this chapter at the market development and regional distribution of biomethane production as well as the value chain. We then analyze factors influencing the different markets and the marketing mix of providers. Our analyses and examples often focus on Germany because it has the largest and most developed biomethane market in the world, but we also consider developments in other European countries and around the world.

A previous version of this chapter has been published in Herbes, C.; Friege, Chr. (Hrsg): Marketing Erneuerbarer Energien. Grundlagen, Geschäftsmodelle, Fallbeispiele, 2015, Springer Gabler.

C. Herbes (✉)
Institute for International Research on Sustainable Management and Renewable Energy, Nuertingen-Geislingen University, Neckarsteige 6-10, 72622 Nuertingen, Germany
e-mail: carsten.herbes@hfwu.de

**Keywords**

Biogas • Biomethane • Green gas • Marketing

# 1    Introduction: Market Development

Biogas is produced through the anaerobic digestion of organic material such as manure, organic waste, or energy crops such as maize. Untreated biogas has a methane content of 50–65 % and can be upgraded to biomethane via a purification process. In the purification process, $CO_2$ and other undesirable ingredients like $H_2S$ are removed, and the gas is further compressed and odorized (Hahn et al. 2014). Biomethane must comply with the standards for natural gas that prevail in the public gas grid, standards that may differ according to the local gas operator.

Most biogas plants around the world use the biogas on site for generating electricity in a combined heat and power (CHP) unit. However, a small but growing number of them—367 out of 17,240 in Europe—(European Biogas Association 2016a, b) upgrade the biogas to biomethane, which is then fed into the public gas grid or used as fuel for natural gas vehicles (NGVs).

In recent years, Europe, and especially Germany, has seen dynamic development in the production and marketing of biomethane. By the end of 2014, 367 biomethane plants were operating in Europe (Fig. 1), 178 of these in Germany (European Biogas Association 2016b). The German plants alone accounted for an annual production of 8.5 TWh of biomethane in 2015 (Dena 2015b). Outside Europe, fewer plants are found. Only in the USA (25), Japan (6), and South Korea (5) does a significant biomethane market exist (Thrän 2014, all figures end of 2012).

The dynamic growth in Germany (Fig. 2) has been fueled by an ambitious target set in 2007 by the federal government, namely, to produce 6 billion ncbm of biomethane by 2020 (Federal Ministry for the Environment, Nature Conservation, Building and Nuclear Safety 2011). To reach this target, the federal government established subsidy programs to stimulate development of the industry. However, with the 2014 reform of the German Renewable Energy Act (REA) came a massive

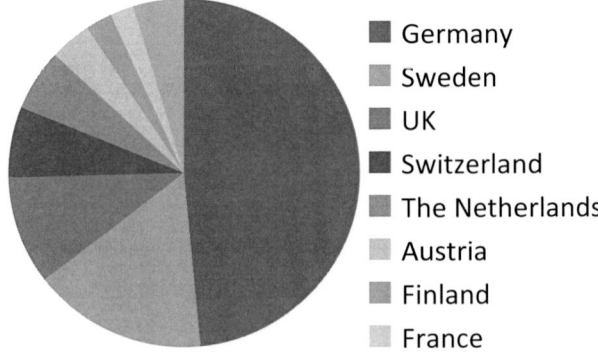

**Fig. 1** Distribution of biomethane plants in Europe, numbers per end of 2014, total 367 (European Biogas Association 2016b)

- Germany
- Sweden
- UK
- Switzerland
- The Netherlands
- Austria
- Finland
- France

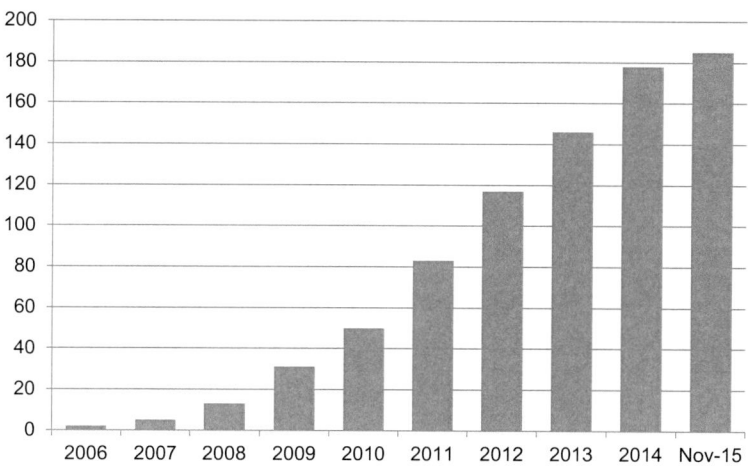

**Fig. 2** Number of biomethane plants in Germany, 2006–2015 (Dena 2015c; Herbes 2015)

lowering of feed-in tariffs (FITs) for biomethane-fueled CHP units, effectively halting the industry's dynamic growth (Dena 2015b).

The French government, on the other hand, has recently announced plans to facilitate construction of 1500 biogas plants over the next 3 years (Ministère de l'Écologie, du Développement durable et de l'Énergie 2014). More specifically, the government is aiming to have 8 TWh from biomethane injected into the grid by 2023 (Ministère de l'environnement, de l'énergie et de la mer 2016). These ambitious goals, supported by favorable FITs, have sparked dynamic growth in the sector: as of June 2016, 21 biomethane plants were injecting gas into the grid (personal calculation based on grid operators' Websites).

In the UK, 50 upgrading plants were in operation as of late 2015 (Zennaro 2015). Further growth in Europe is expected. In fact, biomethane production by 2030 is projected to reach 18–20 billion ncbm/year, meaning nearly 40 % of the biogas produced in Europe will have been upgraded by that time (Green Gas Grids 2013).

Other countries have also identified great potential for their emerging biogas markets. In the USA, for example, it is estimated that biogas upgraded to biomethane could replace up to 56 % of the natural gas consumed in the transportation sector (Milbrandt 2013). Japan started to inject upgraded biogas into the grid as early as 2010 (Osaka Gas 2010); the industry there shows significant potential for growth, especially in biogas derived from waste.

Biomethane markets around the world are driven to a varying extent by state interventions. These range from FITs for biomethane, probably the strongest form of state support, to free markets where explicit support mechanisms do not exist but motivating factors that promote development of the biomethane market do. Figure 3 shows the range of policies various nations have adopted, some of which can work concurrently in the same country. In Germany, for example, there is a FIT (although

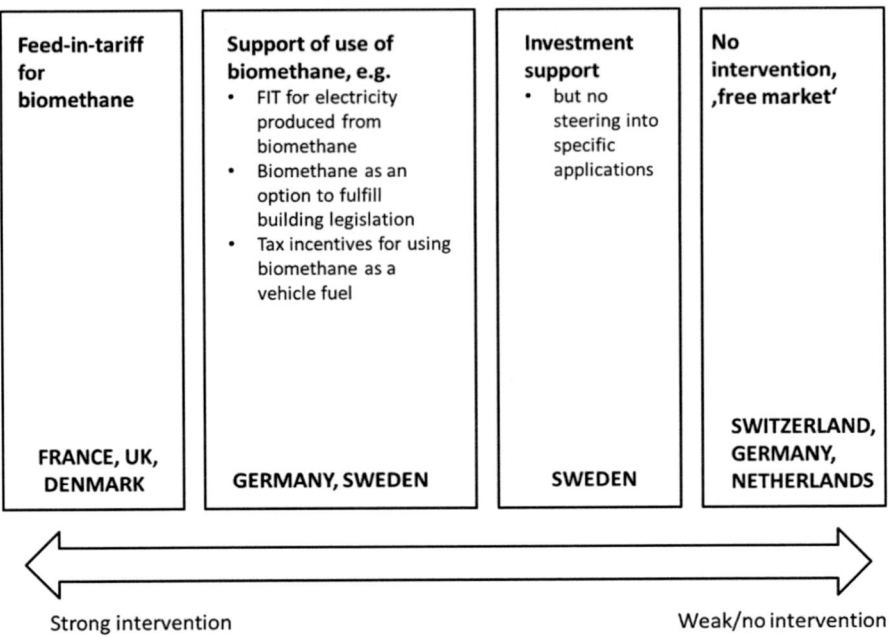

**Fig. 3** Forms of state intervention in biomethane market [author illustration, partly based on Strauch (2014), Larsson et al. (2016), Mathiasson (2016), Schmid (2015), Department of Energy and Climate Change (2013)]

financially no longer attractive) for CHP units using biomethane. At the same time, oil companies selling vehicle fuel must reach a certain quota of renewables in their portfolio, a requirement that drives demand for, among other sources, biomethane and biomethane-based certificates. Similar market-driven support exists in the German construction industry, where new buildings must fulfill certain energy-related standards; one way to meet these standards is to use a biomethane-fueled heating system. Finally, a private market for gas products containing a percentage of biomethane has emerged as households increasingly look to make purchasing decisions that help protect the environment.

Of course, the different support schemes offer investors and operators varying financial incentives to develop biomethane injection plants or to convert existing plants into upgrading plants. FITs provide a high level of security for investors; moreover, they are relatively easy to handle and remove from biogas upgrading plants the need to market their product directly to customers since commercial transactions are handled indirectly through a registry system. Free markets, on the other hand, present investors with greater risks, meaning there is an expectation for greater reward. Risk arises from the fact that after a plant has been built, few options exist for responding to negative developments on the demand side. Plant adaptations and cost reductions offer only marginal flexibility, inasmuch as feed-stock, especially agricultural feedstock, can only be obtained from the local area around the plant, and changes to the gas-generating and upgrading technology are also difficult to realize.

Not only do the different support schemes influence the return and risk of the investor, but when a scheme favors specific application areas, it also determines where and how the biomethane is used. So, for example, favorable FITs exist in Germany for converting biomethane into electricity, a fact that has led to a large proportion of domestically produced biomethane ending up in CHP units. Sweden, on the other hand, offers tax incentives for cars fueled by biomethane, and so the largest share of biomethane serves the automotive sector.

While national markets continue to develop in scope to meet divergent needs, transnational trade of biomethane still faces challenges. Germany has been an exporter of biomethane in recent years, while Switzerland and Sweden have been major importers (Dena 2014b; Schmid 2015; Larsson et al. 2016). Exporters into Germany, however, face in effect a trade tariff since the terms of Germany's REA prohibit CHP units that use foreign biomethane from claiming a FIT. This is just one example demonstrating barriers to international trade of biomethane. To lift these barriers, a European renewable gas registry would be required that documents the flows and attributes of biomethane throughout the value chain from producer to consumer. This would require that criteria and standards for biomethane be made consistent across all European markets (Kovacs 2016).

## 2 Players in the Value Chain

Players in the value chain include both biogas plants that upgrade gas to produce biomethane and those that use biogas on site to produce electricity (Kaltschmitt and Streicher 2009). The latter are generally smaller plants operated by local farmers. The former are larger plants developed and operated by nonagricultural players, such as the biomethane producer NAWARO BioEnergie AG. Their plant in Guestrow, Northern Germany, for example, represents the upper end of the size scale, with a capacity of 5000 ncbm/h.

The average German biogas plant producing electricity on site has a size of ca. 0.5 MWel (German Biogas Association 2015), while the capacity of the average biogas upgrading facility is about five times that size (personal calculations based on German Biogas Association 2015; Dena 2015b).

Plant size has evolved in response to the regulatory thresholds governing the German FIT for electricity produced with biomethane. As a result, many plants in Germany run at capacities between the 350 and 700 ncbm/h thresholds.

These plants operate in a biomethane value chain that has witnessed dynamic development driven by strategic integration moves, both forward and backward. The main steps in this value chain are shown in Fig. 4.

Agricultural businesses, traditionally strong in biomass production, have started activities in biomethane production. Technology providers and Engineering-Procurement-Construction companies like Envitec and the former MT Energie have also been moving forward in the value chain. Envitec even offers contracting

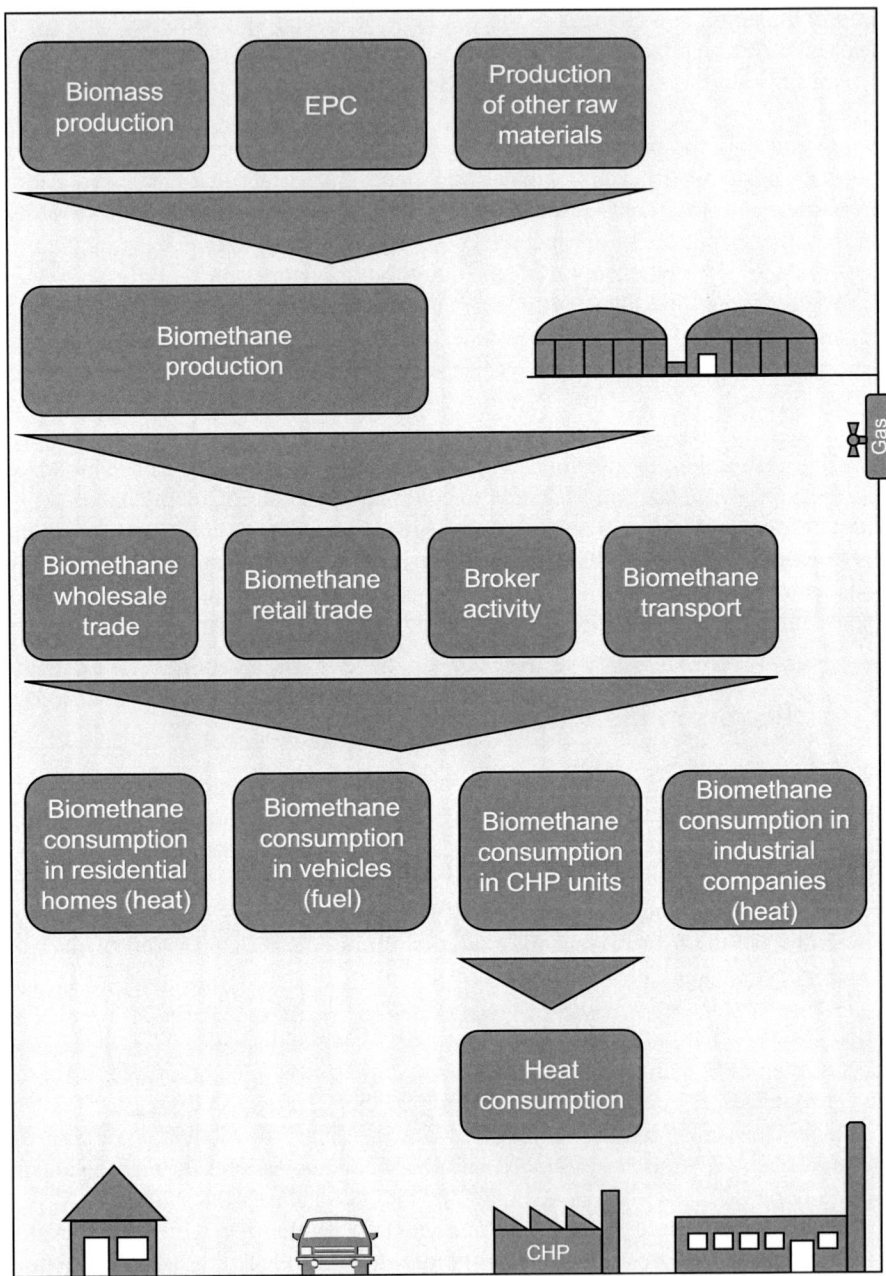

**Fig. 4** Biomethane value chain [author illustration based on Herbes and Hess (2011)]

services in the heat market based on the biomethane it produces in its own plants, thus reaching directly into the end-user market. The classic players in biomethane production are AC Biogas and NAWARO BioEnergie AG, both of which started their businesses in that part of the value chain. AC Biogas, however, has expanded its traditional activities by both backward—producing biomass—and forward integration—providing heat from biomethane.

The wholesale step of the value chain is represented by big utility companies such as E.ON or RWE, which have, however, begun backward integration into biomethane production. Besides these big utilities, new companies such as bmp greengas have established a strong position in the wholesale market. Retail, the next step, is dominated by municipal utilities, the so-called Stadtwerke. Some of these have also expanded their activities into biomethane production. Between the wholesalers and the retailers are the brokers, like Arcanum Energy.

A look at the overall value chain makes it clear that a wide range of players shape the biogas industry, from business activities focused far upstream to those operating all the way downstream. The classic biomethane producers have countered these moves by competitors by starting their own forward and backward integration activities.

# 3    Utilization Paths and Their Influencing Factors

That biomethane can be used in many ways represents a defining characteristic of this market, one that sets it apart from the market for renewable electricity. There are, broadly speaking, four paths or markets that exist for biomethane: using it to produce electricity in CHP units, using it as a substitute for natural gas in NGVs, using it for heating buildings, and using it as a raw material in the chemical industry (Fig. 5).

The presence of four distinct utilization paths creates path dependency in both the biomethane-based end product and in the marketing strategies used by suppliers of that product. Along some paths, the marketing of biomethane is strongly influenced by the prevailing legal framework; for example, the production substrate or the plant size used to generate the biogas may have to conform to categories stipulated by regulations. In Germany, laws such as the REA, the Biomass Ordinance, the Renewable Energy Heat Act, the Biofuel Sustainability Regulation, and others must be considered in any marketing strategy. There are, however, other markets where legislative stipulations play a lesser role and biogas product attributes are best aligned with consumer preferences. But each utilization path poses its own unique challenges, which means that planning for a biomethane plant requires that the target utilization path be considered first. Of course, shifting to other utilization paths once a plant is in operation is possible; however, this may result in less than optimal revenues.

The relative importance of the four utilization paths differs by country. In Germany, the world's largest biomethane market, the gas is used predominantly for firing CHP units, a market whose economic viability depends on the FIT

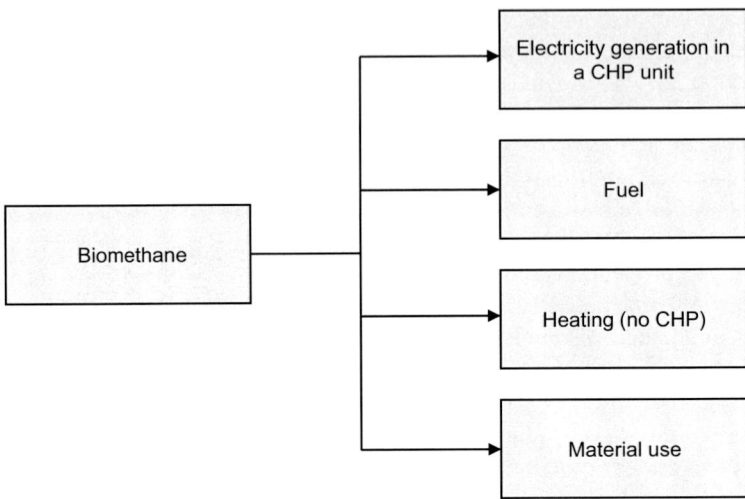

**Fig. 5** Biomethane utilization paths

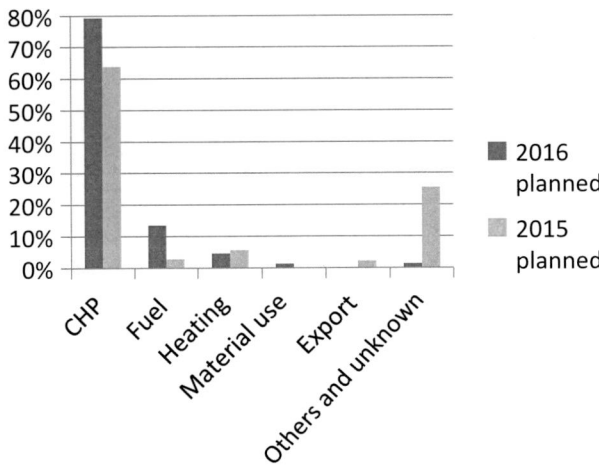

**Fig. 6** Relative importance of four utilization paths in German market, 2015 and 2016 (planned) [author illustration, data based on Dena (2015b, 2014b)]

stipulated by the REA (Fig. 6). The German REA thus has been the main driver behind the German biomethane market. The market for vehicle fuels is the second most important path, but with an expected share of 13 % of the market in 2016, it is much smaller than the market for CHP units. In Sweden, biomethane is used primarily as a vehicle fuel, above all for captive fleets such as buses. Around 54 % of the biogas produced in Sweden is upgraded, and most of it is used as vehicle fuel (Larsson et al. 2016). In Switzerland, on the other hand, while biomethane is used as vehicle fuel, it is used primarily for heating. In fact, Swiss gas distributors import foreign biomethane to satisfy the country's demand (Schmid 2015).

In the following sections, we look at the different paths and their influencing factors in more detail.

## 3.1 Market for Electricity Generation in CHP Units

Using biomethane in CHP plants seems unique to the German market because it is driven by the state-led support system. The most important element of this support system is the FIT for electricity generated from biogas and biomethane (Sorda et al. 2013). The way this system works is that, first, cogeneration units draw biomethane from upgrading plants via the public gas grid; they then feed the electricity they produce into the electricity grid, receiving a fixed FIT from the electricity grid operators. The FIT is determined both by characteristics of the cogeneration unit and by characteristics of the biomethane used. The German REA includes detailed criteria for determining the FIT, including specifications for the substrate (energy crops vs. waste) as well as for the size of the upgrading unit. Driven by the incentives set by the REA, the majority of the feedstock in German biomethane plants is maize silage (Dena 2014b), and many of the upgrading units have a size just below the thresholds set in the REA (350 and 700 ncbm/h in REA 2009, and 700, 1000, and 1400 ncbm/h in REA 2012).

While CHP units have dominated the biomethane market in Germany, they are bound to lose their leading position in the future. With the radical reform of the biogas-related regulation in REA 2014, the FIT for electricity from biomethane has been cut significantly, so that biomethane is in most cases no longer a viable fuel for new CHP units (Herbes et al. 2014a). Businesses that supply heat and use cogeneration as the technical means to do so will probably turn to other options, such as using natural gas or woody biomass.

## 3.2 Heating Market

Biomethane can be used as a substitute for natural gas in heating households and businesses. This is a potentially large market: 48 % of the residential buildings in Germany use gas for heating (BdEW 2014), and in France 50 % of multifamily dwellings and 30 % of the single-family dwellings do so (gas in focus—Observatoire du Gaz 2014). In the UK, 80 % of all residential homes are connected to the gas grid (Department of Energy and Climate Change 2013). Overall, gas accounted for 40 % of the heat produced in the European Union in 2013 (Eurostat 2016). Since heat production in residential homes and other buildings relies heavily on natural gas, the heating market represents a significant opportunity for biomethane.

Many biomethane-based gas products are available in Germany. Herbes et al. counted 170 different products in 2014 with a biomethane content between 1 and

100 % (Herbes et al. 2016). While this path only accounts for a small part of the German biomethane market and less than 1 % of the German market for natural gas (Eberlein 2015), its relative importance for biomethane marketers is expected to rise owing to the difficulties that CHP plants face, as explained in Sect. 3.1.

In Germany, the heating market is partly driven by state-led support schemes. There are regulations (EEWärmeG, MAP, EWärmeG Baden-Württemberg) stipulating the use of renewables in newly constructed buildings. But these regulations allow the developer or owner to decide which form of renewable energy is used. So far, biomethane has not been competitive with solar thermal power, insulation, and other renewable options. This is partly due to the fact that even in new buildings constructed by private households, biomethane must be used in a cogeneration unit to satisfy legal regulations, and the limited demand for heat in a typical residential home makes this option financially unattractive (Loßner et al. 2012). Moreover, the federal Renewable Heat Act applies only to newly constructed buildings, which account for only 0.6 % of existing buildings (EWI/ GWS/Prognos).

But private households also purchase biomethane-based gas products without being incentivized or obligated by legal regulations. These households make choices similar to those made in the market for green electricity: they choose to make purchasing decisions that both protect the environment and support the development of renewable energy. In the Netherlands, households can opt for a green gas product over a pure natural gas product (Eker and van Daalen 2015), and marketers are able to command a price premium to support this choice.

In Switzerland, where heat generation is the predominant utilization path for biomethane, gas distributors have played an important role in shaping the market by offering default green product options. So if a customer of, for example, Energie360° in Zurich does not specify otherwise, the gas product received contains 5 % biomethane (Schmid 2015).

In Sweden, the government supports the use of biogas for heating by granting a full exemption from energy and carbon dioxide taxes (Larsson et al. 2016). In the UK, the Renewable Heat Incentive provides a FIT for biomethane that has been the main driver of biomethane plant development (Adams et al. 2015). In Switzerland, biomethane is exempted from the $CO_2$ levy on natural gas; however, the exemption applies only to domestically produced gas (Die Bundesversammlung—Das Schweizer Parlament 2012).

However, two important barriers—absent from the green electricity market— hamper the development of the biomethane market. First, there are widespread reservations about using biogas (Herbes et al. 2014b; Markard et al. 2016), at least as reflected in the attitudes of German consumers. These reservations center around the large-scale use of energy crops in the biogas industry, a practice that in Germany has drawn public criticism under the tabloid headline "maizification" of the landscape (Herbes et al. 2014a). Whether these reservations would be shared worldwide remains an open question.

However, the second barrier is likely to exist anywhere:consumers have been lulled into thinking that green energy carries little in the way of additional cost.

Price premiums for green electricity products targeting private households have remained low, averaging in 2012 only about 2 % (Mattes 2012). So customers have grown used to purchasing green energy for a price only marginally higher than that of electricity from fossil fuels. They have not had to choose between their pocket-book and their conscience. But biomethane production comes at a cost about double that of natural gas, meaning biomethane retailers face the challenging task of convincing customers to pay a sizeable price premium hitherto unknown to them (Herbes et al. 2016).

## 3.3    Market for Vehicle Fuels

The market for vehicle fuel is potentially interesting for biomethane producers in the short and mid-term. Unlike in the market for electricity generation, renewables such as wind or solar power are not likely to present strong competition to biomethane since electric vehicle technology has progressed far more slowly than the technology for NGVs.

Clearly, the market for biomethane as a vehicle fuel depends on the development of the market for NGVs. As of late 2012, the number of NGVs worldwide stood at nearly 17 million, with Iran, Pakistan, and Argentina representing a combined share of 48 % of the world's NGV fleet. In Europe, Italy stands out with a fleet of roughly 750,000 vehicles, which corresponds to 5 % of the world's fleet. Over the last 10 years, the Asia-Pacific region has recorded a most impressive average annual growth rate of 35 %, but even in Europe the average annual growth rate in the number of NGVs reached 14 % between 2003 and 2012. Only in North America has the growth in NGVs been stagnant over this period (NGV Global 2014).

In Germany, the biggest producer of biomethane, however, the number of NGVs has not grown significantly; with a meager share of 0.2 % of the world's fleet, the market in Germany is generally considered a bust (Rosenstiel et al. 2015). Still, of the 900 compressed natural gas (CNG) filling stations operating in Germany, 350 offer products that contain biomethane, and in many cases, customers can opt for a 100 % biomethane product (Erdgas mobil 2013). As it stands today, biomethane accounts for 20 % of the German CNG market.

The key factors influencing future growth include the delivery infrastructure (number of gas stations), tax benefits and exemptions, vehicle manufacturers' product policies, vehicle conversion costs, changes in consumer perceptions, and coordination among the different players in the market (Rosenstiel et al. 2015).

Apart from these general developments and influencing factors in the market for NGVs, there are unique trends in the different markets for biomethane as a vehicle fuel.

In Germany, two distinct markets exist for biomethane as a vehicle fuel (Geisler 2014). First, consumers can make a conscious decision to buy biomethane instead of natural gas to fuel their vehicles. Biomethane in this case replaces natural gas. In the second market (quota market), companies that sell vehicle fuels can choose to use biomethane to fulfill their quota obligations.

In the first market, the main influencing factors are the number of NGVs, the infrastructure, and consumer preferences. Consumers must choose a biomethane product, and that means in many cases having to drive to a different filling station, since only a third offer biomethane products. The price for biomethane at the filling station is about the same as the price for CNG, so there is no significant negative price influence. However, as in the heating market, negative perceptions of biogas can prevent consumers from choosing a fuel based on biogas.

In the second market, the regulatory framework plays a dominating role. In Germany, the controlling regulations are the German Federal Emission Control Act and the Biofuel Sustainability Regulation. The Emission Control Act obliges companies that sell gasoline or diesel to fulfill an annual biofuel quota, which today stands at 6.25 %. The quota can be fulfilled by blending biomethane with CNG, although alternative renewable fuels like bioethanol and biodiesel compete with biomethane in this application. Moreover, biomethane used to fulfill a quota obligation must meet the requirements of the Biofuel Sustainability Regulation. In particular, companies must demonstrate that by using biomethane they reduce the greenhouse gas emissions of their fuels by a certain percentage (Geisler 2014). In Germany, this figure has stood at 3 % since 2015 and will increase to 7 % in 2020. Market experts estimate that the first quota of 3 % has already been fulfilled with the reduction potentials of currently used biofuels (Geisler 2014). However, the quota of 7 % will be difficult to achieve and could drive the demand for biomethane, especially biomethane generated from waste (Erdgas mobil 2013; Grope and Holzhammer 2012; Geisler 2014), since this has the greatest reduction potential.

That biomethane in the quota market is judged by both fuel companies and customers on its specific greenhouse gas reduction potential changes the nature of competition in this market. Competition is no longer based on the price per amount of energy but on price per amount of greenhouse gas emissions saved. The greenhouse gas reduction potential of other biofuels such as biodiesel affects the competitive landscape, as does the fact that the reduction potential for biomethane differs depending on the feedstock used. This places more stringent requirements on biomethane plants to document the greenhouse gas saving potential of their products.

One negative factor in this market is regulation on the European level that limits the share of food-crop biofuels that qualify as renewables in the transportation sector. The current regulatory limit is 7 % of energy consumption by 2020 (Scarlat et al. 2015).

Blending CNG with biomethane is just one of several options for fuel companies. Alternatively, they can use bioethanol or biodiesel mixed with ethanol and diesel. Thus, the decision to choose biomethane depends on the market development and prices for these alternative biofuels. Overcapacities in biofuel refineries and low prices for used cooking oil have had a negative influence on biomethane prices in the fuel market (Erdgas mobil 2013).

In Sweden, 54 % of the biogas produced in the country is upgraded, and most of it is used as a vehicle fuel. Fully 0.9 TWh of biomethane is sold as vehicle fuel, and Sweden even imports biomethane from other countries to meet its fuel demand.

# 2 Renewable Forms Used in Contracting

Statistics published by the German Association of Heating Energy Suppliers (VfW e. V.) (http://www.energiecontracting.de) provide an overview of the shares attributable to different forms of renewables in the supply of heating energy and electricity in the contracting market. The statistics show that almost 7 % of heating energy in contracting projects across Germany came from renewables in 2014. That corresponds to a connected wattage of 1575 MW for thermal energy. The connected wattage for electrical energy in contracting projects using renewables amounted to around 205 MW in 2014. Solid biomass in the form of timber accounts for by far the largest share of renewables, followed by biogas and wind power (Fig. 1). Photovoltaic (PV) systems have only played a subordinate role to date.

The following sections deal in detail with the individual forms of renewables and their suitability for use in the contracting business.

## 2.1 Photovoltaics (PV)

PV systems only generate electricity. This can either be fed into the grid or used to cover the generator's own needs. A number of business models are available on the market for both options, and these are offered both to commercial and industrial customers and to private households. What all models have in common is the goal of using suitable surfaces at prospective customers to generate solar power and exploiting the hurdle of system financing and operating and maintenance risk on behalf of customers. However, these models only constitute contracting when the power generated in the PV system is also used by customers themselves. Here, generally made of the lease model described in Sect. 1.2.

Until a few years ago, the more attractive model in economic terms was to feed electricity into the grid. This was because of the high rates of compensation and checked cuts in solar power subsidies in recent years on the one hand and

share of
in
atus

w-

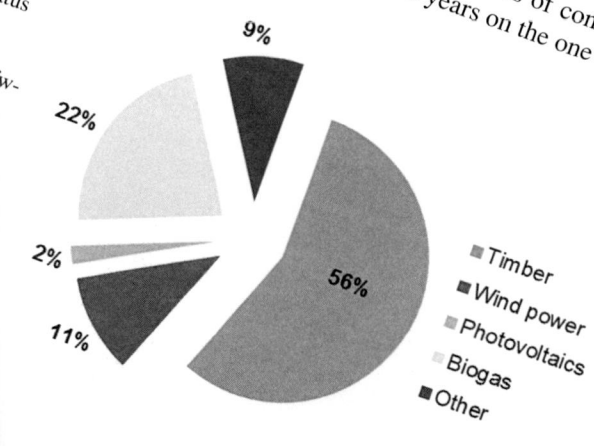

9%

22%

2%

11%

56%

■ Timber
■ Wind power
■ Photovoltaics
░ Biogas
■ Other

Biomethane accounted for 1.2 % of all fuel in Sweden and 11 % of biofuel in 2013 (Larsson et al. 2016). The Swedish government applies a wide set of policy instruments to support the use of biomethane as a vehicle fuel. It grants a full exemption from energy and carbon dioxide tax as well as vehicle tax for 5 years, provides investment support to various parts of the biogas value chain, requires filling stations to sell renewable fuel, and employs a number of additional instruments (Larsson et al. 2016).

## 3.4 Material Use

The chemical industry is a new and still nascent market for biomethane. In principle, there are two ways of using biomethane in this industry. First is a "real" material use, where biomethane replaces natural gas in industrial processes such as the production of plastics. Second, companies in the chemical industry can buy certificates of origin from biomethane producers while continuing to produce plastics based on oil. In both cases, companies can build on the positive environmental characteristics of biomethane and offer their customers "green" plastics, for example, as a packaging material. Given the fact that plastic packaging for fast moving consumer goods such as shampoo or yogurt accounts for only a small fraction of the total cost of the product, yet at the same time oil-based plastics present a well-known threat to the environment, "green packaging" based on biomethane can be an attractive strategy for marketing consumer goods.

German-based BASF, a big multinational player in the chemical industry, has, together with TÜV Süd, already developed and introduced a certification system based on mass balancing to replace natural gas with biomethane in its production processes. This allows BASF to offer its customers raw materials such as polyamide for which biomethane or bionaphta is used instead of fossil-based materials (BASF 2014; Klein and Frietsch 2015).

## 4 Suppliers' Marketing Mix

The four utilization paths have different requirements for biomethane products, different decision criteria, and user economics. Therefore, we differentiate the discussion of the elements of the marketing mix in the following sections accordingly.

## 4.1 Product Policy

When buying biomethane, customers value product attributes differently depending on the utilization path. German customers operating CHP plants buy a biomethane product based on the attributes stipulated in the German REA that are relevant for the FIT they receive. The FIT is calculated based on the size of the biomethane

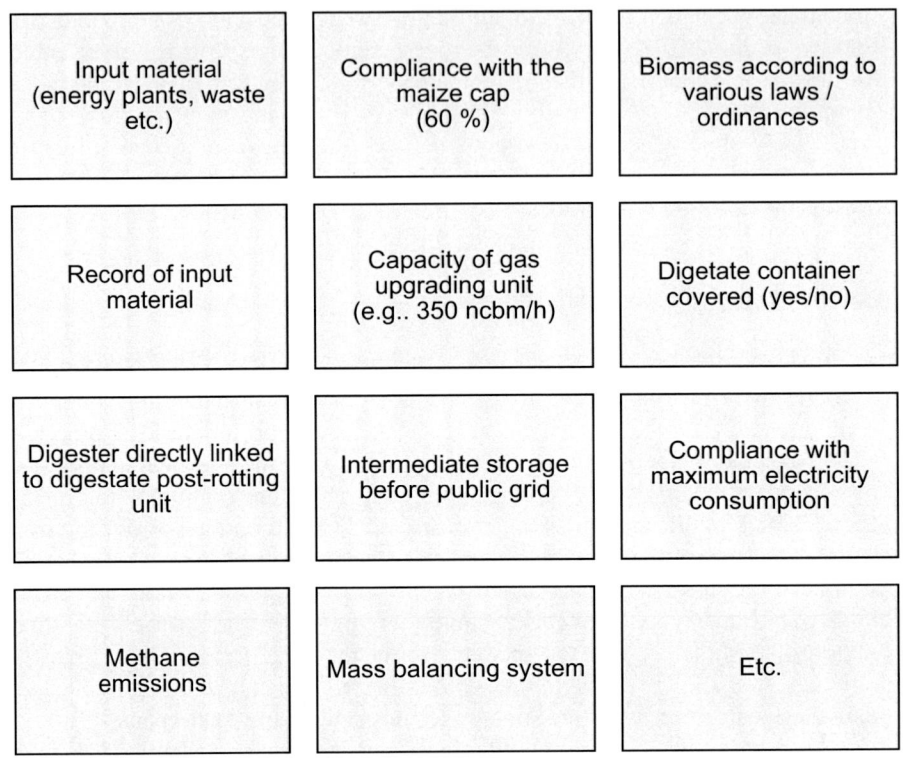

**Fig. 7** Attributes of biomethane products [author illustration, based on Dena (2014a)]

upgrading unit, the year it went into operation, and the kind of input material the plant uses. Since each of these attributes has numerous possible values, many combinations exist, which has led to the introduction of many different biomethane products on the market (Plaas 2014). Figure 7 presents the different attributes of biomethane products that can be combined to produce differentiated products. Currently, the biogas register operated by the German Energy Agency lists more than 100 different biomethane products, each with slightly different product attributes. For customers, these variations are confusing and make it difficult to formulate a business plan for their biomethane-based CHP unit.

Whereas FITs drive the CHP market, it is greenhouse gas emission reduction potentials that drive the market for vehicle fuels. Since the reduction potential depends on the production process and the input material, this market also features a wide variety of products. The greater the reduction potential of a product, the more value that product has for the oil and gas companies purchasing it. The end customer at the filling station, however, is usually unable to detect any differences between products based on reduction potentials, although an advertiser could change that perception.

In the heating market, the product portfolio for private households is not nearly as complex as in the CHP and vehicle fuel markets. Most households do not operate a CHP unit and so are not eligible to receive a FIT. They simply use the biomethane in existing heating systems that run on natural gas. But other attributes can be important for private customers. First, the biomethane content in a product is crucial. Very few products consist of 100 % biomethane; many contain 10 or 30 % biomethane. Biomethane marketers can further differentiate their products based on the input material. Based on the attitudes of German consumers, the use of energy plants, especially maize, for producing biogas is viewed critically, while biogas produced from waste is viewed more favorably (Forsa 2013; Herbes and Ramme 2014). Just as in the market for green electricity, marketers could further benefit from marketing their gas as being "produced locally" or bearing an ecolabel. However, these are only rarely used, at least in Germany (Herbes et al. 2016).

In Switzerland, public utilities have started to integrate a nudging strategy into their product policy. Gas consumers who do not actively opt for a different solution often receive a natural gas product that contains a certain percentage of biomethane, for example, 3 or 10 %. Customers who want a larger share of biomethane can choose from a range of products with up to 100 % biomethane; they can also choose between products with or without ecolabels (Schmid 2015).

## 4.2 Pricing Policy

The pricing policies of suppliers, like product policies, differ greatly across the four different utilization paths. In the CHP market, the largest market in Germany, the user economics are transparent to suppliers. Based on the size and initial date of the CHP operation, the FIT according to the REA is easy to calculate. Investment costs for standard CHPs are also transparent, as is the cost of natural gas as an alternative fuel. If the price for the heat is known as well, the supplier can more or less work out a customer's business plan and set a price for the biomethane accordingly (Herbes and Hess 2011).

Besides the price level, a pricing strategy contains two additional components: the price adjustment mechanism during the contract period and the length of the contract period. While biomethane contracts in the first years of the developing CHP market were often tied to the price of heating oil and then to the price of natural gas, today many contracts have a fixed annual price increase of perhaps 2 %. The length of the contract period varies widely in this market, from a few months to 10 years.

In the fuel market, biomethane producers and traders are rather passive price takers. The price for the biofuel quota is strongly influenced by conditions in the bioethanol and biodiesel markets, both of which are far bigger than the biomethane market. But unlike in the CHP market, in the fuel market there are no long-term contracts.

In the end-user market at the filling station, pricing is a barrier for marketers. While prices for gasoline and diesel are listed in euros per liter, prices for CNG and biomethane are listed in euros per kilogram. Since the energy value per unit varies widely, it is difficult for end customers to compare prices. This has led the biomethane industry to suggest listing prices in euros per the equivalent of one liter of gasoline, a change that would require modifying the law governing the listing of prices (Dena 2013).

For the heating market, a recent study examined the pricing strategy of suppliers for biomethane-based gas products targeted at residential customers (Herbes et al. 2016). The study shows that the price depends mainly on the biomethane content of the product. Research on the willingness to pay of customers from the green electricity market (Herbes et al. 2015) and on the biomethane market (Forsa 2013) suggests that certain attributes like ecolabels, extra climate protection activities, and using waste instead of energy plants could offer room to demand higher prices. However, none of these attributes currently has a significant impact on the price of biomethane products. Moreover, absolute prices as well as price premiums for biomethane products as compared to pure natural gas products differ widely among providers.

Besides the conditions in the different utilization paths or markets, production costs also influence pricing strategies. The costs differ considerably depending on the size of the biogas plant and the upgrading plant as well as on the cost of the input material. For an energy-based plant with a capacity of 500 ncbm/h (raw gas), the costs may fall between 7.8 and 8.4 eurocent/kWh; for a larger plant with a capacity of 2,000 ncbm/h, costs fall between 6.4 and 7.0 eurocent/kWh (Grope and Holzhammer 2012).

## 4.3    Distribution Policy

As part of their marketing strategy, biomethane producers must make a decision on their distribution channels. Apart from the more fundamental decision on the utilization path, they must also decide whether they want to use intermediaries, such as wholesalers or retailers, or market directly to their end customers. This kind of disintermediation is already visible in the market: on its Website, for example, the biomethane producer NAWARO BioEnergiepark Güstrow GmbH addresses CHP operators. Apart from wholesalers and retailers, trading platforms like www.biomethan-markt.de also match suppliers and customers. For the physical distribution of gas, suppliers use the public gas grid.

Another important task for suppliers in distribution is the documentation of the attributes of biomethane for their customers. The CHP market will serve to illustrate. Remember that sellers of electricity on the CHP market must meet regulatory criteria to be eligible to receive a FIT. They must prove that the biomethane used in their CHP units satisfies these criteria. Thus, biomethane producers must provide documentation that CHP operators can use to meet this burden of proof.

The basic approach to such documentation is "mass balancing," that is, the biomethane is not physically transported from the producer to the customer. Instead, the producer injects a certain amount of biomethane into the public grid, and this quantity with its attributes is booked in a registry. In Germany, 80 % of all biomethane upgrading plants use the biogas registry of the German Energy Agency (dena). The customer takes the same amount of gas out of the public grid, marks the corresponding quantity in the registry as used, and receives documentation that can be used to claim a FIT for the electricity produced or to prove to private households that the gas sold to them is really biomethane. The attributes the producer documents in the biogas registry, for example, the size of the upgrading unit or input material, are verified by external auditors, who issue audit reports. Suppliers use the audit reports for their documentation in the biogas registry (Dena 2015a).

## 4.4 Communication Policy

Communication policy, like the other elements of the marketing mix, depends on the utilization path. In the following paragraphs we first look at the arguments suppliers use to win customers and then at the communication channels.

In the CHP market in Germany, the main argument used to be the long-term cost advantage in heat generation over the use of natural gas. This cost advantage was derived from the additional income from the FIT for the electricity produced in the CHP unit. The communication was business to business, mostly through the sales personnel of traders like municipal utilities, but also partly through the Websites of biomethane producers.

However, the heating market is largely business to consumer, making the Websites of marketers vitally important. Generally, it is difficult for biomethane marketers to attract consumer attention. Consider, for example, that 10 % of the users of the German price comparison portal Toptarif in 2013/2014 were specifically looking for green electricity products when shopping for electricity; however, only 1.6 % of gas customers were looking for green gas (Toptarif 2014).

A second barrier for providers of biomethane-based products is the fact that customers looking for an environmentally friendly gas product can choose between products based on biomethane and so-called climate gas products. For the latter, the provider uses natural gas and compensates the $CO_2$ emissions of the product by supporting environmental protection projects abroad or other activities. It is probably safe to assume that not all customers understand the difference between the two product categories, and so not all of those among the aforementioned 1.6 % that switch products will be buying a biomethane-based product. These so-called climate gas products, based on $CO_2$ emission compensation, are available not only on the German market but also in France, Switzerland, and other countries.

Further complicating the marketing of biomethane-based products are the critical views consumers make have toward biogas. Various studies from Germany have shown that biogas is the least popular among the renewable energy technologies (Herbes et al. 2014b) and that there are strong reservations concerning the use of energy plants, especially maize, for producing biogas (Herbes and Ramme 2014; Forsa 2013; Markard et al. 2016).

Providers of biomethane-based products for residential customers communicate the following advantages: positive environmental effects, climate protection, a support of energy transition process, and the building of new renewable energy facilities that reduce dependence on imports. The arguments for buying biomethane are thus similar to those used by suppliers of green electricity (Herbes and Ramme 2014). However, only a minority of the providers in a recent study of biomethane-based products in Germany actively communicate the input material (energy plants vs. waste), and one third did not even disclose the input material upon request (Herbes et al. 2016). One reason for this rather restrained communication could be the suppliers' awareness of the controversial public discourse around energy plants.

## 5    Conclusion

The markets for biomethane are multifaceted and dynamic, varying widely based on four distinct utilization paths and the prevailing regulatory climates. Many markets are still influenced by state-led support schemes in various forms, such as FITs, quota regulations, investment support, tax incentives, and other energy-policy instruments. Germany, boasting the most developed biomethane market in the world, built its dominant position largely on the basis of FIT incentives for electricity produced using biomethane. Now that the FIT has been lowered considerably, other markets such as those for vehicle fuel, heating, and use in industry are expected to gain in importance. Yet another market—one not influenced by regulatory policy—is the private household, where consumers seeking to protect the environment can choose biomethane as a substitute for natural gas. By developing a suitable marketing mix, providers could use these pro-environmental attitudes and preferences of customers to grow the biomethane market and make themselves less dependent on state-led support schemes.

## References

Adams, P., Mezzullo, W. G., & McManus, M. C. (2015). Biomass sustainability criteria: Greenhouse gas accounting issues for biogas and biomethane facilities. *Energy Policy, 87*, 95–109.
BASF. (2014). *BASF: Erstes Polyamid aus dem Massenbilanz-Verfahren im Serieneinsatz. plasticker news.* Accessed June 22, 2016, from http://plasticker.de/news/shownews.php?nr=23396.
BdEW. (2014). *BDEW Studie Wie heizt Deutschland.* Accessed June 22, 2016, from https://www.bdew.de/internet.nsf/id/C44E72D1F65D174EC1257DAC00319783/$file/141212%20BDEW%20Studie%20Wie%20heizt%20Deutschland%20Anhang.pdf.
Dena. (2013). *Transparente Preisinformation für einen Kraftstoffmarkt im Wandel: Lösungsansätze für eine bessere Vergleichbarkeit der Preise alternativer Kraftstoffe.* Berlin. Accessed June 22, 2016, from http://www.dena.de/fileadmin/user_upload/Publikationen/Verkehr/Dokumente/20131007_Positionspapier_web.pdf.
Dena. (2014a). Biogas register Germany—Catalogue of criteria. Dena. Accessed June 22, 2016, from http://www.biogasregister.de/fileadmin/biogasregister/media/englische_Dokumente/Catalogue_of_Criteria_Biogasregister.pdf.

Dena. (2014b). *Branchenbarometer Biomethan. Daten, Fakten und Trends zur Biogaseinspeisung.* 1/2014. Accessed June 22, 2016, from http://www.biogaspartner.de/fileadmin/biogas/documents/Branchenbarometer/Branchenbarometer_Biomethan_I_2014.pdf.

Dena. (2015a). *Biogasregister.de: English Information.* Accessed June 22, 2016, from http://www.biogasregister.de/index.php?id=713.

Dena. (2015b). *Branchenbarometer Biomethan: Daten, Fakten und Trends zu Biomethan 2015.* Berlin. Accessed July 23, 2015.

Dena. (2015c). *Branchenbarometer Biomethan: Daten, Fakten und Trends zu Biomethan 2015.* Accessed May 13, 2016.

Department of Energy and Climate Change. (2013). *The future of heating: Meeting the challenge.* Accessed June 22, 2016, from https://www.gov.uk/government/uploads/system/uploads/attachment_data/file/190149/16_04-DECC-The_Future_of_Heating_Accessible-10.pdf.

Die Bundesversammlung—Das Schweizer Parlament. (2012). *Keine $CO_2$-Abgabe auf importiertem Biogas.* Accessed June 22, 2016, from https://www.parlament.ch/de/ratsbetrieb/suche-curia-vista/geschaeft?AffairId=20123191.

Eberlein, J. (2015). *Biofuels and eco gas—Market overview Germany.* Dena. Accessed June 13, 2016, from http://www.dena.de/fileadmin/user_upload/Veranstaltungen/2015/European_Biomethane_Conference/Jens_Eberlein.pdf.

Eker, S., & van Daalen, E. (2015). A model-based analysis of biomethane production in the Netherlands and the effectiveness of the subsidization policy under uncertainty. *Energy Policy, 82,* 178–196.

Erdgas mobil. (2013). *Entwicklungen im Kraftstoff- und Biokraftstoffmarkt 1. Halbjahr 2013—Eine Analyse der Biokraftstoff-Quotenplattform von erdgas mobil.* Accessed June 22, 2016, from https://www.erdgas-mobil.de/fileadmin/downloads/Presse/Studien_Artikel_Broschueren/Halbjahresinformation_2013_Biokraftstoffmarkt.pdf.

European Biogas Association. (2016a). *17240 biogas plants in Europe.* http://european-biogas.eu/wp-content/uploads/2016/01/Graph-1-Biogas-plants.png.

European Biogas Association. (2016b). *367 biomethane AD plants in Europe.* http://european-biogas.eu/wp-content/uploads/2016/01/Graph-2-Biomethane-plants1.png.

Eurostat. (2016). *Electricity and heat statistics.* Accessed June 22, 2016, from http://ec.europa.eu/eurostat/statistics-explained/index.php/Electricity_and_heat_statistics#Derived_heat_production.

EWI/GWS/Prognos. Energieszenarien für ein Energiekonzept der Bundesregierung.

Federal Ministry for the Environment, Nature Conservation, Building and Nuclear Safety. (2011). *Das Integrierte Energie- und Klimaschutzprogramm (IEKP).* Accessed May 13, 2016, from http://www.bmub.bund.de/detailansicht/artikel/das-integrierte-energie-und-klimaschutzprogramm-iekp/.

Forsa. (2013). *Vorstellungen und Erwartungen der Verbraucher in Bezug auf Biogasangebote.* Berlin: Bericht.

Gas in focus—Observatoire du Gaz. (2014). *Types of energy used for heating by type of housing.* Accessed June 22, 2016, from http://www.gasinfocus.com/wp-content/uploads/2014/09/E18.xlsx.

Geisler, R. (2014, Juni). *Biomethan auf Spur? Neue Entwicklungen im Kraftstoffmarkt.Vortrag bei der dena biogaspartnerschaft am 24.* Berlin. Accessed June 22, 2016, from http://www.dena.de/fileadmin/user_upload/Veranstaltungen/2014/Biogaspartner_Podium/09_Robin_Geisler.pdf.

German Biogas Association. (2015). *Biogas segment statistics 2014/2015.* Accessed June 22, 2016, from http://www.biogas.org/edcom/webfvb.nsf/id/DE_Branchenzahlen/$file/15-11-19_Biogasindustryfigures_2014-2015_english.pdf.

Green Gas Grids. (2013). *Proposal for a European biomethane roadmap: WP 3.* Accessed May 13, 2016, from http://european-biogas.eu/wp-content/uploads/2013/11/GGG-Biomethane-roadmap-final.pdf.

Grope, J., & Holzhammer, U. (2012). *Ökonomische Analyse der Nutzungsmöglichkeiten von Biomethan: Biomethanverwertung in Kraft-Wärme-Kopplung, als Kraftstoff und als Beimischprodukt im Wärmemarkt: Vortrag beim VDI Wissensforum.* Accessed June

22, 2016, from https://www.dbfz.de/fileadmin/user_upload/Vortraege/Extern/2012-06-28_%
C3%96konomische_Analyse_der_Nutzungsm%C3%B6glichkeiten_von_Biomethan.pdf.

Hahn, H., Krautkremer, B., Hartmann, K., & Wachendorf, M. (2014). Review of concepts for a
demand-driven biogas supply for flexible power generation. *Renewable and Sustainable
Energy Reviews, 29*, 383–393.

Herbes, C. (2015). Marketing von Biomethan. In C. Herbes & C. Friege (Eds.), *Marketing
Erneuerbarer Energien: Grundlagen, Geschäftsmodelle, Fallbeispiele* (pp. 183–201).
Wiesbaden: Springer Gabler.

Herbes, C., Braun, L., & Rube, D. (2016). Pricing of biomethane products targeted at private
households in Germany—Product attributes and providers' pricing strategies. *Energies, 9*(4),
252.

Herbes, C., Friege, C., Baldo, D., & Mueller, K.-M. (2015). Willingness to pay lip service?
Applying a neuroscience-based method to WTP for green electricity. *Energy Policy, 87*,
562–572.

Herbes, C., & Hess, F. (2011). Herausforderungen in Marketing und Vertrieb von Biomethan—ein
neuer Markt entsteht. In *Tagungsband 5. Rostocker Bioenergieforum*, 95–110.

Herbes, C., & Ramme, I. (2014). Online marketing of green electricity in Germany—A content
analysis of providers' websites. *Energy Policy, 66*, 257–266.

Herbes, C., Jirka, E., Braun, J. P., & Pukall, K. (2014). Der gesellschaftliche Diskurs um den
„Maisdeckel" vor und nach der Novelle des Erneuerbare-Energien-Gesetzes (EEG) 2012.
*GAIA, 23*(2), 100–108.

Herbes, C., Pustišek, A., McKenna, R., & Balussou, D. (2014). Überraschende Diskrepanz bei
Biogas: lokal akzeptiert, global umstritten. *Energiewirtschaftliche Tagesfragen, 2014*(5),
53–56.

Kaltschmitt, M., & Streicher, W. (2009). Energie aus Biomasse. In S. Kaltschmitt, M. Kaltschmitt,
& W. Streicher (Eds.), *Regenerative Energien in Österreich* (pp. 339–532). Wiesbaden:
Vieweg + Teubner.

Klein, D., & Frietsch, S. (2015). Identical product properties—Chemical products based on
renewable resources through use of the mass balance approach. *FAPU—European Polyure-
thane Journal* (January 2015), 3–4.

Kovacs, A. (2016). *BIOSURF project serving the evolution of the European biomethane market.*
Accessed June 22, 2016, from http://www.biosurf.eu/wordpress/wp-content/uploads/2015/06/
09_Kovacs-Malmo-12May2016.pptx.

Larsson, M., Grönkvist, S., & Alvfors, P. (2016). Upgraded biogas for transport in Sweden—
Effects of policy instruments on production, infrastructure deployment and vehicle sales.
*Journal of Cleaner Production, 112*, 3774–3784.

Loßner, M., Gawel, E., & Herbes, C. (2012). Einsatz von Biomethan in Neubauten nach
EEWärmeG—eine Hemmnis- und Wirtschaftlichkeitsanalyse. *Zeitschrift für
Energiewirtschaft, 36*(4), 267–283.

Markard, J., Wirth, S., & Truffer, B. (2016). Institutional dynamics and technology legitimacy—A
framework and a case study on biogas technology. *Research Policy, 45*(1), 330–344.

Mathiasson, A. (2016). *Present status and future projects of biomethane in Sweden (Scandinavia).*
Accessed June 22, 2016, from http://www.biosurf.eu/wordpress/wp-content/uploads/2015/06/
03_EBASwedenmay2016.pptx.

Mattes, A. (2012). Grüner Strom: Verbraucher sind bereit, für Investitionen in erneuerbare
Energien zu zahlen. *DIW-Wochenbericht, 79*(7), 2–9.

Milbrandt, A. (2013). *Biogas potential in the United States (Fact Sheet), energy analysis, NREL
(National Renewable Energy Laboratory).* Accessed June 22, 2016, from http://www.nrel.gov/
docs/fy14osti/60178.pdf.

Ministère de l'Écologie, du Développement durable et de l'Énergie (2014). *La transition énergé
tique pour la croissance verte.* Accessed June 22, 2016, from http://www.developpement-
durable.gouv.fr/-La-transition-energetique-pour-la-.html.

Ministère de l'environnement, de l'énergie et de la mer (2016). *Arrêté du 24 avril 2016 relatif aux objectifs du développement des énergies renouvelables.*

NGV Global. (2014). *Natural gas vehicles statistics.* Accessed June 22, 2016, from http://www. iangv.org/current-ngv-stats/.

Osaka Gas. (2010). *Biogas introduced to gas utility's pipeline.* Accessed June 22, 2016, from https://www.osakagas.co.jp/en/whatsnew/1209735_11885.html.

Plaas, B. (2014). *Getrennte Bilanzierung—Auswirkungen auf den Biomethanmarkt: Vortrag im Rahmen von „Zukunft Biomethan—der Auftakt", Berlin, 24. Juni 2014.* Accessed June 22, 2016, from http://www.dena.de/fileadmin/user_upload/Veranstaltungen/2014/ Biogaspartner_Podium/03_Bjoern_Plaas.pdf.

von Rosenstiel, D. P., Heuermann, D. F., & Hüsig, S. (2015). Why has the introduction of natural gas vehicles failed in Germany?—Lessons on the role of market failure in markets for alternative fuel vehicles. *Energy Policy, 78,* 91–101.

Scarlat, N., Dallemand, J.-F., Monforti-Ferrario, F., Banja, M., & Motola, V. (2015). Renewable energy policy framework and bioenergy contribution in the European Union—An overview from National Renewable Energy Action Plans and Progress Reports. *Renewable and Sustainable Energy Reviews, 51,* 969–985.

Schmid, M. (2015). *Biomethane for heating purposes: Case study Switzerland.* Dena. Accessed June 22, 2016, from http://www.dena.de/fileadmin/user_upload/Veranstaltungen/2015/Euro pean_Biomethane_Conference/Michael_Schmid.pdf.

Sorda, G., Sunak, Y., & Madlener, R. (2013). An agent-based spatial simulation to evaluate the promotion of electricity from agricultural biogas plants in Germany. *Ecological Economics, 89,* 43–60.

Strauch, S. (2014). *Biomethane markets and policies, presentation on the European workshop on biomethane—Markets, value chains and applications.* Accessed June 22, 2016, from http:// european-biogas.eu/wp-content/uploads/2014/03/6_Sabine-Strauch_Biomethane-markets-2.pdf.

Thrän, D. (2014). *Biomethane: Status and factors affecting market development and trade: A joint study by IEA bioenergy Task 40 and Task 37.* Accessed 13 May 2016 from http://www. bioenergytrade.org/downloads/t40-t37-biomethane-2014.pdf.

Toptarif (2014). *Verbraucher zeigen kaum Interesse an Ökogas.* Accessed June 22, 2016, from http://www.toptarif.de/presse/presse-announcement/article/613.

Zennaro, K. (2015). *Biomethane Roadmap in the UK.* Accessed June 22, 2016, from http://www. biosurf.eu/wordpress/wp-content/uploads/2015/06/UK-Roadmap_REA.pptx.

**Carsten Herbes** is a professor of International Management and Renewable Energy at Nuertingen-Geislingen University (NGU) and Director of the ‚Institute for International Research on Sustainable Management and Renewable Energy'. Before joining NGU he worked for ten years with a leading management consulting firm in Europe and Asia. Subsequently he joined a bioenergy company where he finally became CFO. He obtained a Master's degree in business administration from Mannheim University and a Ph.D. from the University of Frankfurt (Oder). His research topics include marketing, acceptance and cost of renewable energy. He acts as advisor on sustainability issues to various companies and associations and is a frequent speaker at national and international conferences.

# Renewable Energies in the Contracting Market

Ralf Klöpfer and Ulrich Kliemczak

**Abstract**

Contracting is an innovative service model in which tasks relating to the supply of energy and other utilities are assigned to a contractor. This produces a number of benefits for prospective customers. Many varieties of contracting have become established on the market. These mainly differ in terms of the scope of services offered by the service provider. Renewable energies ("renewables") are deployed in contracting concepts, particularly when it comes to supplying heating energy. That said, the different forms of renewables are not equally suited to this purpose and therefore offer varying potential for deployment. Various subsidy programmes are available to further increase renewables' share of energy consumption and, thus, successfully promote the so-called energy turnaround. However, the use of renewables also involves numerous challenges and risks. Contracting provides prospective users with the opportunity to overcome these hurdles and use renewables in their energy supply. As a general rule, it is the customer who decides whether renewables will be used in the context of a contracting solution by stipulating the requirements it has in its utility solution. Furthermore, lawmakers also play a key role by setting mandatory requirements for the use of renewables or by determining the relevant subsidy framework. Usually, renewables can only be used when they come

A previous version of this chapter has been published in Herbes, C.; Friege, Chr. (Hrsg): Marketing Erneuerbarer Energien. Grundlagen, Geschäftsmodelle, Fallbeispiele, 2015, Springer Gabler.

R. Klöpfer (✉)
MVV Energie AG, Mannheim, Germany
e-mail: ralf.kloepfer@mvv.de

U. Kliemczak
MVV EnergySolutions GmbH, Mannheim, Germany
e-mail: u.kliemczak@mvv.de

with inexpensive supply concepts or enable customers to comply with legal requirements.

**Keywords**
Contracting • Renewable energies • MVV • Contractor • Contracting models

# 1 Basic Principles of Contracting

## 1.1 Definition and Benefits of Contracting

The term "contracting" refers to the conclusion of a contract in which tasks involved in supplying energy or other utilities are assigned by the client (the "contractee") to a service provider (the "contractor"). Within the contractual framework, the contractor assumes responsibility for a specified period for supplying the contractee's property or production site with the energy and utilities it requires, such as heating energy, cooling energy, electricity, compressed air, water or nitrogen. In other models, the contractee assigns to the contractor only those operations management tasks that are involved in supplying energy and utilities. In this case, the contractee may have the necessary plant technology provided additionally by the contractor or elsewhere.

For the contractee, the assignment of individual tasks to a specialist service provider offers numerous benefits. These may include:

- Drawing on the service provider's expertise when it comes to plant design and planning, procurement, plant construction, operations management and optimisation;
- Ensuring that costs remain plannable;
- Avoiding proprietary investments in energy and utility supply measures;
- Ensuring high plant availability rates;
- Working with the contractor as a partner for all-round optimisation and efficiency enhancement;
- Reducing primary energy consumption;
- Outsourcing economic and technological risks as well as planning and investment risks;
- Reducing the carbon footprint;
- Assigning operations management personnel to contractor, where applicable.

In return, the contractee commits itself to work together with the contractor for several years. Depending on the contracting model selected, contract terms of between 5 and 15 years are customary. To avoid any disadvantages arising for the contractee over the long contractual term, it is important to select a competent contractor, one who views the contract as a relationship between partners and is willing and able to react flexibly to any changes in the relevant legislation and market conditions.

**Table 1** Value chain elements covered by various contracting models

| Energy supply contracting | Project development, analysis, consulting, concept design | Planning | Financing | Construction | Operation / operations management | Energy and utilities supply |
|---|---|---|---|---|---|---|
| Savings contracting | Project development, analysis, consulting, concept design | Planning | Financing | Construction | Operation / operations management | Energy and utilities supply |
| Financing contracting | Project development, analysis, consulting, concept design | Planning | Financing | Construction | Operation / operations management | Energy and utilities supply |
| Technical plant management | Project development analysis, consulting concept design | Planning | Financing | Construction | Operation / operations management | Energy and utilities supply |

Tasks performed by contractor
Tasks performed by contractee

## 1.2 Contracting Models

Depending on the services to be provided by the contractor, a distinction is made between various contracting models which are referred to on the market with their own designations, some of which nevertheless differ from provider to provider. The four main contracting models as described in the industry standard DIN 8930-5 are briefly outlined in what follows. An overview of the value chain elements covered by the various contracting models is presented in Table 1.

### 1.2.1 Energy Supply Contracting

In this contracting model, the contractor assumes all tasks involved in supplying energy and utilities to the customer, including plant operations, optimisation, maintenance and repairs, on the basis of a long-term contract. The contractor thus bears all of the economic, legal and technical risks involved in supplying the customer. As a general rule, plant operation also involves assigning ownership of the respective plants to the contractor. If the contract provides for the modernisation or construction of energy and utility supply plants, then the contractor plans, builds and finances these at its own risk and at its own expense. As well as bearing responsibility for the plant, the contractor also sees to the procurement of the energy to be used and sells the required useful energy and utilities to the contractee. These services are usually settled with a basic charge and consumption-related prices. Accounting for a share of around 84 % (http://www.energiecontracting.de/6-verband/wir-ueber-uns/vfw-in-zahlen.php) of the overall contracting market, this contracting model is by far the most widespread contract form.

### 1.2.2 Savings Contracting (Energy Savings Contracting)

In savings contracting, which accounts for around 9 % (http://www. energiecontracting.de/6-verband/wir-ueber-uns/vfw-in-zahlen.php) of the market, the contractor identifies energy savings potential at the customer and contractually guarantees specific energy savings. The contractor plans, finances and implements the optimisation measures impacting energy consumption at the energy and utility supply plants or downstream distribution and utilisation plants. As a general rule, the contractee retains ownership of the plants themselves, while the contractor sees to plant operation and optimisation. The contractor finances its expenses by participating in the energy cost savings achieved. When it comes to documenting annual energy savings, it is important that the energy requirements reference basis should be jointly determined at the beginning of the contractual relationship. This is because annual energy requirements at production plants in particular, but also in normal real estate, fluctuate from year to year, and very significantly so in some cases.

### 1.2.3  Technical Plant Management (Operations Management Contracting)

In technical plant management contracting, the contractor merely assumes responsibility for operating and optimising the energy and utility supply plants. The contractee retains ownership of and legal responsibility for the plants and also finances them.

Furthermore, the contractor may also provide advisory and planning services for optimisation or modernisation measures. This contracting model has a market share of around 4 % (http://www.energiecontracting.de/6-verband/wir-ueber-uns/vfw-in-zahlen.php).

### 1.2.4  Financing Contracting

Projects in which the contractor plans, finances and builds the energy and utility supply plants are referred to as financing contracting. The contractee retains responsibility for operations and related risks. With a market share of 3 % (http://www.energiecontracting.de/6-verband/wir-ueber-uns/vfw-in-zahlen.php), this model plays a subordinate role since it does not allow for key contracting benefits resulting from the assumption of responsibility for the plants and supplies to take effect.

One special form of financing contracting that should be mentioned at this point is the so-called lease model. This special contracting model has proven its worth in recent years as a means of offering customers economically interesting solutions for efficient, decentralised proprietary electricity generation. Here, the contractor plans, finances and builds the proprietary electricity generation plant and leases this to the contractee. As the owner and operator of the plant, the contractee benefits from the so-called proprietary electricity privilege and is thus exempted from the levy charged under the German Renewable Energy Sources Act (EEG) for the electricity it generates itself. If so desired by the customer, the contractor can also take over all of the operations management for the proprietary electricity generation plant and thus offer the customer the additional benefits of technical facility management.

## 2      Renewable Forms Used in Contracting

Statistics published by the German Association of Heating Energy Suppliers (VfW e.V.) (http://www.energiecontracting.de) provide an overview of the shares attributable to different forms of renewables in the supply of heating energy and electricity in the contracting market. The statistics show that almost 7 % of heating energy in contracting projects across Germany came from renewables in 2014. That corresponds to a connected wattage of 1575 MW for thermal energy. The connected wattage for electrical energy in contracting projects using renewables amounted to around 205 MW in 2014. Solid biomass in the form of timber accounts for by far the largest share of renewables, followed by biogas and wind power (Fig. 1). Photovoltaic (PV) systems have only played a subordinate role to date.

The following sections deal in detail with the individual forms of renewables and their suitability for use in the contracting business.

### 2.1      Photovoltaics (PV)

PV systems only generate electricity. This can either be fed into the grid or used to cover the generator's own needs. A number of business models are available on the market for both options, and these are offered both to commercial and industrial customers and to private households. What all models have in common is the goal of using suitable surfaces at prospective customers to generate solar power and eliminating the hurdle of system financing and operating and maintenance risk on behalf of customers. However, these models only constitute contracting when the electricity generated in the PV system is also used by customers themselves. Here, use is generally made of the lease model described in Sect. 1.2.

Until just a few years ago, the more attractive model in economic terms was to feed electricity into the grid. This was because of the high rates of compensation paid. The marked cuts in solar power subsidies in recent years on the one hand and

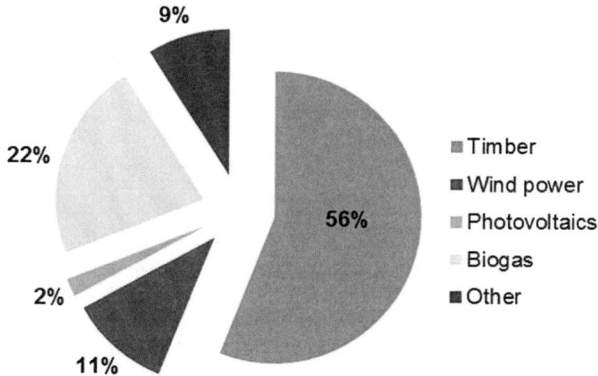

**Fig. 1** Percentage share of renewable fuels used in contracting market: status 2014 (http://www.energiecontracting.de/6-verband/wir-ueber-uns/vfw-in-zahlen.php)

the sharp rise in electricity procurement prices (mainly owing to the levy charged under the German EEG) on the other hand mean that new PV systems are often only economically viable when the electricity generated is primarily used for proprietary needs, thus reducing the volume of electricity procured externally.

As a general rule, the use of PV systems remains highly dependent on the legal framework. Alongside the volume of solar power subsidies, this also includes levies charged under the EEG for electricity resulting from proprietary generation. However, electricity generation costs associated with PV systems are expected to decrease further in the medium term, thereby also reducing their dependence on subsidies.

## 2.2    Solar Heating

With solar heating systems there is no alternative to using the heating energy generated in the immediate vicinity of the system. This means that solar heating systems are basically suited for supplying heating energy in contracting solutions. To achieve sufficiently high supply reliability irrespective of weather conditions, however, such systems must always be combined with other heating energy generation systems. This results in more complex systems technology than for PVs and means that contracting solutions are mainly relevant for municipal and industrial customers. However, solar heating finds only very limited application among these customers. This is due to the significantly higher specific heating energy production costs resulting from high volumes of investment when compared with conventional fuels such as natural gas. Not only that, fewer subsidy options exist for this approach (market incentive programme).

One key motivation for using solar heating systems is provided by the requirements of the German Renewable Energy Heating Energy Act (EEWärmeG), which calls for a share of heating and cooling energy to be generated from renewables. However, these requirements only apply to public and residential buildings.

## 2.3    Biogas

When it comes using biogas,[1] a distinction can be made between the two following deployment options:

a) Deployment as biomethane (bio-natural gas) that is fed directly into the natural gas grid and can theoretically be used to generate electricity and heating energy at any supply point;

---

[1]Sewage, landfill and mine gas constitute special forms of biogas that will not be considered separately here. Generally speaking, these fuels can be used in contracting in the same way as biogas.

b) Deployment as biogas for direct use at a supply plant in the direct vicinity of the biogas plant (electricity and heating energy).

Biomethane generation requires greater technical input to achieve the quality needed for the gas to be fed into the natural gas grid. This model nevertheless offers the advantage that customers not directly in the vicinity of the biogas plant can also be supplied with electricity and heating energy from biogas-powered energy supply plants.

Both models are basically suitable for contracting solutions, with the supply of heating energy to the customer generally forming the basis for the arrangement. The electricity is fed into the grid and paid for in line with the provisions of the EEG. Given significantly higher fuel prices for biogas compared with natural gas, the feed-in compensation is also a prerequisite for offering economically viable biogas-based supply solutions. The subsidy cuts introduced in the current amendment to the EEG legislation (2014) nevertheless mean that biogas-based supply solutions will in future be economically viable only in special cases. Such solutions will therefore hardly offer any further potential for use in contracting solutions.


## 2.4    Solid Biomass

The use of solid biomass in the form of timber and waste timber to generate energy has gained enormously in significance in recent years. While virtually no more fuel volumes are available on the waste timber market in Germany, depending on the quality and region involved, timber fuels still hold potential for further expanding the generation of energy from biomass.

The available plant technology permits the use of a very wide range of different fuel qualities and offers a suitable incineration technology for nearly all kinds of timber fuel. Energy generation then involves either the combined generation of electricity and heating energy or the generation of heating energy alone. Particularly when the heating energy is used to heat residential and public buildings, timber-powered boiler plants offer suitable potential to meet the requirements of the EEWärmeG legislation.

Given its wide-ranging flexibility in different heating levels and heating media, timber is very well suited for use as a fuel in contracting solutions. However, such solutions are restricted almost exclusively to industrial or municipal customers and large residential buildings. The expertise required to generate energy from solid biomass and experience in operating complex plants can be offered to customers as an attractive added value, as can plant financing and fuel procurement. Not only that, the contractor also assumes responsibility for assuring the quality of the fuel as one of the key prerequisites for ensuring adequate plant availability levels. That said, additional potential for deploying biomass will be determined not least by developments in wood pellet and wood chip prices, as well as in prices for conventional fuels such as natural gas and heating oil.

## 2.5    Biogenic Residues

One special form of solid biomass that should be mentioned at this point is the possibility of generating energy from biogenic residues. These are mainly incurred in the food and beverage industry and include grain mill waste, such as husks, residues from oil mills and grape marc. Given the different volumes and qualities of residues arising at individual production locations, these fuels are chiefly suited for use in the generation and consumption of heating energy at the respective location and thus offer a good basis for contracting solutions. Generating heating energy from these very different residues generally requires customised firing technologies. These are available on the market from various plant manufacturers. However, companies have significantly less experience with operating this kind of plant than timber-powered biomass plants. Particularly for this kind of plant concept, contracting can offer clear benefits to customers because experienced contractors have the expertise needed to plan and operate such plants and manage the risks associated with using biogenic residues.

Whether economically viable concepts can be developed for generating energy from these kinds of residues, however, depends on the heating requirements at the given production site. To a very great extent, it also depends on whether other options are available for using the residues (material recovery/recycling), as well as on the relationship between the technical input required to market the product and the potential revenues.

## 2.6    Geothermal Energy

In Germany, geothermal energy is used above all to generate heating and cooling energy. As a general rule, it is therefore also suitable for use in contracting solutions. With installed capacity of around 4.2 GW of thermal energy output for heating energy generation, Germany already holds the fifth position in international rankings (http://www.geothermie.de/aktuelles/geothermie-in-zahlen.html). Given geological conditions in Germany, however, the generation of electricity from geothermal sources currently still only plays a subordinate role.

The current political framework enables larger-scale geothermal plants to be operated viably in many areas of Germany (http://de.wikipedia.org/wiki/Geothermie). That said, these plants also carry a number of risks and have in the past resulted in sometimes considerable damage to buildings in their immediate vicinity. Despite its promotion in the market incentive programme, the full costs involved mean that the use of near-surface geothermal energy to heat or cool buildings by means of heat pumps can only compete with solutions based on conventional fuels in individual cases that offer ideal conditions. Geothermal energy is nevertheless already being put to widespread use. The requirements of the EEWärmeG offer potential for using geothermal energy pump plants above all in private and municipal contexts. In an industrial context, however, geothermal energy pumps and plants generally play a very minor role. The key focus here is usually on using industrial waste heat sources.

Given these factors, providers of contracting solutions for geothermal plants do operate in the market, but the share of the contracting market attributable to such plants is currently still low.

## 2.7 Wind Power

Wind turbines only generate electricity. For conventional turbines with capacities greater than 100 kW, the electricity generated is usually fed into the grid and not earmarked for use by any specific customer. This means that traditional wind turbines that feed electricity into the grid do not offer a basis for contracting solutions.

On the other hand, smaller wind turbines with capacities less than 100 kW have played an increasingly significant role for several years now when it comes to using wind power. These mostly involve vertical turbines that are very quiet and already start up at low wind speeds. Given their size, these turbines even offer interesting potential for use in a private context. The number of these small wind turbines is growing rapidly in wind-rich regions. In Germany, for example, an estimated total of 10,000 such turbines had already been installed by 2010 (http://www.klein-windkraftanlagen.com). It nevertheless remains to be seen whether these plants will also play a more major role in contracting solutions.

In principle, if the electricity is put to proprietary use, then the contracting models available for wind turbines are similar to those for PV systems. Because companies are usually not able to develop wind power projects with just their own resources, suitable service offerings by contractors may provide a good incentive for using wind power. According to an investigation carried out by EnergieAgentur.NRW, the reluctance to draw on wind power contracting is largely attributable to the political and planning law framework, which is subject to frequent changes (EnergieAgentur.NRW, "Windenergie-Contracting in NRW", May 2015).

## 2.8 Hydropower

Hydropower plants represent another solution; such plants generate only electricity and generally feed it into the grid. They therefore do not offer a basis for contracting solutions. One exception relates to small-scale hydropower plants that can also be used for proprietary electricity generation. In this case, contracting solutions are conceivable in principle. However, the potential for this kind of plant in Germany must be questioned and depends, among other factors, on the availability of corresponding water rights.

# 3    Using Renewables in Contracting: Key Factors and Challenges

Cores aspects of Germany's energy turnaround includes raising renewables' share of energy consumption and the associated expansion in the decentralised energy supply. To support the successful implementation of this turnaround, the government has introduced a number of subsidies aimed not only at expanding the use of renewables but also at enhancing energy efficiency and saving energy. These measures are intended to create a basis for making the use of renewables economically viable.

Despite the availability of subsidies for renewables, however, companies often face a number of hurdles that must be overcome when implementing corresponding projects. These include:

- Lack of trust in economic viability of renewables,
- Lack of relevant expertise,
- Lack of access to fuels potentially required,
- Financing,
- Supply of suitably qualified operations staff,
- Technical, approval law and economic risks,
- Organisational input.

Contracting provides potential users with the possibility of overcoming these hurdles and thus smoothes the way for companies to use renewables (Sect. 1.1).

That said, the decision as to whether renewables can be used to satisfy a supply requirement is usually not taken by the contractor. The role of the contractor is rather to design a customised plant concept that enables the customer to be securely supplied with energy and utilities in line with its requirements. As a general rule, customers give priority to inexpensive energy and utility supply solutions. When devising plant concepts, the contractor assesses whether renewables can also be viably used within the relevant legal framework.

Alongside cost factors, there are further criteria depending on the customer group that significantly influence the selection of the relevant plant concept and use of renewables. Distinctions can be made, for example, between industrial, real estate and municipal customers. These are presented in Table 2.

**Table 2** Decision-making criteria when selecting plant concept or renewables

| Customer group | Decision criteria |
|---|---|
| Industrial customers | Energy and utility requirements, carbon footprint, costs, risks, plant complexity, availability |
| Real estate customers | Energy pass (appreciation in property value), requirements of EEWärmeG legislation, costs |
| Municipal customers | Energy pass (appreciation in property value), requirements of EEWärmeG legislation, costs, climate protection targets |

While real estate and municipal customers draw on renewables chiefly to meet legal requirements (EEWärmeG), the decisive factor for industrial customers is the cost of their supply of energy and utilities. Particularly when they compete on an international level, companies tend to focus on ensuring low energy and utility prices. In this case, renewables are only selected when, drawing on all subsidy options, they make it possible to implement inexpensive supply concepts.

For contractors, financing requirements in particular play a key role in determining prices for supplying energy and utilities. These depend primarily on the way in which the necessary investments are to be financed. The prospective contractee may have access to more favourable financing options than the contractor. However, the contractor may be able to offer its customers added value compared with in-house solutions. Viewed as a whole, therefore, these would not be cheaper for the customer.

One challenge when using renewables relates to the structuring of price adjustment clauses in contracting agreements in cases where the electricity is subsidised under the EEG. Here, fuel price increases may only be compensated for via revenues from heating energy since the EEG compensation paid for electricity remains fixed for the whole subsidy period. Particularly when the EEG compensation accounts for a high share of total revenues, the resultant disproportionate increases in heating energy prices may be difficult to communicate to customers. That said, customers face this disadvantage regardless of whether they opt for a contracting solution.

With regard to the legal framework, risks relate above all to amendments in the relevant subsidy laws. However, owing to the protection of the status quo still applicable for plants already in operation, these risks are limited to the period within which legislative amendments are prepared and adopted. During these periods, investment decisions are usually not taken and projects already planned are implemented faster to enable existing subsidy regulations to be drawn on. Should future legislative amendments nevertheless infringe on the protection of the status quo, then the resultant insecurity among investors would have unforeseeable consequences for the further expansion in renewables and their use in the contracting market.

## 4    Renewables Contracting in Practice: Select Examples

### 4.1    Solar Heating

The construction of a new gastronomy business in 2010 also involved installing an energy centre to supply the restaurant with heating energy. The municipal council required the building to meet so-called green building standards and be fully supplied with heating energy from renewables in the context of an energy supply contracting solution. The tasks performed by the contractor included designing, planning, building and financing the necessary plant technology and supplying the restaurant with heating energy.

**Table 3** Select project data of contracting concept for energy supply to a gastronomy business

| Project data | Figure |
|---|---|
| Plant investment | € 194 k (€ 144 k subsidies) |
| Contractual term | 15 years |
| Installed heating energy capacity | 100 kW timber; 60 m$^2$ collector surface |
| Pellet requirements | 20 t/a |
| Heating energy turnover | 100 MWh (26 MWh solar) |
| $CO_2$ savings | 55 t/a |
| Invoicing | Basic charge and volume-dependent price |

Because of the high volume of warm water required and the minimum supply temperature of 65 °C, a CPC vacuum tube collector system directly linked to the heating buffer storage was selected for the base load supply. Medium and peak loads are covered by a wood pellet boiler. Select project data are summarised in Table 3.

The project was implemented by MVV EnergySolutions in 2010 and exemplifies a classic contracting approach to supplying a customer with renewables. In this case, the decision to work with renewables was determined by the requirements of the municipal council. As a result, the economic viability of the solution was not compared with that of conventional fuels.

## 4.2   Biogas (Biomethane)

The biogas project described here was based on an invitation for tenders issued by a public sector organisation that involved taking over the entire heating energy supply in the context of an energy supply contracting solution. The contractor was required to plan the supply concept, install and finance the necessary plant technology and ensure the heating energy supply to the customer's properties. The building for the energy centre would be provided by the state government.

The exclusion criteria set by the customer for the supply concept included compliance with EEWärmeG and a primary energy factor of $\leq 0.5$ for the heating energy supply. For logistical reasons, the customer excluded the use of solid biomass, as a result of which the primary energy factor requirements could only be met by using biomethane.

Consistent with the conditions set by the customer and subsidy options, MVV EnergySolutions submitted a concept in which the heating energy would be supplied by two biomethane combined heat and power (CHP) plants with an electrical capacity of 750 kW each and two natural-gas-powered boilers.

The CHP plants would be installed 1 year apart to enable both plants to be assessed as individual plants under the current requirements of the EEG. This would guarantee higher EEG compensation and make it possible to offer a more favourable heating energy price.

**Table 4** Select project data of contracting concept for energy supply to public sector organisation

| Project data | Figure |
|---|---|
| Planned plant investment | € 2.9 million |
| Planned contractual term | 15 years |
| Installed heating energy capacity | 9.6 MW |
| Installed electricity capacity | 1.5 MW |
| Natural gas input | 6 GWh/a |
| Biomethane input | 23 GWh/a |
| Electricity generation (EEG) | 8,626 MWh/a |
| Heating energy turnover | 14,000 MWh/a |
| $CO_2$ reduction potential | 3,200 t/a |
| Invoicing | Basic charge and volume-dependent price |

The electricity generated by the CHP plants would be fully fed into the grid for general supply and would receive constant compensation over a 20-year period pursuant to current EEG provisions.

In their final state, the CHP plants would have provided almost 60 % of the desired heating energy. Select project data are presented in Table 4.

This project was not implemented, but it nevertheless illustrates a classic contracting approach to supplying heating energy to a customer on a renewable basis. Here, too, the use of biomethane was determined by customer requirements.

## 4.3    Solid Biomass

Based on a need to convert its existing energy centre owing to the expiry of the relevant approvals, in 2009 a public sector organisation issued an invitation for tenders for the supply of heating energy to its properties in the context of an energy supply contracting solution. In addition to heating energy in the form of heating water, the invitation for tenders also provided for the supply of permeate and concentrate and the operation and maintenance of the local hot water grid. The necessary operations staff for the heating plant would be provided by the customer by way of a personnel agency agreement. Furthermore, with regard to the renewal of the heating energy generation plants, the customer wanted the contractor to meet the legal requirements in force since January 2009 with respect to the use of renewables upon construction of additional buildings (EEWärmeG). The contractor would also be required to ensure the necessary planning and financing, the conversion work and operations management at the energy centre.

Consistent with the conditions contained in the enquiry, MVV EnergySolutions tendered and implemented a concept based on solid biomass. The technical concept involved the following main measures:

- Dismantling of existing gas/oil boilers (fossil fuels),
- Construction of two timber-powered boilers of 10 $MW_t$ each based on solid biomass (timber) to secure base load supply,
- Construction of a timber store

**Table 5**  Select project data of contracting concept for energy supply to public sector organisation

| Project data | Figure |
|---|---|
| Planned plant investment | € 12 million |
| Planned contractual term | 20 years |
| Installed heating energy capacity (timber) | 20 MW |
| Installed heating energy capacity (gas/oil) | 51 MW |
| Timber input | 132 GWh/a |
| Gas/oil input | 3 GWh/a |
| Heating energy input | 95,000 MWh/a |
| $CO_2$ savings | 24,000 t/a |
| Invoicing | Basic charge and volume-dependent price |
| Annual energy cost savings for customer | 20 % |

- Installation of a new gas/oil boiler (16 $MW_t$) to cover peak loads,
- Continued use of an existing gas/oil boiler as a reserve,
- Installation of a heating energy storage facility,
- Gradual renovation of heating energy grid.

The biomass boiler was dimensioned in such a way as to facilitate high operating hours and full utilisation while at the same time enabling scheduled inspection work to be performed without incurring additional expenses for expensive secondary fuel (gas or extra light heating oil). Following the complete conversion of the energy centre, 95 % of heating energy is now generated using timber as a fuel, while the remaining amount is covered by light heating oil or natural gas. Select project data are presented in Table 5.

The concept implemented not only enabled the requirements of the EEWärmeG to be met in full. It has also resulted in a less expensive supply of heating energy than previously. In this case, the contractee benefited not only from an ecological perspective but also in economic terms from a sustainable supply concept based on renewables. However, this state of affairs was only possible because the contractor was able to secure corresponding volumes of timber on favourable terms upon conclusion of the contract. Not only that, the plant technology planned and installed was capable of working at high availability levels with the fuel quality thereby deployed.

## 4.4    Biogenic Residues

In response to a customer request, a concept for generating heating energy from grain husks at a new production site was compiled for a grain mill operator in 2014. This concept was to form the basis of a contracting offer.

The technical concept involved making use of the grain husks incurred as a waste product during the grinding process. Previously, husks had been used as an additive to fodder. The processing effort (grinding) required for this is relatively high, however, and the customer faced difficulties in selling the grain husks in the quantity incurred (around 20–25 t/d).

**Table 6** Select project data for investigation of grain husk incineration as basis for supplying steam to mill operator

| Project data | 1 t/h steam | 2 t/h steam |
|---|---|---|
| Required plant investment | € 1.0 million | € 1.5 million |
| Natural gas costs saved | € 235 k/a | € 470 k/a |
| Grain husk fuel costs including ash disposal | € 78 k/a | € 155 k/a |
| Operating costs | € 70 k/a | € 77 k/a |
| Grain husk input | 1,260 t/a | 2,530 t/a |
| Steam requirement | 6,500 t/a | 13,000 t/a |
| Annual savings for customer | € −35 k | € 60 k |

The grain mill also requires steam, which was currently to be provided from a natural-gas-powered boiler. During the first stage of expansion, steam requirements would remain constant at 1 t/h. Following further expansion in the production plant, this figure would rise to 2 t/h.

The task now involved reviewing whether the grain husks could be used to generate steam under the conditions outlined earlier and whether an economically viable concept could be developed on this basis. In terms of fuel costs, the volume of costs stated for the grain husks should correspond to the revenues previously generated by the customer from using the husks as a fodder additive minus the processing costs thereby avoided. No subsidies would be available for heating energy generation in this case.

As a result of the investigations, the customer was shown that, assuming sufficiently high steam requirements, the supply of steam from grain husks was less expensive than using natural gas. The principal project data are presented in Table 6.

To be economically viable, however, the use of grain husks presupposes a minimum steam requirement of 2 t/h, a figure that would not be reached in the first production phase. For this reason, the project was initially not pursued any further. The concept nevertheless shows that under ideal conditions, the supply of heating energy using regenerative fuels is also possible and that this approach holds potential for contracting solutions.

---

## 5 Conclusion

Expanding the use of renewables is a key prerequisite for successfully implementing the energy turnaround. Despite government subsidies, potential users of renewables nevertheless face a number of challenges and risks. As an innovative service model, contracting represents one way of assisting customers to overcome existing hurdles and facilitate the use of renewables in their supply of energy and utilities.

A variety of contracting models have become established in the market. These offer customers a high degree of flexibility in terms of having their supply of energy and utilities structured by a service provider. The associated benefits for the customer mean that contracting is an important instrument that can contribute to the success of the energy turnaround.

As a general rule, however, it is not the contractor who decides whether renewables will be used in contracting solutions. This decision is rather taken by customers, who stipulate the requirements they have in their utility supply, and lawmakers, who make the use of renewables mandatory and structure renewables subsidies. Energy and utility prices are often the decisive criterion for customers when selecting contractors. This means that opportunities for renewables only arise when they can be put to economically viable use in inexpensive supply concepts. The provision of a reliable and suitable subsidy framework by lawmakers still plays a key role in this respect and will continue to do so in the medium term.

**Ralf Klöpfer** has been a member of the Executive Board of MVV Energie AG in Mannheim since 2013. He is responsible for the sales division, to which the contracting service provider MVV EnergySolutions GmbH is also allocated. Before joining MVV Energie AG, he worked as Managing Director at enevio GmbH. Prior to that, he held various senior positions at EnBW in Karlsruhe for more than 15 years, most recently as Spokesman of the Management at EnBW Vertrieb GmbH. He studied electrical technology, majoring in energy technology, at the University of Stuttgart.

**Ulrich Kliemczak** is an employee of MVV EnergySolutions GmbH. As a planning engineer and project developer, he has dealt with a wide variety of contracting projects for more than 10 years now. As well as compiling supply concepts and handling negotiations with customers, he has also assumed technical project management for various projects and accompanied and been responsible for planning, building and launching operations at supply plants. Before joining MVV EnergySolutions GmbH, he worked for two years as a planning engineer at MVV Energie AG. He studied environmental process engineering at Technische Universität Bergakademie Freiberg, subsequently obtaining a doctorate there in the field of energy process technology.

# Renewable Energy in the Marketing of Tourism Companies

Susanne Gervers

**Abstract**

The specific situation of tourism companies causes difficulties with regard to sustainability. After all, their job is to sell perfectly staged counterworlds to everyday life. In essence, the touristic experience invokes "the crossing of boundaries"—spatially, socially, but also morally. The postulate of so-called sustainable tourism—to "fully" take into consideration future economic, social, and ecological requirements—resembles the squaring of a circle, at least for those companies that attempt to create a "coherent overall picture" to satisfy the demands of their customers. For that reason, tour operators face specific difficulties when it comes to the adequate integration of positive approaches to marketing renewable energy in their client-focused service packages. These positive approaches certainly exist in the tourism value chain and can be used to satisfy the minimum requirements for "sustainable" tourism, the criterion of the Global Sustainable Tourism Council. The relevance of this difficulty can even be demonstrated for a leading corporation in the field of tourism, Studiosus Reisen München GmbH. This case study provides food for thought. What is needed in the future to sustainably reduce tourism's frequently evoked so-called green gap?

**Keywords**

Sustainable tourism • Sustainable marketing • Tour operator management • Touristic experience • Green gap in tourism

A previous version of this chapter has been published in Herbes, C.; Friege, Chr. (Hrsg): Marketing Erneuerbarer Energien. Grundlagen, Geschäftsmodelle, Fallbeispiele, 2015, Springer Gabler.

S. Gervers (✉)
Hochschule für Wirtschaft und Umwelt Nürtingen-Geislingen, Geislingen an der Steige, Germany
e-mail: susanne.gervers@hfwu.de

# 1    Introduction

On the occasion of the first fvw Online Marketing Day on 7 May 2014 in Cologne, the best online marketing campaigns of tourism companies were honored. It was noteworthy that none of the top 12 candidates even hinted at the topics of renewable energy, climate change, or sustainability (see fvw, 09.05.2014, pp. 28ff.). Marketing takes its cues primarily from the wishes of customers or from the ideas of companies about these wishes. What do tourism companies think about the wishes of their customers? Apparently customers do not want to be reminded of climate change and other problems; in tourism, the perfectly staged alternative to our daily routine plays a specific role, or, as the head of a travel agency pointedly formulated: "We want to sell the best weeks of the year and not optimal solutions to crises and problems" (fvw, 27.02.2014, p. 13).

It is definitely not the case that climate change and sustainability are ignored by the industry. Products such as the green rail card by Deutsche Bahn, which features $CO_2$ neutrality, are considered pioneers. The same is true of new products of the sharing economy in the area of transport operations or lodging. However, ecologically correct relaxation at home, let alone the trend to avoid travel, is less likely to satisfy the deep-seated wishes of the guest. It may be the case that $CO_2$ emissions can be compensated via so-called climate donations, but not the intercultural exchange or the experience of the unknown and unfamiliar (and, thus, of one's identity). It does not exist without mobility across borders, interpreted in multiple ways. And this may even imply that the standards and accomplishments of our society, for example concerning climate protection and sustainability, are forgotten during supposed the best weeks of the year.

"When making the booking decision, sustainability is irrelevant" (fvw, 14.03.2014, p. 74). It is more important to let the guest experience sustainability on site, based on the principles of storytelling. It was a major conclusion at a number of panel discussions at the ITB Congress 2014 in Berlin (fvw, 14.03.2014, pp. 72–74) that the luxury segment offers the greatest potential for "sustainable" tourism. To experience people in their local setting, for example energy generation in a village in Calabria, is more likely to make the traveler aware of structures and problems and to foster engagement and responsibility through personal contact. What is the meaning of "sustainable" tourism, what is its official definition, and what actors are involved along the value chain in tourism? What companies and their marketing activities are of the greatest relevance for studying the topic of "sustainable energy in tourism marketing"?

# 2    Tourism Companies and Sustainable Marketing

Meffert et al. (2012, p. 15) stress the sales focus of practitioners in the field of tourism, even though marketing is currently seen in a much broader and generic context. The definition of the American Marketing Association, which was officially adopted in July 2013,[1] explicitly makes reference to society at large when entering into exchange relations. This should sound particularly familiar to tourism companies since networks of various kinds are the main reference points for their daily work. Networking is indispensable for tourism professionals since they are required to work together with a number of different companies, societal groups, and even directly with their customers in the course of providing their services. For that reason, Pechlaner et al. (2011) appropriately point out the importance of competent cooperation and networking, especially in regions and destinations.

What exactly is *sustainable marketing*, and what is its precise meaning in the tourism industry? Pomering et al. (2011, p. 959) describe sustainable marketing as a development stage with lower target orientation than "sustainability management." To clearly highlight the aim of sustainability and to accept it without a doubt, a "holistic perspective" is required, in the opinion of El Dief and Font (2010, p. 159); this is where they observe the decisive criterion for differentiation from green marketing and *greenwashing*, a term frequently cited in the tourism industry. Such a holistic perspective requires the systematic pursuit of sustainability at all levels of the corporation, not only in marketing. And indeed, internationally accepted criteria with clear guidelines for action are in place for the tourism industry. They allow categorization and assessment (GSTC 2012, 2013) and help in meeting the target.

So what is meant by *sustainable tourism*? The environmental agency of the United Nations (UN) developed the following formulation jointly with its special agency, the World Tourism Organization (UNWTO):

Definition Start
   **Tourism that takes full account of its current and future economic, social and environmental impacts, addressing the needs of visitors, the industry, the environment and host communities** (UNEP/WTO 2005, p. 12)
   Definition Stop

The resulting demands on tourism marketing, namely, to take full account of environmental aspects in addition to all social and economic considerations while simultaneously satisfying the needs of customers, appears rather unrealistic against the backdrop of the previously described industry setting. At its essence, the tourism experience involves crossing boundaries—spatially, socially, but also from a moral perspective. In all areas of multidisciplinary tourism research, boundaries and borderline experience play a role; Pomering et al. (2011, p. 957) explicitly refer to the importance of the marketing activities of tourism companies. Boundaries give structure to a journey and our sense of time, and they allow us to

---

[1]"Marketing is the activity, set of institutions, and processes for creating, communicating, delivering, and exchanging offerings that have value for customers, clients, partners, and society at large." (AMA)

leave the familiarity of the daily routine behind. The traveler—this is revealed by contributions in tourism psychology—enters a different world:

> The various forms of otherness consumed in tourism seem able (and are often purposely produced) to satisfy desires that are hidden or otherwise repressed in tourists' everyday lives. (Picard and Di Giovine 2014, p. 23)

The touristic experience is characterized by abandoning the daily routine, by "consuming" otherness and unfamiliar settings, and, thus, also by reassurance of one's own identity. What types of travel and what types of companies are most informative about the topic of renewable energies in the marketing of tourism companies? Not all activities of a traveler are considered to be part of tourism. *Tourism* is defined as the temporary departure from the customary center of one's life. For the international statistical measurement of the social, cultural, and economic phenomenon of tourism, the following official definition by the UNWTO[2] is used:

> Definition Start
> **A visitor is a traveler taking a trip to a main destination outside his/her usual environment, for less than a year, for any main purpose (business, leisure or other personal purpose) other than to be employed by a resident entity in the country or place visited. These trips taken by visitors qualify as tourism trips. Tourism refers to the activity of visitors.** (UN/UNWTO 2010, p. 10)
> Definition Stop

Based on this definition, travel without a purpose (leisure travel) and travel that has a specific aim (such as business travel) are not distinguished; both are considered to be tourism as long as additional requirements are met, such as the limitation to 1 year. However, travel that takes place without a clearly defined purpose, such as the classical vacation, which serves the vague aim of recreation, is more likely to reveal a "coherent overall picture" of the phenomenon of tourism. In the business travel segment, meanwhile, the marketing of renewable energy is prominently positioned by the companies involved. The concept of green meetings, the planning, organization, and implementation of so-called environmentally appropriate events is increasingly developing into a flagship of companies and Germany as a travel destination. While business travel also belongs to the tourism industry, it must be differentiated from tourism in the narrow sense, which is defined by the travelers' extraordinary experiences and the importance of the unfamiliar for personal identity.

So what types of companies should be studied, what segments most likely represent a coherent overall perspective on travel and tourism, and who were the guests of these companies? Tour operators undertake organizational, informational, distributional, and social tasks in the source and target regions. Depending on the

---

[2]The status of the UNWTO is that of a specialized agency of the United Nations. It is headquartered in Madrid.

type and occasion of the journey, the tourism service chain includes different types of service providers, but normally transportation and lodging companies. The tour operators play a central role, since they combine the various offerings into a comprehensive bundle of services, which they distribute independently. Even though the share of so-called classical package tours has been steadily declining for many years, while the independent gathering of information and the booking of partial services online by travelers has been increasing at the same rate, tours compiled by tour operators remain the most important type of organized vacation at 42 % in the year 2013[3] (FUR 2014, p. 4). To obtain a consistent overall perspective on travel and tourism, we should first take a closer look at the offerings of the organizers, more precisely, the offerings of the providers in the segment of leisure travel. In the tourism industry, a distinction is made between tourism (with operators at the center) and business travel.

The operators deal with numerous additional tourism companies in their offerings. Do these potential cooperation partners provide favorable conditions for a "green" product of the operator? What is the importance of sustainability targets such as an increase in the share of renewable energy for these companies? A fragmented picture currently emerges among operators in the field of tourism concerning the topics of climate change and renewable energy, which can be outlined as follows:

## 2.1 Digression: Service Providers

### 2.1.1 Transport Providers

- With its strategy DB 2020, **Deutsche Bahn AG** positions itself as a "green" company and actively promotes its role as an "environmental leader": it is planned that by the year 2050 all trains of Deutsche Bahn AG will operate with 100 % electricity from regenerative sources.
- While the statement that long-distance coaches are the most environmentally friendly means of transportation after trains is well accepted among tourism professionals, this aspect is not found in the marketing materials. The providers and their associations instead focus on a possible image change, now that new and attractive target audiences have (re)discovered the bus as a means of transportation, following the liberalization of long-distance bus services in 2013. Speakers at the annual industry meeting in Cologne, organized by RDA **International Coach Tourism Federation**, primarily pointed out that coaches are an inexpensive and safe means of transportation. The fact that it is an "environmentally friendly" alternative to car, plane, or train was only a secondary consideration (RDA 2013). The German Tourism Association, the umbrella organization of the tourism industry in Germany, meanwhile thinks that a

---

[3]In 2005, this share was still 48 % (FUR 2014, p. 4).

change in image is possible since the long-distance coach fits well with "new travel behavior": "environmentally conscious, more frequent and shorter travel" (DTV).

– The adjustment process in light of the changing environmental situation is particularly difficult for air carriers: to counter increasing costs and competitive pressures, they still focus on technological solutions ("fuel efficiency") but currently do not adjust their business models. This is even true of **Deutsche Lufthansa AG,** a company with exemplary activities in the UN Global Compact. A joint public relations campaign aims at creating awareness about the "high ecological efficiency of German air carriers" (Deutsche Lufthansa AG) among people living in Germany. Roland Conrady, head of the ITB Berlin Congress and president of the scientific association DGT,[4] initially an air carrier manager himself, sums it up in a presentation to numerous industry representatives:

Major change in the transport sector seems unavoidable in the future. (Conrady, 15.05.2014)

### 2.1.2   Hospitality Industry

– Measures to conserve energy, even including **zero-emission hotels**, are also an important topic in the hospitality sector owing to the high relevance of fixed costs. Initially, the large and globally active hotel chains addressed the issue in their marketing initiatives, for example the hotel company Hilton, which has been in business since 1919 and currently operates 4661 hotels (Hilton 2016a). The reduction of $CO_2$ emissions was considered to be an important strategic aim to cut costs and was pursued in cooperation with numerous partners. The documentation and development of new management approaches aiming at a reduction of $CO_2$ is seen as an important marketing topic: Hilton, for instance, claims to participate in specific projects that advance change in other industries as well. The company also states that it bought carbon credits from a steel producer from South Korea to offset carbon emissions caused by Hilton customers' meetings and events (Hilton 2016b). However, this should not be seen as purely positive, since—similar to the compensation of $CO_2$ emissions in air travel—this apparently leads to a negative habituation effect instead of the necessary change in consumption patterns needed in the long run (see subsequent discussion on destinations). In business-to-business marketing, meanwhile, purely economic criteria are listed, and not only by the German hotel and restaurant association DEHOGA (2012) but also, for example, by the much smaller marketing cooperation Green Hotels in the USA (Green Hotels 2016).

---

[4]Deutsche Gesellschaft für Tourismuswissenschaft e.V.

Others may be too aggressive in highlighting the changing awareness and support of ecological change, for example, Boutique Hotel Stadthalle Wien, according to its own assessment the first city hotel with a zero global energy balance:

> In the course of a year our boutique hotel creates the same amount of energy that we require to run it. For this, we only use renewable energy sources like solar and photovoltaic panels, ground water heat pumps and even three wind turbines. A calculation that is guaranteed to pay off! (Hotel Stadthalle 2016, 2014b).

However, the company, which has been awarded a number of prizes for environmental awareness and innovation over the previous years, must concede that the required permits for the three wind turbines have not yet been granted by the city. At any rate, this seems hard to imagine in the middle of a metropolitan area such as Vienna. At the front desk, a fact sheet is available upon request that includes the sentence: "We continue to hope that we will obtain the permits" (Hotel Stadthalle 2014c), while the text on its Website is formulated in the present tense:

> Our new building is not only the perfect addition to the existing, thoughtfully renovated period townhouse, but maintains a zero-energy balance as well. (Hotel Stadthalle 2014b).

Boutiquehotel Stadthalle very successfully works with this approach and is co-founder of the marketing cooperation Sleep Green Hotels (Sleep Green 2016a, b). According to the hotel, the guests, international city tourists, but also business travelers, value these activities and make a conscious choice:

> We are aware of the fact that everybody can support the environment. For that reason, our guests sleep with a clear conscience at our place. (Hotel Stadthalle 2014a).

A sentence on the bill informs readers that by staying "at the zero-energy balance hotel," the traveler has helped the environment (Hotel Stadthalle 2014d).

### 2.1.3   Destinations

– For important source countries such as Germany, earthbound travel is gaining in importance. The manifold reasons supporting this development include the so-called demographic change, an increasing aging of the traveling population, a growing attractiveness of domestic offerings, and, recently, the complex security situation in international destinations. The changes surely do not stem from travelers' heightened environmental awareness or a new focus in the marketing activities of tourism companies. A large number of existing offerings do not point in that direction: particularly new entrants in the heavily contested market for holiday travel continue to use all means to attract vacationers with a love for travel from the rich source countries.
– As an example, new destinations such as Costa Rica attempt to position themselves in the important German source market as a trendy ecological travel destination (CST 2014). This may seem strange, considering the $CO_2$ balance

of long-haul travel. While there is a theoretical possibility of voluntarily compensating for personal $CO_2$ emissions, for example via atmosfair or myclimate if booking travel from TUI, the effects are nothing special. To achieve something in practice, a certain element of compulsion is obviously needed, for example, a surcharge on ticket prices. But the possibility of providing a compensation payment should also be viewed critically as travelers may get the impression that no behavioral change is needed and unnecessary long-haul travel is subject to greenwashing (Schmücker 2011, p. 140).

–　Climate change and travel that is low in $CO_2$ emissions pose massive challenges for a globalized tourism industry. At ITB Berlin, the world's largest tourism convention, the picture is mixed: At the ITB Congress as "the leading travel industry think tank," (ITB 2016) only a few, but clearly critical, statements were made, for example, a keynote by the Swedish climate researcher Gössling at CSR Day (ITB 2016, p. 40). The official partner country of 2016, the Maldives, meanwhile used slogans such as "…the sunny side of life" or "where the weather is a dream" (Maledives 2016) and remained correspondingly guarded on the topic of climate change during numerous appearances, including a panel discussion in the context of the CSR Day. At any rate, environmentally conscious travel is frequently considered to be part of the luxury segment in the tourism industry, as discussed in the introduction (Sect. 1). This seems rather fitting.

This fragmented description at the level of operators reveals different strategies and behavioral patterns of the providers in the tourism sector: a broad spectrum of possible business partners is available to the independent organizer who manages the travel package as a coherent unit and interacts with guests. When compiling an offer, to what extent do tour operators take into consideration the bundling of various components and the positive contributions of the providers? Organizing travel is among the most creative activities in the broad field of tourism: to what extent do organizers make use of their scope, to what degree do they address the issue of renewable energy in their interactions with the guest? Once it has been clarified that, from the perspective of the guests, tour operators are largely responsible for the overall presentation of the tourism offering, the type of organizer and the evaluation criteria need to be identified.

The customer preference for an environmentally conscious offering does exist in principle—if no additional burdens are created: "Sustainability should not cost anything" (Conrady 2014, p. 36). At the same time, 22 % of customers are considered to have an interest in sustainability and represent an attractive market segment for tourism companies (Conrady 2014, p. 5). Tour operators frequently know too little about their target audiences or about the available options for targeted market cultivation. It also appears that there are significant research gaps. Specifically, a theoretical framework is missing, which allows the structuring of issues such as successful interaction with customers concerning sustainability in tourism, the identification of organizers or types of companies that could be in charge of this so-called educational task, and, finally, the identification of values and visions that are likely to be of relevance in the future.

## 3     Uniform Criteria for Sustainable Tourism

Dörnberg et al. (2013, p. 13) estimate the number of tour operators[5] in Germany at 1500, while the operator ranking of fvw (Dossier, 13.12.2013) lists 57 noteworthy market participants for Germany. Little is known about the way they operate and their distinguishing features. Worth mentioning at any rate is the broad range of those companies in Germany: from small, albeit very professional, partnerships or capital companies with 2000 travelers a year to vertically integrated groups, such as TUI AG with TUI Deutschland and 7.5 million participants in 2013 (Ibid., p. 5). While these companies also differ with respect to their portfolios and their quality policy, some of them, despite all their differences, come pretty close to their own aspirations of assuming a leading role in the field of sustainability. The individual companies are quite different and include TUI AG, Studiosus Reisen München GmbH, and forum anders reisen, a marketing cooperation of smaller tour operators.

While a direct comparison of these three companies might be very appealing, the contribution of this paper is to clarify with the help of a corporate example how difficult it is to actually "implement" the individual aims in applied business practice. Considering the title of this contribution, the specific question must be: To what degree is the topic of renewable energy present in the marketing of the selected company? To what degree do the people in charge explicitly address the issue? Is there any indirect coverage of the topic of climate change? Tour operators are in a difficult position in this regard since, on the one hand, they are very close to the needs and wishes of their clients while, on the other hand, they are far away and have very little opportunity to influence them. According to Dörnberg et al. (2013, p. 228):

> During vacation travel, the customer is subjected to such sensory overload that he will normally not be able to identify the factor in this complex service package which dominates as a brand or image.

Prior to compiling the actual bundle of services, during the offer and information phase, the organizers ought to look out for any opportunity to understand their customers so that they can take all appropriate steps in planning and structuring their Websites.

For the systematic incorporation of so-called green topics into the customer dialogue, hotels and tour operators can utilize the criteria of the Global Sustainable Tourism Council. They were developed in 2012 and are widely recognized at the

---

[5]Tour operations as the main source of income, excluding those companies that organize travel as a sideline job, only occasionally or without commercial purpose.

international level. These criteria serve as global reference values for minimum requirements and have been continually refined since 2007 in a cooperative process that includes a total of 27 organizations, including the UNWTO and TUI AG: "The minimum that any tourism business should aspire to reach" (GSTC 2012). These Global Sustainable Tourism Criteria for Hotels and Tour Operators not only formulate internationally recognized standards but also provide important pointers concerning concrete implementation. With regard to the relevance of the topic of renewable energy or the problem of climate change in corporate marketing, the following **criteria and indicators** ought to be present in the Web presence of the tour operator:

> **D1.3** Energy consumption is measured, sources are indicated, and measures are adopted to minimize overall consumption and encourage the use of renewable energy.
>
> IN-D1.3.a Total energy consumed, per tourist-specific activity (e.g., guest-night, tourists), per source. Percentage of total energy used that is a renewable versus nonrenewable fuel. . ...
>
> **D2.1** Greenhouse gas emissions from all sources controlled by the organization are measured, procedures are implemented to minimize them, and offsetting remaining emissions are encouraged.
>
> IN-D2.1.a Total direct and indirect greenhouse gas emissions are calculated as far as practical. The carbon footprint (emissions less offsets) per tourist activity or guest-night is monitored. . ...
>
> **D2.2** The organization encourages its customers, staff, and suppliers to reduce transportation-related greenhouse gas emissions.
>
> IN-D2.2.a Customers, staff, and suppliers are aware of practical measures/opportunities to reduce transport-related greenhouse gas emissions (GSTC 2013).

Since the topic of renewable energy also has strong societal relevance through a regional focus and the development of local self-supporting communities, the following point is also important:

> **B1** The organization actively supports initiatives for local infrastructure and social community development. . . (ibid.).

Does the company selected satisfy these minimum requirements of the Global Sustainable Tourism Council, and do the responsible parties directly or indirectly address the topic of renewable energy in their Web presence? The following section presents the results of the Web analysis and makes the transition to a critical reflection and categorization of these results.

## 4    Corporate Example

TUI AG, Studiosus Reisen München GmbH, and forum anders reisen consider themselves to be leaders in the field of sustainability. Among these three very different companies, the example of Studiosus stands out. It is one of only a handful of tourism companies—the only tour operator in Germany until 2014—that actively support the Global Compact of the UN and provide visible documentation of these activities. In contrast to TUI Deutschland and forum anders reisen, Studiosus is a pure operator of study tours, which implies certain demands on the quality of the travel program and its organization by tour guides that are predominantly employed by the company. The impeccable reputation of the company with respect to its commitment to sustainability is also derived from the positive attitude toward a solid medium-sized company in a very volatile and at times even shady environment. Study tour operators in general, and especially Studiosus in particular, appear to be far removed from any suspicion of greenwashing.

## 4.1  Studiosus Reisen München GmbH

Studiosus is a family-owned medium-sized company, founded back in 1954 in Munich. It is currently managed by the second generation and continues to enjoy great success in the marketplace. Studiosus is the largest study tour operator in Germany. It is characterized by innovative concepts ("extra tours") as well as a credible positioning regarding sustainability. Studiosus has been a member of the UN Global Compact since 2007 and is additionally seen in the industry as a benchmark for the implementation of sustainability measures. Consequently, it has received numerous awards and prizes over the years. In 2013, Studiosus received the CSR prize of the federal government in the category of medium-sized companies (50–499 employees). For the first time, this prize was awarded to "exemplary and innovative companies, which pursue a path of structuring their entire business activity in such a way that social, ecological and economic sustainability is achieved" (Studiosus 2014h).

Jointly with Marco Polo, Studiosus Reisen ranks 11th in the operator rankings of fvw (Dossier, 13.12.2013, p. 5) and has sales of € 233 million and 90,620 customers. According to information obtained from the company, approximately 83 % (Studiosus 2014g) of the company's study tours are sold via travel agencies in Germany, Austria, and Switzerland. The products of Studiosus are study tours of high quality with specialized tour guides, who are trained predominantly by Studiosus.

However, an analysis of the Webpages of Studiosus reveals that the problems of climate change and other aspects of renewable energy are described rather cautiously at the product level.

– The search term "climate change" provides only one entry on the homepage: in the year 2008, Studiosus received an award from GEO SAISON for the tour "The Alps and climate change" (Studiosus 2014b). This tour is no longer offered.
– The search term "renewable energy" similarly provides one entry: in a press release, Studiosus highlights a new offering, a study tour to Calabria with the title "Italy's wild tip of the boot." Included is the sentence:

Furthermore, the Studiosus tour manager also points out social topics of the region: What is the role of alternative energy and environmental protection in the South of Italy? And how big is the influence of the Mafia in Calabria? (Studiosus 2014d)

The catalogue does not allow these types of search terms. "Calabria" yields the following result:

- The tour offering initially contains no suggestions concerning "alternative energy and environmental protection." Furthermore, it involves air travel (8 days), and on the relevant pages of the catalogue, no data about the $CO_2$ balance of this trip are provided (Studiosus 2014e). While no information is given concerning possible compensating payments for air travel, travel by bus and train is designated as climate neutral "via $CO_2$ compensation" (Studiosus 2014e).
- The detailed description of the trip is not easily located (under "print function"), but it contains no references to renewable energy. During the ridge walk on the Monte Tiriolo, which is described as challenging, the guests can look out at the end a 360-degree panorama . . .

. . . and observe wind turbines in the distance. What is the role of alternative energy and environmental protection in the South of Italy? Ask your tour manager! In the village, the weaving looms rattle just like they did in the old days. (Studiosus 2014c)

Following Müller and Mezzasalma (Müller 2007, pp. 169ff), minimum requirements are in place for the marketing mix of tour operators with an ecological orientation:

1. Resource-efficient, low-emission means of transportation are selected (PRODUCT).
   - Studiosus makes this choice available but does not reference it in the description of the travel offering or in the detailed description of the tour. The information is only provided in the general travel directions. At this point, the decision in favor of the tour has factually been made.
2. The company supports products that are environmentally friendly, for example, by favoring these products in their offering or by a $CO_2$ compensation, which is included in the price (PRICE).
   - At € 1440 per person (if applicable plus surcharge for air travel) for 8 days full of adventures and support from a tour manager, which is equivalent to a daily price of €180, this study trip is well priced for the organizer, but at the same time not unreasonably expensive from the perspective of the guest. There is no incentive to select earthbound travel as the standard version and no $CO_2$ compensation is included.
3. Distribution only plays a minor role, for example, by producing the catalogues in an environmentally conscious manner. However, McKercher et al. (2014) demonstrate for the case of Hong Kong that both employees at the service desk as

well as distribution managers tend to completely ignore the problems of climate change. This is most likely due to the fact that the higher fees imply a preference for selling expensive long-distance travel[6] (DISTRIBUTION).

- If Studiosus sells 83 % (Studiosus 2014g) of the tours via travel agencies and additional tours via nontraditional outlets, their influence is reduced considerably.

4. The company provides its guests with decision tools that are easy to follow, for example, environmental certificates, the $CO_2$ balance of the tour, reports about the environmental situation at the destination, or suggestions for behavior (COMMUNICATION).

- Travel, particularly study tours, is a product that requires a large amount of information. While Studiosus is very transparent about its operations and products, it does not directly address the problematic issue of climate change when presenting individual tours. This is done much later. It would be an interesting question to check whether this approach serves the aim of sustainability and to what extent this is pursued systematically. An operator of study tours would most likely also be able to "educate" his guests without bothering them.

The search term "green energy" provides advice about ecologically sound ways of traveling to a destination and the possible offset of greenhouse gas emissions under "journey to destination." Information is also provided about the fact that airplanes cause the most harm to the climate among all means of transportation (Studiosus 2014a). The required compensation amount was already calculated and it would be easy to include it in the booking. Alternatively, it would be just as easy to calculate the individual values for each trip as well as the needed compensation payments with only a few clicks. On this occasion, Studiosus stresses that all business travel of its own employees is also calculated and compensated. All $CO_2$ compensations of the company and its customers finance the construction of biogas plants in southern India. This is done through a charitable organization (Studiosus Foundation e.V.) and in cooperation with the Swiss climate protection agency myclimate and provides high visibility for the company's efforts at climate protection.

The search terms "environmental protection," "ecology," and "sustainability" produce no hits in the packages section. Instead, fairly comprehensive details about ecologically responsible program planning are contained in the general information about the company. Studiosus organizes air travel with minimum stays of 3 nights, and the length of stays in the destination area exceeds the customary value by 25 % according to the company. Given the demands of older target audiences for study tours, however, this appears to mostly make economic sense.

Conclusion Start

---

[6]Travel agencies act as agents and receive a fee of 10 % on average.

The relevant criteria of the Global Sustainable Tourism Council were only partially satisfied: criteria D1.3 and D2.1 with indicators IN-D1.3.a and IN-D2.1. a (Sect. 3) were not satisfied, criterion D2.2 with indicator IN-D2.2.a (Sect. 3) only very marginally; criterion B.1 meanwhile, which relates to social sustainability initiatives, such as the use of climate donations to build biogas installations in the south of India, was obviously fulfilled. This criterion perfectly matches the self-image and market positioning of Studiosus:

"We consider it our task to build bridges across inner and outer boundaries with the aim of creating true intercultural understanding" (Studiosus 2014f).

Conclusion Stop

## 5    A Brief Summary

Very little research exists on tour operators; for that reason they are considered a type of "black box" in the tourism industry. Do they remain consciously or unconsciously vague on the issue of bridging the "green gap" in tourism? While numerous studies, including the annual travel analysis "Reiseanalyse" (FUR 2014, p. 6), stress the relevance of so-called green topics for customers of tourism companies, there appears to be little willingness to pay more or to make voluntary $CO_2$ compensation payments (Conrady 2014, pp. 5, 36). Individual participants along the service chain of the tourism industry, as well as customers, always see the responsibility as lying elsewhere or, when in doubt, with political leaders. Tour operators, as they even argue themselves during specialist conferences, also have an "educational role" to play. Due to their exposed position in the service chain and the proximity to the needs and wishes of their clients, this is certainly accurate.

So what is missing? Tour operators shape encounters and much more, but they are practically invisible when those encounters actually happen; only local people and circumstances matter. During and after a trip, they are normally contacted only in case of a complaint. Tour operators need to inquire about the topics and means needed to enter into a more intense discussion with their customers about suitable and ethically acceptable methods. This would require considerable effort, however, including openness, creativity, and entrepreneurial spirit.

## References

American Marketing Association. (2014). *Definition of marketing*. Accessed June 20, 2014, from https://www.ama.org/AboutAMA/Pages/Definition-of-Marketing.aspx, quoted as: AMA.

Boutiquehotel Stadthalle Wien. (2014a). *Das weltweit 1. Stadthotel mit Null-Energie-Bilanz.* Accessed September 09, 2014, from http://www.hotelstadthalle.at/null-energie-bilanz, quoted as: Hotel Stadthalle, 2014b.

Boutiquehotel Stadthalle Wien. (2014b). *Fact sheet.* Unpublished manuscript, quoted as: Hotel Stadthalle, 2014c [On-site request of the author, 26 August2014].

Boutiquehotel Stadthalle Wien. (2014c). Bill Nr. 63338, 27 August 2014. Unpublished manuscript, quoted as Hotel Stadthalle, 2014d.

Boutiquehotel Stadthalle Wien. (2016a). *Das Weltweit 1. Stadthotel mit Null-Energie-Bilanz*. Accessed April 28, 2016, from https://www.hotelstadthalle.at/nachhaltigkeit/null-energie-bilanz-hotel/, quoted as: Hotel Stadthalle.

Boutiquehotel Stadthalle Wien. (2016b). *Das umweltbewusste Hotel mitten in Wien*. Accessed September 09, 2014, from http://www.hotelstadthalle.at/nachhaltigkeit, quoted as: Hotel Stadthalle, 2014a.

Conrady, R. (2014, May 15). Corporate Social Responsibility in der touristischen Wertschöpfungskette: Status Quo, Trends, Herausforderungen, Keynote zur Jubiläumstagung „Verantwortliche Gestaltung der Zukunft—Innovative Ansätze für die Destinations- und Standortentwicklung" des Lehrstuhls Tourismus/Zentrum für Entrepreneurship der Katholischen Universität Eichstätt-Ingolstadt in Ingolstadt. Presentation published at: Accessed June 21, 2014, from http://www.ku.de/fileadmin/150306/css/Jubil%C3%A4umstagung/Conrady.pdf

Deutsche Bahn AG. (2014). *Erneuerbare Energien*. Accessed June 21, 2014, from http://www.deutschebahn.com/de/nachhaltigkeit/oekologie/klimaschutz/Erneuerbare_Energien/, quoted as: DB.

Deutsche Lufthansa AG. (2014). *Die-Vier-Liter-Flieger: Gemeinsame Kampagne der deutschen Luftfahrt*. Accessed June 21, 2014, from http://www.lufthansagroup.com/de/themen/die-vier-liter-flieger.html

Deutscher Hotel- und Gaststättenverband e.V. (2014). *Energiesparen leicht gemacht*. Accessed June 22, 2014, from http://www.dehoga-bundesverband.de/fileadmin/Inhaltsbilder/Publikationen/ Broschuere_Energiesparen_ leicht_ gemach t_Okt_2012_final.pdf, quoted as: DEHOGA 2012.

Deutscher Tourismusverband e.V. (2014). *Der Fernbus rollt durch Deutschland*. Accessed June 21, 2014, from http://www.deutschertourismusverband.de/themen/bustouristik.html, quoted as: DTV.

von Dörnberg, A., Freyer, W., & Sülberg, W. (2013). *Reiseveranstalter-Management. Funktionen, Strukturen, Management*. München: Oldenbourg.

El Dief, M., & Font, X. (2010). The determinants of hotels 'marketing managers' green marketing behavior. *Journal of Sustainable Tourism, 18*, 157–174. forum anders reisen (2014). Reiseperlen. Accessed June 23, 2014, from http://konradinheckel.tpk6.de/smart2/pub/reiseperlen-2014

FUR Forschungsgemeinschaft Urlaub und Reisen e.V. (2014). ReiseAnalyse 2014. Erste Ausgewählte Ergebnisse der 44. Reiseanalyse zur ITB 2014. Accessed June 14, 2014, from http://www.fur.de/fileadmin/user_upload/RA_Zentrale_Ergebnisse/RA2014_ErsteErgebnisse_DE.PDF, quoted as: FUR.

fvw-magazin 10/14. (2014, May 09). fvw-magazin 06/14 (2014, March 14). fvw-magazin 05/14 (2014, February 27).fvw -magazin (2013, December 12). Dossier "Deutsche Veranstalter 2013", quoted as: Dossier.

Global Sustainable Tourism Council, Washington (2012, February 23). *Global sustainable tourism criteria for hotels and tour operators, Version 2*. Accessed June 14, 2014, from http://www.gstcouncil.org/images/pdf/gstc-hto-indicators_v2.0_10dec13%20.pdf, quoted as: GSTC 2012.

Global Sustainable Tourism Council, Washington. (2013, December 10). *Global sustainable tourism criteria for hotels and tour operators—suggested performance indicators*. Draft Version 2.0. Accessed June 14, 2014, from http://www.gstcouncil.org/images/pdf/ global%20sustainable%20tourism%20criteria%20h-to%20version%202_final.pdf, quoted as: GSTC 2013.

Green Hotels Association. (2016). *Why should hotels be green?* Accessed April 28, 2016, from http://www.greenhotels.com/

Hilton Worldwide (2016a). *Carbon factsheet*. Accessed April 28, 2016, from http://cr.hiltonworldwide.com/downloads/, quoted as: Hilton, 2016a.

Hilton Worldwide. (2016b, March 31). *Corporate fact sheet*. Accessed April 28, 2016, from http://www.hiltonworldwide.com/about/, quoted as: Hilton, 2016b.

Instituto Costarricense de Turismo. (2014). *Nachhaltigkeit CST*. Accessed June 22, 2014, from http://www.visitcostarica.com/ict/paginas/sostenibilidad.asp?tab=4, quoted as: CST.

ITB Berlin. (2016). Der ITB Kongress 09.- 11. 03. 2016. Accessed May 05, 2016, from http://www.itb-kongress.de/Programm/ProgrammDownload/

Maledives. (2016). ...the sunny side of life. Visit Maldives Year 2016. Accessed May 05, 2016, from http://www.visitmaldives.com/de/

McKercher, B., Mak, B., & Wong, S. (2014). Does climate change matter to the travel trade? *Journal of Sustainable Tourism, 22*, 685–704.

Meffert, H., Burmann, Chr., & Kirchgeorg, M. (2012). *Marketing. Grundlagen marktorientierter Unternehmensführung. Konzepte—Instrumente—Praxisbeispiele* (11th revised and expanded ed.). Wiesbaden: Gabler.

Müller, H. (2007): *Tourismus und Ökologie. Wechselwirkungen und Handlungsfelder* (3rd revised ed.). München: Oldenbourg.

Pechlaner, H., Fischer, E., & Bachinger, M. (Eds.). (2011). *Kooperative Kernkompetenzen. Management von Netzwerken in Regionen und Destinationen*. Wiesbaden: Gabler.

Picard, D. & Di Giovine, M. (2014) Introduction: Through other worlds. In: D. Picard & M. Di Giovine (Eds.), *Tourism and the power of otherness. Seductions of difference*. Bristol: Channel View.

Pomering, A., Noble, G., & Johnson, L. (2011). Conceptualizing a contemporary marketing mix for sustainable tourism. *Journal of Sustainable Tourism, 19*, 953–969.

RDA Workshop Touristik Service GmbH. (2013) Fernlinienbusverkehr in Deutschland—Erste Bestandsaufnahme der Betreiber und Portale auf dem RDA-Workshop 2013. Accessed June 21, 2014, from http://www.rdaworkshop.de/presse/detailseite-pressemeldungen/article/fernlinienbusverkehr-in-deutschland-erste-bestandsaufnahme-der-betreiber-und-portale-auf-dem-rda-w-1//104.html, quoted as: RDA.

Schmücker, D. J. (2011). Freiwillige Kompensation von Flugreisenemissionen als nachfragerinduzierte Anpassungsstrategie—ein empirischer Anbietervergleich. *Zeitschrift für Tourismuswissenschaft, 3*, 139–149.

Sleep Green Hotels. (2016a). Boutique-Hotel Stadthalle***, Wien. Accessed 28 April 2016, from http://www.sleepgreenhotels.com/de/hotels/sleep-green-hotels/oesterreich/boutiquehotel-stadthalle.html, quoted as: Sleep Green, 2016a.

Sleep Green Hotels. (2016b). *Warum "Sleep green Hotels"?* Accessed April 28, 2016, from http://www.sleepgreenhotels.com/de/mitglied-werden.html, quoted as: Sleep Green, 2016b.

Studiosus Reisen München GmbH. (2014a). *Ausgleich der Treibhausgas-Emissionen der Flüge*. Accessed June 21, 2014, from http://www.studiosus.com/Informationen/WichtigeInformationen/Saison_2014/Anreise, quoted as: Studiosus, 2014a.

Studiosus Reisen München GmbH. (2014b). *Auszeichnungen und Preise für das ökologische Engagement*. Accessed June 23, 2014, from http://www.studiosus.com/Ueber-Studiosus/Unternehmensprofil/Auszeichnungen, quoted as Studiosus, 2014b.

Studiosus Reisen München GmbH. (2014c). Druckansicht [Detailansicht WanderStudienreise Kalabrien]. Accessed June 23, 2014, from http://www.studiosus.com/pdf/0469.pdf?opsid=1689521, quoted as: Studiosus, 2014c.

Studiosus Reisen München GmbH (2014d, April 17). *Italiens wilde Stiefelspitze entdecken: Kalabrien ist neu im Wanderprogramm von Studiosus*. Accessed June 23, 2014, from http://www.studiosus.com/Presse/Pressemitteilungen/Italiens-wilde-Stiefelspitze-entdecken-Kalabrien-ist-neu-im-Wanderprogramm-von-Studiosus, quoted as: Studiosus, 2014d.

Studiosus Reisen München GmbH (2014e). *Kalabrien—Italiens wilde Stiefelspitze*. WanderStudienreise Italien. Accessed June 21, 2014, from http://www.studiosus.com/Reiseangebote/Reisefinder/%28ops_id%29/1689521/%28Reise%29/Kalabrien/?f=b, quoted as: Studiosus, 2014e.

Studiosus Reisen München GmbH. (2014f). *Unternehmensleitbild.* Accessed June 29, 2014, from http://www.studiosus.com/Ueber-Studiosus/Unternehmensleitbild, quoted as: Studiosus, 2014f.

Studiosus Reisen München GmbH. (2014g). *Mittelständisches Unternehmen mit Tradition.* Accessed June 23, 2014, from http://www.studiosus.com/Ueber-Studiosus/Unternehmen sprofil/Daten-Fakten, quoted as: Studiosus, 2014g.

Studiosus Reisen München GmbH. (2014h). *Studiosus gewinnt begehrte Auszeichnung der Bundesregierung.* Accessed June 21, 2014, from http://www.studiosus.com/Ueber-Studiosus/ Unternehmensprofil/Auszeichnungen, quoted as: Studiosus, 2014h.

TUI AG. (2013, September 12). *AG notiert erneut mit Bestplatzierung im Dow Jones Sustainability Nachhaltigkeitsindex.* Accessed June 23, 2014, from https://www.tui-group. com/de/presse/presseinformationen/archiv/2013/20130912_Dow_Jones_Sustainability_Index.

United Nations Environment Programme/World Tourism Organization. (2005). *Making tourism more sustainable. A guide for policy makers.* Accessed June 14, 2014, from http://www.unep. fr/shared/publications/pdf/DTIx0592xPA-TourismPolicyEN.pdf, quoted as: UNEP/WTO.

United Nations Environment Programme/World Tourism Organization. (2012). *Tourism in the Green Economy.* Background Report. Accessed June 20, 2014, from http://www.unep.org/ greeneconomy/Portals/88/documents/ger/ger_final_dec_2011/Tourism%20in%20the%20green_ economy%20unwto_unep.pdf, quoted as: UNEP/UNWTO.

United Nations, Department of Economic and Social Affairs, Statistics Division/World Tourism Organization. (2010). International Recommendations for Tourism Statistics 2008, Studies in Methods, Series M, No. 83/Rev. 1. Accessed June 14, 2014, from http://unstats.un.org/unsd/ publication/Seriesm/SeriesM_83rev1e.pdf#page=21, quoted as: UN/UNWTO.

World Tourism Organization (UNWTO). (2014). *Understanding tourism: Basic glossary.* Accessed June 14, 2014, from http://media.unwto.org/en/content/understanding-tourism-basic-glossary, quoted as: UNWTO.

**Prof. Dr. Susanne Gervers** has been professor of tourism management at Hochschule für Wirtschaft und Umwelt Nürtingen-Geislingen since 2010. As the founder and managing director of a tour operator, she has 11 years of applied business experience, following a first exposure to the field of tourism during her student years. In Bamberg, but mainly in Heidelberg, York/England, and Hamburg, she studied political science, public law, history, and sociology. She worked at Universität der Bundeswehr in Hamburg and completed her doctorate in economics and social sciences at Universität Lüneburg supported by a stipend of the Friedrich-Naumann-Foundation. In her role as university professor, she studies questions of teaching methodology and creativity research as well as new forms of tourism in the sharing economy or the foundations of a theory of tourism. In addition, she researches problems of ethics and the management of innovations.

# From Energy Supplier to Capacity Manager: New Business Models in Green and Decentralized Energy Markets

Ben Schlemmermeier and Björn Drechsler

**Abstract**

This chapter concentrates on the description of new business models in green and decentralized energy markets and derives recommendations for action for energy suppliers—particularly municipal utilities, of which there are around 1000 companies in Germany. The chapter focus is outlined in the introduction and clarified further by providing the authors' vision concerning the future of the energy sector. In what follows, the increasing importance of renewable and decentralized energy production and the resulting necessity to further develop the design of the power market in Germany—with reference to the current energy policy—are described. Based on a short description of the main capabilities of the electrical power system, the future core tasks, challenges, and business models of energy companies are derived and explained with reference to current market and regulatory developments. The chapter ends with a description of strategic and organizational requirements needed by energy companies to implement the new business models. In the summary, the main aspects and success factors of viable business models for energy companies are again highlighted.

**Keywords**

Business model • Renewable energy • Electricity market design • Capacity management

A previous version of this chapter has been published in Herbes, C.; Friege, Chr. (Hrsg): Marketing Erneuerbarer Energien. Grundlagen, Geschäftsmodelle, Fallbeispiele, 2015, Springer Gabler.

B. Schlemmermeier (✉) • B. Drechsler (✉)
LBD-Beratungsgesellschaft mbH, Mollstraße 32, 10249 Berlin, Germany
e-mail: ben.schlemmermeier@lbd.de; bjoern.drechsler@lbd.de

# 1      Introduction

## 1.1      The Energy Market Is a Political Market

The German energy sector crosses the bridge of the *Energiewende* ("energy transition"): the way back is blocked, since the earnings of many utilities are declining significantly. Going forward, the bridge has not yet been completed, as the legal and regulatory reforms that are turning the energy sector into a market for energy and, thus, into a business are far from complete: the Renewable Energy Act (*Erneuerbare-Energien-Gesetz, EEG*), Combined Heat and Power Generation Act (*Kraft-Wärme-Kopplungs-Gesetz, KWKG*), design of the electricity market, grid usage charges, smart metering, energy efficiency, and emissions trading. While the EEG 2014 had as its main focus the improved integration of renewable energy into the energy market, the next revision (EEG 3.0), and thereby the switch to tendering procedures, is already on the way (BMWi 2014a, p. 3). At the same time, the design of the power market is being developed further and the entire system of grid usage charges, levies, and fees in the pricing of electricity for the end user is being challenged.

Since the beginning of market liberalization in Germany in 1998, there has never been such a high degree of uncertainty in the energy sector concerning further market developments and regulatory initiatives. In addition to the developments in the regulatory framework for the German energy transition, climate protection targets, innovations in energy technology, information and communication technologies (digitization), and changing customer needs are the main drivers behind the transformation of the energy sector. In this uncertain market environment, numerous energy suppliers, especially municipal utilities, face the challenge of protecting the profitability of their existing business while at the same time developing and implementing business models that assure a successful future.

## 1.2      Situation in the "Old World" of Energy Retailers

In the "Old World," an energy supplier predominantly buys and sells electricity and gas. This requires the ability to develop electricity and gas products, to handle them operatively, and to sell them. Most of the innovative potential in the development of this business model and product was accomplished long ago. Green electricity rates, predominantly on the basis of certificates of origin, are a standard offering of most utilities. New green electricity labels, which certify the sustainability of the provider in addition to the product (TÜV Süd 2014), can again help to achieve greater differentiation. These days, only electricity products that market renewable electricity regionally continue to have a unique selling point. However, to market electricity from sources that qualify under the Renewable Energy Act as green electricity products, the relevant statutory instrument from EEG 2014 still needs to be implemented. Other new and innovative offerings are based on the value-oriented integration of private capacities into the market or the promise of delivering (green) power to customers that comes from clearly identifiable sources from a defined pool of installations. Business models like these are often described as community power or sharing electricity. They depend largely on intelligent metering systems and digitization of commercial processes and represent a specific form of capacity management or virtual power plants (Sect. 4.3).

The traditional market for simple electricity and gas products will continue to shrink owing to the growing trend toward decentralization. And since electricity and gas discounters will not be able to raise prices to cover their costs of customer acquisition and negative margins, they will leave the market. Pricing differences and incentives to change suppliers will consequently be reduced. For the "Old World" of energy retailing, this implies the need to radically simplify products, communications, and customer management, including processes and information technology (IT). To cope with the increasing digitization of everyday life and the wishes of customers and to reduce costs, the traditional electricity and gas business of the future will need to be processed online. Product management must become quantitative in nature: margins and market shares need to be managed in a targeted fashion. The main message is: be simple, efficient, and profitable.

## 1.3    Chapter Structure

This chapter begins with a brief overview of the current and future relevance of decentralized and renewable energy generation in Germany—as an obvious sign of the irreversible transformation of the entire energy sector (Sect. 3). The necessity of further developing the design of the energy market and the core tasks of the future are analyzed in Sect. 4.

Based on the capabilities of the power system presented in the previous section, the most important fields of action for energy companies are highlighted and business models for the so-called New World are derived and described in Sect. 5.

In Sect. 6 we cover the strategic and organizational requirements needed to develop and implement new business models. In the conclusion, the main aspects of this chapter are summarized in the form of propositions, and success factors for future business models of municipal utilities are stated.

Before we tackle these issues, we start in Sect. 2 with our vision of the energy sector of the future (Schlemmermeier 2012, p. 42). This view of the future serves as a framework for the contents of the entire chapter.

## 2    Our Vision: From Energy Supplier to Capacity Manager

**Climate Protection as Undisputed Societal Consensus** An energy sector that is free of $CO_2$ by 2050: The reaction to climate change is currently the major propellant of change for the energy industry. Germany's $CO_2$ reduction target of more than 80 % by 2050 compared to 1990 (BMUB 2014) implies that the use of fossil fuels needs to be reduced consistently and that the demand for energy must be covered a lot more efficiently and almost completely from renewable energy sources.

**Decentralized Production and Storage Change Customer Behavior** The availability of new technologies opens the door to decentralized electrical power generation (which is also known as distributed generation). The costs of decentralized generation decreases while the costs of conventional large power plants increases. The lack of constant availability of renewable electricity generated by wind turbines and solar photovoltaic (PV) systems must be compensated via constantly available gas-fired production units, preferably via the cogeneration of heat and electricity (combined heat and power, CHP). Small CHP units will become efficiency leaders in energy usage and emissions and cost leaders in investments and operations.

Increasing availability and declining cost of technologies for a decentralized energy supply will increase the desire of consumers to provide not only their own heating energy but also their own electricity. Given sufficient purchasing power and financial incentives, this opens up enormous growth potential. Private customers of the future will take care of their own energy demand: They will produce electricity and heat via cogeneration, install solar PV systems on their roofs, and have a battery in the basement or an electric vehicle in the garage. Instead of putting money in a savings account, they will look for shrewd investments. At the same time, customers will want to be sure that they can always obtain electricity if their own production is insufficient to cover their demand (darkness, no wind). If possible, they will also want to turn energy production into a business. If they produce excess energy, they will want to be able to sell it.

Industrial clients of the future will also partially cover their own electricity demands but will continue to rely on the capacities provided by the grid. They will flexibly adjust the quantity and timing of their demand to the supply of electrical power (demand-side management).

**Further Development of Electricity Market Design** When discussing the backup needed in addition to the fluctuating supply by renewables, storage technologies are frequently mentioned as the missing link. This is only half the story. In the future, a complex system of flexible capacities will be needed: generation facilities, storage units, and demand response resources, as well as distribution and transmission systems. Competition in innovation and efficiency will determine what technology to use for any specific purpose. This competition will be driven by a new market segment: the capacity market. When capacities are scarce and especially storage facilities are expensive, there will be significant incentives for business and industrial consumers to accept compensation for load shifting or peak shaving via demand response measures. The future will no longer be characterized by a competition of various types of energy or fuels but by a competition of capacities and system services. This is a major opportunity for European economies. Instead of importing raw materials, value will be created domestically (building production facilities, developing the grid, system services, marketing). This value creation will enhance the economic robustness of Europe.

**From Energy Supplier to Capacity Manager: A New Business Model** In an energy world dominated by renewable energy sources, especially wind and sun, the marginal cost of electricity and heating will tend toward zero. In other words, there will no longer be any significant variable costs for providing electrical and thermal power. This enhances the importance of fixed costs and will ultimately lead to the disappearance of an energy price in favor of a "flat rate," which is already familiar from the telecommunications industry.

In the future, two major types of services will be in demand in the energy market: End users will expect the reliable provision of energy, while their own decentralized production facilities will be integrated as efficiently as possible into the overall system. As this process unfolds, customers will partially adjust the timing of their energy demand to the supply of energy. The independent system operator will demand flexible capacities, which will synchronize physical supply and physical demand (load) in the overall system. Energy providers will become capacity managers and the complexity of their products will increase rapidly. They will generate contribution margins from revenues for services provided to end users and for services in the entire system.

**What Will Matter?** Intelligent metering systems (often called smart meters) are a key component of the energy transition. They collect data, are an important communications unit, and provide the interface for the management of supply (decentralized generation), storage and demand (load management). The management of complex processes, growing data volumes, and sophisticated IT systems will become the decisive factor for future business success.

To increase competition in innovation and efficiency, a framework that supports market outcomes will be essential. Thus the decoupling of the functions of grid operation on the one hand and ancillary and balancing services on the other hand is necessary. Only those services that can be provided efficiently in a natural monopoly need to be regulated as a monopoly.

The potential of electric mobility for the energy sector is underestimated. The state of development in storage technologies is currently the limiting factor. As soon as that problem is solved, a revolution in the automobile industry will come up. At that point, thousands of megawatts of flexible storage capacity will be connected to the grid and constitute a major part of the overall energy system.

**The Disappearance of the Dinosaur** Dinosaurs are extinct. The personal computer has displaced the mainframe, smartphones have replaced cell phones. The Internet replaces the national libraries of the world. Social networks are a democratic movement. The world is being increasingly decentralized, yet interconnected. This trend will not end with the energy sector. It will force adaptation and displacement on many technologies and companies. Back to the metaphors: Dinosaurs represent nuclear power plants and the mainframe corresponds to the coal plant. The Internet, smartphones, and social networks represent decentralized generation, intelligent metering systems, and virtual power plants.

**Conclusions** Visions like these serve as the foundation for the development, positioning, aims, and strategies for business in the "New World" (Sect. 6.1). They also make the valuable contribution of engaging all involved parties in the development process from the very beginning (Sect. 6.3). It is thus essential that all energy suppliers and utilities develop their own vision about the future of the energy sector.

## 3 The Energy Sector Will Become Decentralized and Renewable

### 3.1 The Decentralization of Electrical Power Generation

The energy transition in Germany and innovations in technology drive the decentralized generation of electrical power and vice versa. The centralized power system, which consists of approximately 450 conventional power plants with a capacity of approximately 95 GW that feed into the high voltage or extra-high voltage grid (BNetzA 2014a), is complemented by about 1.44 million decentralized generation units, which had a capacity of roughly 85 GW at the end of 2013 (BMWi 2014b, p. 7; BSW-Solar 2014, p. 1; BWE 2013, p. 2).

But this is only an interim description of the paradigm change that is under way in the electricity market. In a system of coexistence of centralized and decentralized systems, the decentralized capacities will gradually gain the upper hand. This follows logically from the German targets of strengthening electricity production from renewables and exiting nuclear power as well as the growing societal resistance against centralized power generation based on the environmentally damaging use of fossil fuels that began in the twentieth century.

The grid development plan, which is drafted annually by the four transmission system operators in Germany in cooperation with the federal states, the public, and the federal network agency, must specifically also incorporate this rapid increase of decentralized generation in addition to the availability and operation of large conventional power plants. The scenario analysis of the grid development plan therefore provides a solid foundation for the assessment of future trends in decentralized generation in Germany. Based on the main scenario B for the grid development plan 2014 (BNetzA 2013, p. 2) it is thus likely that the number of decentralized capacities will grow up to a total of approximately three million units in the next 10 years and up to four million within the next 20 years. Already today— and this will be true even more so in the future—it is absolutely essential to efficiently integrate millions of decentralized units into the energy market and electrical power systems.

## 3.2 Developments and Challenges for Renewable and CHP Installations

As compensation rates continue to decline, producers of renewable energy are increasingly challenged to optimize their own consumption and to efficiently market any excess quantities. Operators of existing and new units that generate electrical power via cogeneration (CHP) are forced in the current market situation to equip their installations with a greater degree of flexibility and an electricity-driven operation to open up urgently needed value-added potential. Especially for wind turbines or solar power it will no longer be sufficient to simply pick the location with the highest yield at the lowest available cost. Instead, self-consumption and competencies in the energy business field and in regulatory matters will become increasingly relevant in the future. The same is true of the management of all other decentralized capacities, where a high forecast quality and great operational flexibility must be achieved (Sect. 4).

## 3.3 The Markets for Electricity and Heating Will Continue to Converge

As cogeneration (CHP) will continue to expand and as excess renewable electrical power will be utilized to generate heating energy via heat pumps and electric heating systems, the markets for electrical power, gas, and heating will continue to coalesce even more strongly. Approximately 55 % of energy consumption by the German end user relates to the provision of space heating, warm water, and process heat (BMWi 2014c, Table 7). The heating supply in Germany thus holds massive potential for energy efficiency and cost reduction and opens up a large number of technological and economic possibilities to develop new business models and products for energy providers.

In the German household sector alone, approximately 21 million heat-generating facilities are currently in operation. Of these, approximately 65 % of all oil and approximately 67 % of all gas heaters, 9.5 million units in total, are older than 17 years as of today. Approximately four million of these units are even older than 24 years (Shell und BDH 2013, p. 27). Despite declining heating requirements as a consequence of efficiency gains, most of the old heating installations need to be replaced step by step. Currently, the annual rate of modernization only stands at about 3 % (BDH 2014), which means that it will take about 33 years before currently existing systems are completely turned over once. In sum, there is an enormous modernization backlog in the heating sector, which opens up a massive market potential for the installation of modern condensing boilers (especially gas-fired), solar thermal energy, cogeneration units, biomass boilers, heat pumps, and electric heating systems.

The future of the energy sector is clearly mapped out. Numerous energy service companies, newcomers, start-ups, and industry outsiders are already actively and speedily capturing future business potential with innovative business models and sufficient capital. It remains to be seen what role the established energy suppliers—and especially the municipal utilities—will play in this "New World" and whether their traditionally close relations with customers and local craft businesses can be used as trump cards while simultaneously mastering the challenges of the "Old World".

## 4  Advancing the Electricity Market Design and Core Tasks for the Future

### 4.1  Energy System Capabilities

To consistently, reliably, and efficiently cover end users' energy demands in the future, the integration of the demand side with the fluctuating supply of renewable energy will be needed since this segment is increasingly dominating the energy sector. Increasing the flexibility of supply and demand and their synchronization will be the key tasks in the electricity market of the future. The growing need for flexibility must be guaranteed through a technology mix of flexible generation, electric storage, and demand response resources and with the help of correspondingly liquid wholesale and control energy markets. Figure 1 demonstrates these complex relationships with reference to the most important capabilities of the energy system.

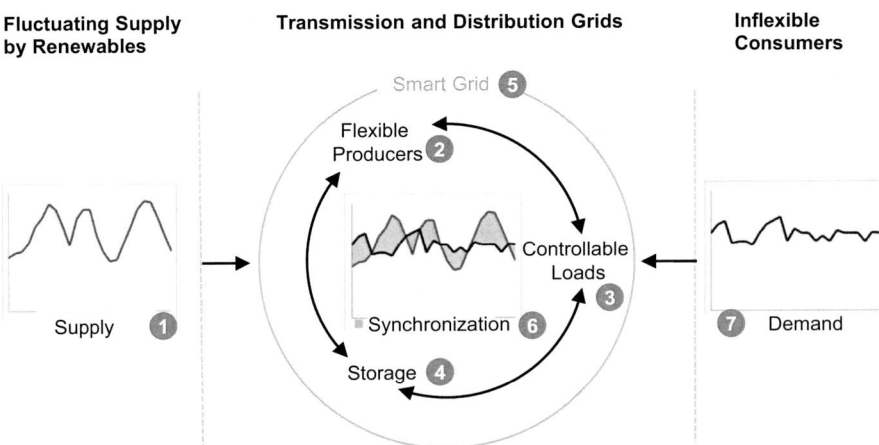

**Fig. 1**  Synchronizing supply and demand of electricity as main future challenge

The most important capabilities of the energy system of the future (also called "capacities") are:

1. Fluctuating supply by renewables (wind turbines, solar PV);
2. Flexible producers (available and controllable generation);
3. Controllable loads (demand response resources);
4. Electric storage (stationary batteries, electric vehicles);
5. Smart grid (transmission and distribution);
6. Synchronization of supply and demand (capacity management/virtual power plants);
7. Inflexible consumers (mainly energy efficiency measures).

Based on these capacities of the power system, numerous market and business opportunities for different customer segments and stages along the value chain exist that will be considered in greater detail in Sect. 5. But these market and business opportunities cannot and should not be captured without adequate consideration of the aims of the current energy policy, namely, the secure, cost-efficient, and environmentally friendly energy supply. For that reason it is mandatory to continuously develop the electricity market design along these lines.

## 4.2    Continuous Development of the Electricity Market Design

### 4.2.1    Current State of Conventional Power Plants

As a consequence of the energy transition in Germany and the decentralization of the power sector, the relative importance of the various technologies available for the generation of electrical power is changing. Driven by the massive buildup of renewable generation, the intensity of competition in the wholesale power market has increased enormously in recent years, including predatory pricing based on marginal cost. The existing market model thus not only puts economic pressure on existing power plants[1] but also prevents necessary investments in the modernization of power plants, which are needed to maintain the reliability of supply and to achieve climate protection targets. It is therefore the prevailing opinion among politicians and utilities in Germany that the current situation with conventional power plants calls for a reform of the electricity market. Still open, however, are the concrete contours of such a market reform [for a more detailed perspective see Schlemmermeier (2014)].

---

[1] By September 2014, the Federal Network Agency had already received requests for the temporary or terminal closure of power plants totaling approximately 13 GW (BNetzA 2014b).

### 4.2.2    Market and Regulatory Components for the Future

Fundamentally, future electricity market design should contain the following market and regulatory components:

1. Capacity mechanism to obtain renewable electrical power,
2. Capacity mechanism to assure reliability of supply,
3. Energy-only market to optimize capacity utilization,
4. Market for control energy and ancillary services.

The existing high reliability of supply in Germany must be assured via a capacity mechanism that offers a financial incentive for providing sufficient electric capacities to meet demand. This is the only way to generate the income streams required to keep existing power plants on the market and to encourage investment in new facilities. To keep costs for the consumers as low as possible, a capacity market needs to foster competition in innovation and efficiency and limit windfall profits for the market participants. At the same time, the energy-only market loses nothing of its importance for the energy sector since it provides the necessary incentives for efficiency and the corresponding utilization of those capacities in the energy system that have the lowest marginal costs.

The generation of electricity from renewables constitutes a separate market segment since renewable energies and fossil-based power plants produce different products. For that reason, they are not complete substitutes but, rather, complements. Hence, an autonomous capacity mechanism for the purchase of renewable electrical power must be developed that allows for price competition on the basis of administered quantities. This capacity mechanism must continue to provide incentives for innovation in the various technologies, while the roll-out needs to be oriented on efficiency criteria for the entire system. And finally, all ancillary and balancing services—and not only the primary control, secondary control, and minute reserve—should be given a market price in order to adequately compensate the participants for this important contribution to guaranteeing the reliability of supply.

A fundamental prerequisite to assure the reliability of supply and the transformation of the energy sector is an adequate power grid. Future grid development must be structured as a process of weighing up the efficiency of installing capacities close to the load versus building new power lines over long distances. Thus, owing to the high impact on costs and the environment, the development of transmission and distribution grids needs to be restrict to what is absolute necessary.

### 4.2.3    State of Debate About the Future Design of the Electricity Market

Concrete suggestions concerning future market design and the improved integration of the different market segments for electricity from renewable and fossil energy sources, as well as for the energy (energy-only market), power (capacity market), and ancillary and balancing services (control energy market) have been around for a number of years. As an example, in 2012 the concept of the focused capacity market was proposed by Öko-Institut, Raue LLP, and LBD-Beratungsgesellschaft (Öko-Institut et al. 2012). The pros and cons of this and of additional proposals concerning different capacity mechanisms have been discussed intensely among experts and politicians in Germany. The most discussed models are briefly presented in the following tables.

### 4.2.4 Description of Five Main Models for the Design of the Future Power Market with Capacity Mechanisms in Germany

In the following tables five main models for the design of the future power market with capacity mechanisms in Germany are briefly described (Tables 1, 2, 3, 4, and 5). As a consequence of the systematic supply shortage in connection with the high price volatility in the energy-only market, neither the grid reserve nor the strategic

**Table 1** Grid reserve

| Brief description | The aim of the grid reserve is to guarantee the secure operation of the power grid. It was introduced in Germany in reaction to transmission bottlenecks in the north–south direction. Components include a regulation on reserve power plants *(Reservekraftwerksverordnung, ResKV)* and a regulation on interruptible loads *(Verordnung zu abschaltbaren Lasten, AbLaV)*.<br>The grid reserve factually works like a strategic reserve but lacks a transparent pricing mechanism. It can only be a transitory instrument, which supports the discussion about fundamental market reform without jeopardizing the reliability of supply. |
|---|---|
| Provision of power | Systemically relevant installations.<br>"Insurance function" that is external to the electricity market. |
| Compensation payment | Reimbursement of cost |

**Table 2** Strategic reserve (BMUB 2013; Consentec 2012)

| Brief description | The further development of the grid reserve results in the strategic reserve, which is utilized in the case of a supply scarcity, in other words, if demand cannot be satisfied by the energy-only market.<br>The aim is to maintain scarcity and, thus, a higher level of electricity prices and margins. This is expected to assure profitable investments and the economic operation of power plants.<br>The amount of strategic reserve must be fixed administratively. The strategic reserve may not be used in the wholesale markets. In essence, a kind of insurance is provided with this approach. |
|---|---|
| Provision of power | Only power plants. The demand for capacity is set by the administration.<br>"Insurance function" that is external to the electricity market. |
| Compensation payment | Capacity charge for maintaining capacity. Payment at highest spot market price (e.g., 3000 Euro/MWh). |

**Table 3** Comprehensive, centralized capacity market (EWI 2012)

| | |
|---|---|
| Brief description | The concept of the comprehensive capacity market is based on the demand for a capacity that is set by the administration. It is acquired via an auction. The power plants are used in the energy-only market and the control energy markets. |
| Provision of Power | The entire market consisting of power plants, controllable loads, and storage. The amount of capacity is determined administratively. |
| Compensation Payment | Fixed capacity charge to provide output over amortization period of capacity (e.g., 15 years). |

**Table 4** Focused, centralized capacity market (Öko-Institut et al. 2012)

| | |
|---|---|
| Brief description | The concept of the focused capacity market is also based on an administratively determined amount of capacity, which is obtained via an auction. |
| | The main difference with the comprehensive capacity market lies in the redistribution effect between consumers and companies as well as among companies. Also included is the option to take into consideration regional scarcity. |
| | The focused capacity market aims at market segmentation and the corresponding reduction of costs for the consumer by reducing windfall profits of the providers of electrical power. |
| Provision of power | Only installations that are at risk of being shut down or new facilities (highly flexible and environmentally friendly generation, controllable loads, storage—output determined administratively). |
| Compensation payment | Fixed capacity charge to provide output over amortization period of capacity (e.g., 15 years for new facilities, 5 years for existing plants). |

**Table 5** Comprehensive, decentralized capacity market (BDEW 2014; Enervis 2014)

| | |
|---|---|
| Brief description | This is one of the various models that focus on demand. In those models, the consumer independently determines the degree of supply reliability that he expects and is willing to obtain from his supplier. Here the degree of reliability of supply is not set administratively but instead determined by the demand for certificates of supply reliability. |
| Provision of power | Overall market consisting of power plants, controllable loads, and storage. Demand determines the range of services offered. |
| Compensation payment | Capacity charge to provide service over the contract term with the customer (e.g., 2 years). Energy charge is equal to market price. |

reserve is suitable for removing the current uncertainty and investment constraints of market participants. Because of the very short-term price signals and the high volatility of the capacity charge, the decentralized capacity market is similarly unable to sufficiently incentivize investments. Still, the decentralized capacity market—just like the focused and comprehensive capacity market—maintains a competitive intensity in the wholesale markets. But only the focused and comprehensive capacity markets are truly capable of providing investment incentives. The different models not only have different repercussions for the business fields of the power plant operators and utilities but also for the degree of reliability of supply and the cost of electricity to consumers. From the various perspectives of politics, corporations, business fields, and consumers, it therefore follows that different models will be preferred. When designing electricity market reforms, politicians must therefore assess the different models on the basis of their effects on the reliability of supply, climate protection, intensity of competition in the energy market, and distribution of costs between the energy industry and consumers.

As for the Renewable Energy Act 3.0, a possible concept was proposed, for example, in a study of the Öko-Institut, commissioned by Agora Energiewende in 2014. In summary, this concept suggests an even greater integration of renewable energy plants compared to the current situation, which also implies that these installations need to accept a greater degree of electricity price risk. At the same time, they should be allowed to collect capacity premiums to refinance their investments if they are structured in a way that supports the overall system (Öko-Institut 2014, p. 1). In some sense, this model advances and puts into concrete terms the concept of the focused capacity market for renewable energy.

## 4.3    Capacity Management as One of the Key Business Models of the Future

Independent of the future electricity market design, the demand for flexibility will be one of the main value drivers for both producers and consumers of electrical power. The operators of available generation units, demand response resources, and storage facilities have numerous marketing options, which poses different challenges concerning technical parameters and the operation of the flexible capacities as well as processes for optimization (in front of and behind the meter) and for marketing. The intelligent management of the provision of power from different capacities as well as the permanent optimization of energy sales and purchases in the various markets with the aim of creating the highest value or the lowest cost will be the task of capacity managers (Fig. 2).

The foundations for the business model of the capacity manager are as follows:

- Decentralized metering and control systems to capture data and to trigger commands at the metering point in real time;
- Automated IT systems that are suitable for a mass market and communicate with the facilities and other market participants;

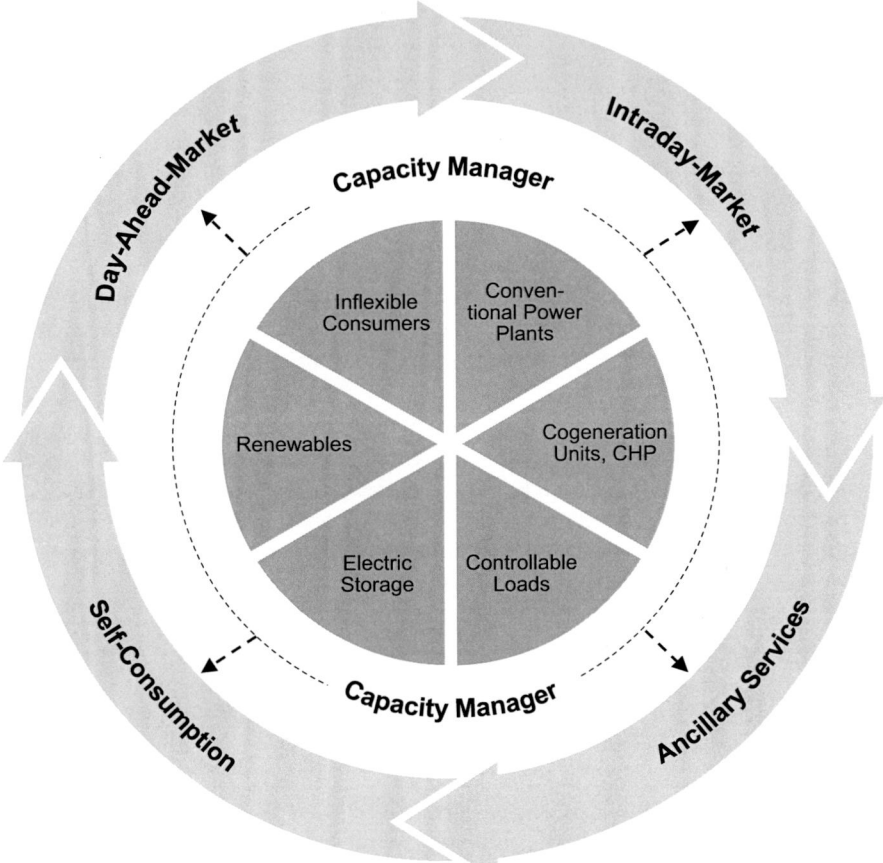

**Fig. 2** Principle of value optimization of capacities in different market segments by capacity manager

- Complex, high-performance IT systems to evaluate and analyze various kinds of forecasts and metering data, to structure portfolios, and to generate switch and control commands.

Mastering highly complex processes, massive amounts of data, and powerful IT systems will thus be a decisive success factor for capacity managers. We call the combination of these functionalities the *factory of the capacity manager* (Fig. 3). It is not mandatory that capacity managers own all capacities or that all elements of his so-called factory are furnished by them. Depending on regulatory developments, certain parts of the factory could very well be under the control of

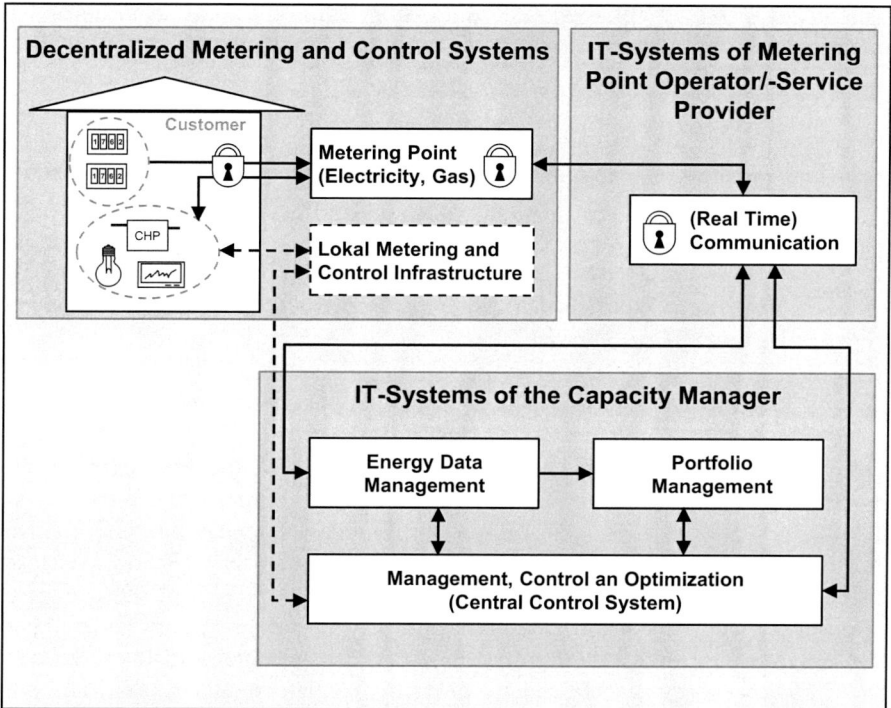

**Fig. 3** Outline of the "factory of the capacity manager"

the metering point operator or the metering service provider. The metering systems and relevant processes (at present, specifically settlement and customer change processes, in the future possibly also switching/controlling) are subject to regulation. This poses additional challenges with regard to the competencies of the capacity manager.

The business model of the capacity manager will develop in several stages and different versions in the market. In its first development stage it has already been observable for a number of years in the German energy market: for example, at companies that directly market renewable electricity in line with the Renewable Energy Act of 2014[2] and at companies that bundle renewables, small CHP units,

---

[2]According to Sect. 5.9 of the Renewable Energy Act of 2014, direct marketing refers to the selling of electrical power from renewable energy or from mine gas to third parties, except where the electrical power is used in direct proximity to the installations and not passed through the grid.

emergency power generators, and controllable loads in so-called virtual power plants and sell it on the wholesale and control energy markets.

In addition to the business model of the capacity manager, numerous other business models, including various products and services, exist that can advance the energy transition. We summarize these opportunities under the term *energy transition business opportunities* and present an overview in the following section.

## 5    Business Models for Energy Suppliers and Utilities in the "New World"

### 5.1    Structuring and Description of Future Business Models

#### 5.1.1    Development of a Vision for the Future of the Energy Market

Energy suppliers in general and municipal utilities in particular need to answer a number of questions: Which business opportunities—today and in the next 5 years—will present enough potential for the so-called New World? What are their capabilities in these business opportunities? And which of these business opportunities do they want to tackle and which abandon? Among other things, this requires a vision for the future that describes expectations about the development of the energy market and its political, regulatory, and social framework conditions. In addition, it will be important to structure the multitude of possible business opportunities and to provide a qualitative and quantitative description.

#### 5.1.2    Structuring and Description of Future Business Models

Business opportunities can be described with regard to their technological attributes (e.g., available and renewable power generation, controllable loads or demand response resources, battery storage and electric mobility) or from a customer perspective (such as energy efficiency, self-consumption, reduction of energy costs, comfort, and return on investment).

In our opinion, future business models will be best structured along the capabilities of the power system, which are described in Sect. 4.1. Using this structure, Figure 4 provides a brief overview of the numerous business opportunities presented by the energy transition for selected customer segments.

The depth and complexity of the analysis of potential business opportunities can be reduced significantly by focusing on specific customer segments (for example, only households and commercial consumers in the home market of the municipal utility) and capabilities (for example, no offshore wind energy). Following the

| Capabilities | Private consumers | Commercial consumers | Key accounts |
|---|---|---|---|
| Renewable energy (fluctuating supply) | | Photovoltaics | |
| | Solar thermal energy | | Wind energy |
| Decentralized, flexible generation | | Condensing boilers | |
| | | Heat pumps | |
| | | CHP units | |
| Controllable loads and consumers | | Demand side management (load shifting etc.) | |
| Storage and electric vehicles | | Stationary battery storage | |
| | | Electric vehicles and loading infrastructure | |
| Capacity management | | Virtual power plants | |
| Inflexible consumers/ energy efficiency | Smart home | Energy management | |
| | | Modernization of heating and cooling systems, energy efficiency measures | |
| | | Analysis and advisory services | |

**Fig. 4** Overview of multitude of energy transition business opportunities (excluding grid operation)

structure presented in Fig. 4, energy transition business opportunities are defined in three dimensions and therefore require detailed and comprehensive planning (Fig. 5):

– Capabilities in the energy system that can be provided via various technologies,
– Customer and market segments where value added can be created,
– Stages in the value chain to implement the business model or marketing the products and services.

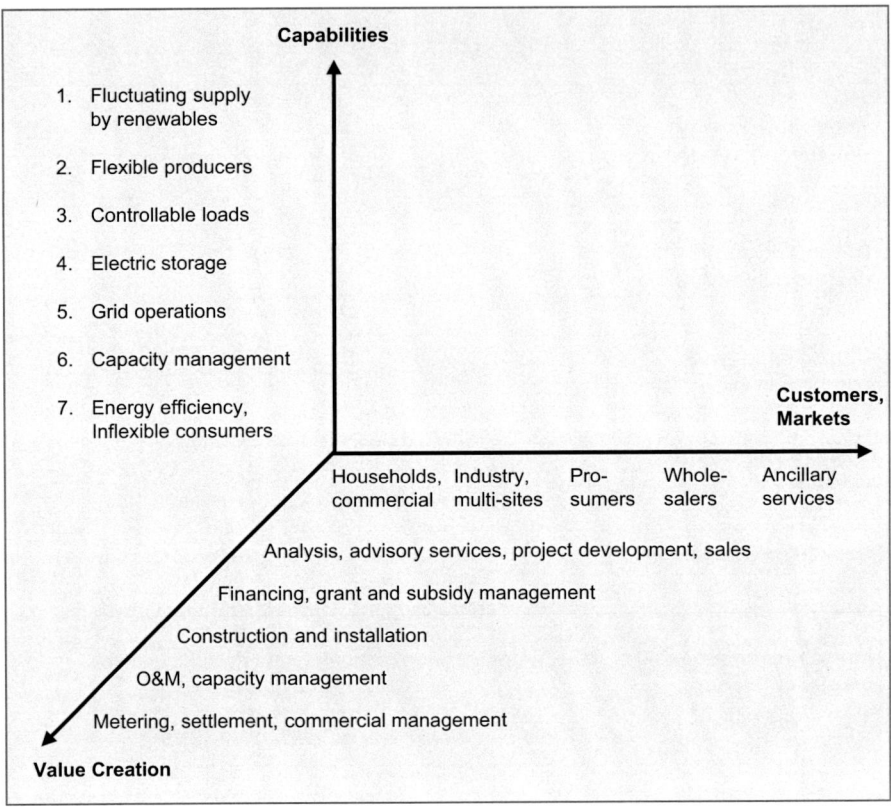

**Fig. 5** Description of business opportunities for energy transition in three dimensions

### 5.1.3  Value Chain of Energy Transition Business Opportunities

Figure 6 displays the value chain for the energy transition business opportunities and outlines the needed competencies. Assuming the necessary clarity concerning the future development of markets and customer demands, energy companies need to identify those stages in the value chain where they want to be active as well as the stages where they decide or are forced to rely on the competencies of partners and market services.

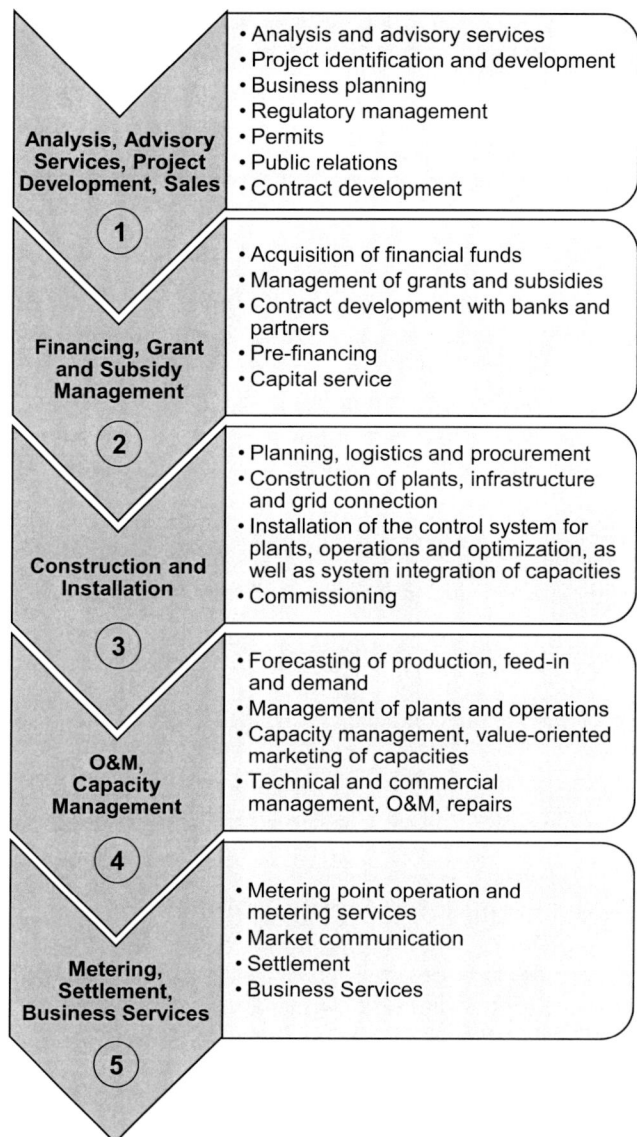

**Fig. 6** Description of value chain for energy transition business opportunities

### 5.1.4 Criteria for Assessing and Selecting the most Promising Business Opportunities

The assessment and selection of the most promising business opportunities from the perspective of an energy supplier must be based on a set of predefined criteria. Potential business models should target a robust and varied business area, allow the use of economies of scale and synergies, and at the same time facilitate a speedy and flexible adjustment to changing framework conditions.

Useful criteria to prioritize business opportunities therefore include the following:

– Customer needs and market potential for possible business models, both local and supraregional;
– Potential value contribution and return or contribution margins;
– Sustainability and susceptibility to regulatory change;
– Feasibility concerning timing, competencies, processes, and organization;
– Company's share in value chain and significance for the company.

Once the possible energy transition business opportunities have been assessed and evaluated and the market and business potential analyzed, energy companies must be in a position to answer the following questions concerning their operations:

– What are the future markets and business potentials with reference to the respective customer segments?
– Which business models are of interest and how can their value contribution be measured?
– How competitive are the new technologies and costs?
– How and by when can a new business model, product, or service be introduced to the market? What are the competencies needed at the company level?

The answers to these questions will then be used to develop the strategy and organize the implementation of the business models for the future (Sect. 6).

## 5.2    Selected Examples of Business Models

In this section we describe possible models for two business opportunities that follow from the energy transition. In Germany they were already available on the market in 2013/2014 and can be profitable in the current regulatory environment.

### 5.2.1    Self-Consumption of Solar Energy for Households

Depending on the specific conditions, a PV system with the right dimensions can cover more than 30 % of the electricity demand of a typical family living in a single-family home in Central Europe. If battery storage, which is becoming more reliable and less expensive every year, is employed, this consumption share can be increased to more than 70 %. Self-consumption of affordable solar energy is thus very attractive for most homeowners, and many have already taken action. Nonetheless, there is still huge market potential among private and commercial customers who have remained passive up to this point and can be motivated by lucrative offers. Customers usually have the choice between ownership of the solar power system by financing the investment independently and a full-service

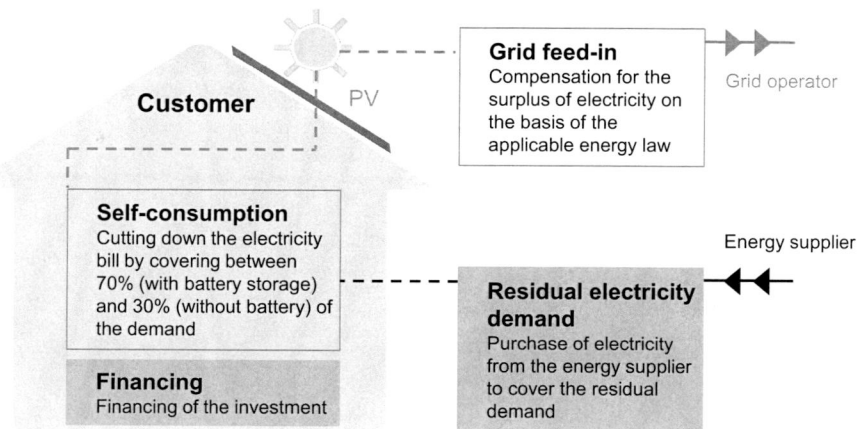

**Fig. 7** Principle of self-consumption of solar energy (with or without battery storage)

offering, which involves the rental of a rooftop by the energy company. Figure 7 describes the self-consumption of solar energy in principle. In the German energy sector, important pioneers in this field are, among others, the companies greenergetic GmbH, rhenag Rheinische Energie AG, and STAWAG AG [for a comprehensive overview see Schorsch (2014), p. 10]. Globally there are many more important players in the steadily growing market for solar PVs. For example, there is a huge market for solar power in the USA, with several companies providing solar solutions for homeowners and businesses.

Self-consumption via decentralized power systems are primarily a topic for single-family and two-family homes. Possible models for multi-family homes, where the tenants consume the electricity produced on site either by the real estate owner or the energy supplier (keyword »tenant electricity«) are considerably more challenging to implement in the current regulatory environment. In Germany, first projects are already tested in the market and increasingly more energy suppliers and housing companies are considering »tenant electricity« as a viable business opportunity. The German-based LichtBlick SE with PV systems and the municipal utility Augsburg Energie GmbH with mini CHP units are two pioneers that have started to accumulate experience in this field (Focht 2014, p. 19).

### 5.2.2 Model for "Tenant Electricity" from a Mini CHP Unit

A suitable business model to supply a multifamily home with electricity and heat from a mini CHP unit can be structured as follows:

- The energy supplier is the owner and operator of the mini CHP unit in a multifamily home.
- The produced heating energy is sold to the real estate owner or the housing company, which charges the tenants for the delivery of heating energy as an ancillary cost in the usual way.

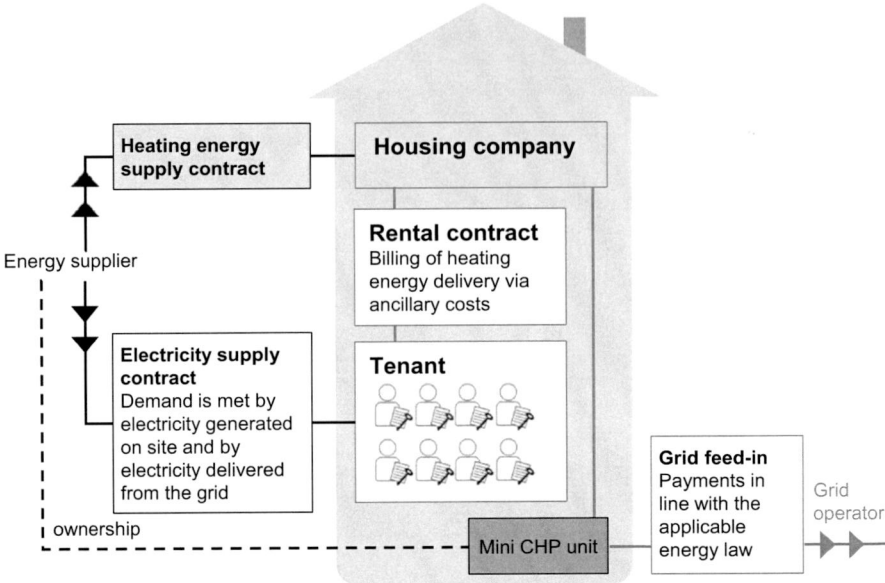

**Fig. 8** Principle of a business model for self-consumption of multifamily home based on a mini CHP unit

– The energy supplier offers a special tariff for "tenant electricity" to the tenants, which guarantees the delivery of electricity generated on site (as much as possible) and electricity from the grid (as little as possible) at a cheaper rate than the local standard tariff.
– The grid operator compensates the energy supplier for the electricity produced by the CHP unit and for the grid feed-in of the excess energy that is not used on site according to the applicable law.

Against the backdrop of increasing requirements on the energy efficiency of existing and new buildings, a growing number of housing companies are becoming interested in efficient energy supplies provided by mini CHP units. Offering electricity from a mini CHP unit to private or commercial tenants of a building can significantly increase the profitability of the investment. Figure 8 shows an example of a business model in which electricity and heating energy are provided for a multifamily home from a mini CHP unit in a simplified and schematic fashion.

### 5.2.3   Factors That Influence the Economic Efficiency of Energy Transition Business Models

The ultimate economic efficiency of business models such as the one presented earlier will always depend on the current regulatory framework and market environment as well as the specific situation of the company implementing the concrete model. In Germany, the Renewable Energy Act and the Combined Heat and Power

| Relevant aspects for developing energy transition business opportunities | |
|---|---|
| Market | Customer segments, target area, market potential |
| Customer needs | Reducing energy costs, ecology, profitability, autarky, reliability of supply |
| Regulation | Opportunities and requirements from energy laws and regulation |
| Financing | Financing, grants and subsidies, profitability and amortization |
| Business models | Sale, lease, contracting, self-consumption, tenant electricity, energy management... |
| Technology | Manufacturer, quality, technological innovation, prices, cost |
| Value creation (Make or buy) | Delivery, installation, operation, management, capacity management, maintenance, service |
| Profitability | Cost and revenue of the utility |

**Fig. 9** Relevant aspects for developing business opportunities for energy transition

Generation Act heavily influence the economic efficiency of business models involving solar PV and CHP units the most. At the same time, it is absolutely essential for any economic assessment to know whether business models that are mainly based on the self-consumption of electricity by the end user are offered in the home market (potential reduction of electricity sales to existing customers) or outside the home market (potentially new customers). And finally, the allocation of the roles of owner, operator, and consumer between energy supplier, owners of the real estate, and end users is extremely relevant for the viability and ultimate success of these business models. Figure 9 summarizes graphically most of the aspects that are relevant to energy companies when developing business plans based on the energy transition.

### 5.2.4 Challenges for Municipal Utilities

Even though numerous obstacles stand in the way, municipal utilities must deal with models of self-consumption when looking at energy transition business opportunities. Providing products and services for self-consumption from solar PV systems is a good starting point for market entry since it is relatively simple. Numerous municipal utilities have already included similar models in their strategy development for the field of decentralized generation. But many issues remain, especially concerning the regulatory framework. Additionally, municipal utilities are partially cannibalizing their home market while at the same time securing their existing electricity delivery contracts. Simply ignoring the trend toward increasing self-consumption is certainly not an option. If municipal utilities do not offer any products for self-consumption, competitors or customers will. Similarly, the possible cannibalization effect is offset by numerous advantages, such as long-term customer retention via service and maintenance of assets, an increase in gas sales, or a growing market for energy management or business services. Marketing and operations for energy-related products and services of the "New World" that are of high quality and have a significant impact on the profit of a company will remain very challenging (Schlemmermeier 2013, S.32).

## 5.3    Conclusions

The market and for self-consumption and similar products and services in the "New World" is still at an early stage. Technologies continue to develop and prices on installations will continue to drop. The producers of CHP facilities are pushing into the market for smaller units. The manufacturers and installers of PV systems are offering increasingly attractive solutions for self-consumption—with and without battery storage—to their customers. Even the manufacturers of stationary batteries are beginning to offer not only hardware but also energy-related services and software solutions to their customers for maximizing the benefits of their installations. This increases the requirement to not only optimize these assets with regard to the needs of customers and providers but also to integrate them efficiently into power systems and energy markets. This market and systems integration, which combines numerous technologies with the creation of value is the core aspect of capacity management (Sect. 4.3). As the number of these types of business models in the market increases, customer interest will grow, and vice versa.

## 6    Strategy Development and Organization for Future Business Models

### 6.1    Development of Positioning, Aims, and Strategies

A major foundation for developing business models for the "New World" is the description of positioning, aims, and strategies for the new business field. This requires, among other things, making decisions about the following issues:

– Market positioning, intended market share; and revenue;
– Elements of the product and service portfolio so as to create (added) value for customers as well as energy companies;
– Development of specific competencies at energy companies and cooperation with service providers and partners;
– Sales organization and processes for the distribution of products and services for the energy transition.

Aims and strategies must be developed consistently on the basis of the individual long-term vision of the specific energy company. Also, aims and strategies must be assessed regularly with regard to current market developments and regulatory initiatives and adjusted if necessary.

## 6.2    Needs of Municipal Utilities in Germany in Current Market Environment

In the current, highly regulated energy sector of Central Europe, market reforms are an important prerequisite for the creation of business opportunities that promise profitable growth and significant sales volume. The existing uncertainty about the future regulatory framework makes it harder to develop energy transition business opportunities. Against this backdrop it seems advisable, especially for municipal utilities in Germany, to follow a dual strategy when developing business models for the "New World." On the one hand, those energy transition business opportunities that are profitable in the current market and regulatory environment need to be integrated into the existing product portfolio. On the other hand, it is necessary already to develop business that can become profitable only in the reformed market and regulatory environment of the coming years.

The specific requirements of the municipal utilities also play an important role in developing a strategy and business model. These include the following:

– Generating profitable growth with new business opportunities to compensate for lost earnings in other segments;
– Maintenance of the customer portfolio (margins, market shares) by offering new services, even as the self-consumption of customers grows;
– Participation in the energy transition in order to strengthen the positioning of the company and to satisfy the demands of the municipal owners or shareholders;
– Efficiency (cost, resources, knowledge) and success in developing new and profitable business opportunities;
– Maintaining independence, which also relates to external perceptions of the company.

Market developments, changes in the regulatory framework, and growing expectations of owners require substantial modifications of the business models of Germany's municipal utilities. A specific challenge is the enormous growth in the complexity of the business and the correspondingly higher demand on corporate resources. Municipal utilities thus face the strategic challenge of deciding which services they want to provide independently and which services they want to offer in cooperation with other municipal utilities, strategic partners, or independent service providers.

## 6.3    New Business Models Require New Structures and Processes

Growing complexity, intense competition, and enormous price-reduction pressures, coupled with increasing digitization, lead to the fact that more and more highly specialized companies are entering the value chain of energy-related products and services. Therefore, margins at the various stages of the value chain will remain under pressure. Technological progress and frequent changes in the regulatory framework conditions of the energy market meanwhile imply that the product lifetime of most business opportunities is becoming ever shorter. Thus utilities tend to shy away from capital-intensive business models owing to the high risk associated with the amortization of relevant expenditures.

Since the "New World" of the energy sector differs in numerous ways from the "Old World," the development of new business models will frequently require fresh thinking across the entire company. This holds for the corporate culture and the organizational structure as well as for the motivation, willingness to innovate, knowledge, and self-conception of employees. Especially with regard to the frequently used argument of the "Old World" that new business models would jeopardize existing business areas of many energy companies (keyword: models of self-consumption, Sect. 5.2), the idea of the "New World" as a new and independent organizational unit in the corporate structure is a noteworthy one. This makes sense since the two worlds will continue to coexist in the energy industry for many years, until finally the "New World" will completely overshadow the "Old."

Currently in Germany, many municipal utilities still fail to include in their portfolios significant amounts of products that allow profitable growth in the age of energy transition. The development of such energy transition business opportunities and the implementation of new operational processes should therefore be at the forefront of strategic corporate development (Schorsch und Schlemmermeier 2014, S.21).

## 6.4    Process Model for the Development, Production, and Marketing of Energy-Related Products and Services

Compared to the "Old World" of retailing energy, the "New World" of energy transition business opportunities will require a new process and management model, which in our opinion consists of three pillars (Fig. 10):

1. Product development: business and product development as well as product management;
2. Production and operations: production and creation of consumer solutions;
3. Sales: distribution of consumer solutions.

In what follows, we will discuss these three pillars of the process model in more detail.

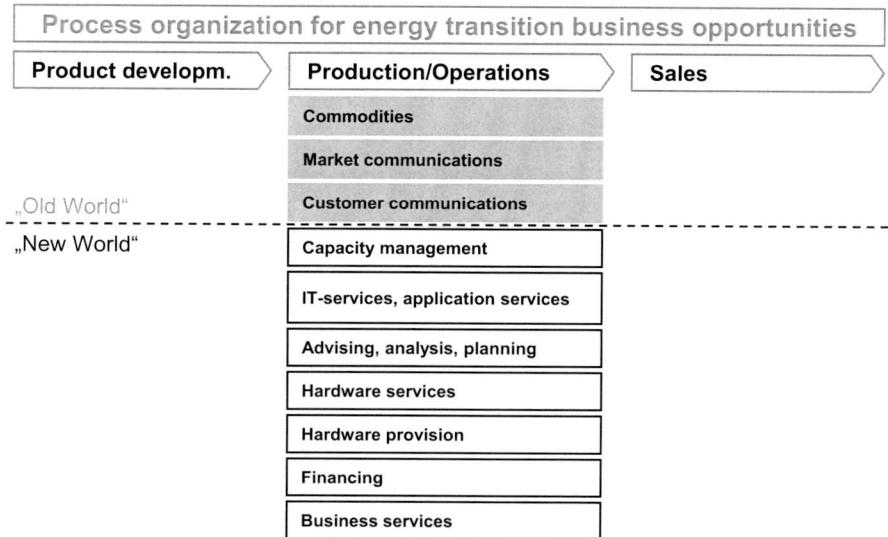

**Fig. 10** The three-pillar-process model for implementation of business opportunities for energy transition

**1. Product Development** The core task of the development unit is product management, which includes responsibility for products, prices, communications, and sales. If the right products are the key to success (profitable growth), product development must become the nucleus of this process model. The development unit should thus take responsibility for the following activities:

– Product range and quality,
– Design of customer contacts in the production processes,
– Efficiency and cost considerations for the provision of services.

Owing to the multitude and complexity of future business opportunities, product development will increasingly become an interdisciplinary task requiring interdisciplinary teams. At a minimum, employees in the field of product development must demonstrate competencies and qualifications in the following areas:

– Understanding markets and regulations;
– Understanding customer needs and client communications;
– Development of technical solutions;
– Modeling of quantities, costs, and prices;
– Pricing and contract development, risk analysis, and management, which includes developing the requirements for operational processes and sales performance.

**2. Production and Operations** The products for the "New World" are produced in the operating unit based on the guidelines provided by the development team. The employees in the operating unit are thus responsible for the quality and cost of the production process as well as the incorporation of suppliers of materials and third-party services. The major challenge is to arrive at the correct decisions, jointly with the product development team, concerning in-sourcing and outsourcing of the various tasks and services against the background of aims and strategies, available competencies at the company, and economic requirements. At the same time it is absolutely crucial, especially in the case of technically complex energy transition business opportunities, to involve craftsmen, that is, installers of solar power and CHP systems, from the outset in the development and value creation process.

**3. Sales** Employees in the sales unit are responsible for product placement and distribution. The sales force is best structured along customer segments and geographical regions and needs to be supported by product specialists. The complexity of many products in the "New World" is so high that no all employees can be expected to represent all products up to the point of contract closing. However, it is precisely this complexity that makes it important for a majority of (potential) customers to have to deal with only one corporate representative who can take care of all their needs and coordinate all activities in the background.

The reality of energy retailing in the "Old World" demonstrates that in Germany new customers will not switch to another delivery contract on their own, despite significant pricing differences of more than 100 Euro per annum. Instead they need to be convinced via direct marketing or sales partnerships. This reality is even more relevant for the sale of difficult-to-explain products of the "New World." Nonetheless, strong competitive pressures and the need to lower costs, combined with a trend toward digitization, will mean that, going forward, the highest possible share of sales activities for products and services of the "New World" needs to be organized online.

## 6.5    Conclusions

To tackle the big challenges of the energy transition, energy companies must have a division that takes responsibility for the development, management, and economic results of new business activities and products. This division can be understood as a functional unit of the company that in the classical sense of marketing is responsible for developing and managing product and pricing policies as well as communication and selling activities. The abilities to innovate and "understanding customers" are the key factors that enable profitable growth in the days of the energy transition. Whether the products of the "New World" can indeed be sold and how this selling process can be implemented are subject to constant learning. Both issues need to be clarified before the newly developed structures can be further expanded in a meaningful way.

# 7 Summary: Main Points Concerning the Future of Energy Companies

**Energy Transition and Customer Needs Change the Business of Energy Companies** The business of energy supply will change in fundamental ways. The drivers of change are the energy transition and customer needs. Customers decide what they want and what offerings are suitable for them. The customers of the future will want to be autonomous and self-sufficient and to participate in the energy transition. This is true for homeowners in the countryside as well as for tenants in urban areas. They all should be involved and want to be involved. In cities, housing companies will hold a key position regarding the energy transition.

**Traditional "Old World" Market Shrinks Owing to Increasing Decentralization** In the "Old World," energy suppliers buy and sell electricity and gas. The sales margins that can currently be achieved exceed the amount that can be justified on the basis of value creation. The traditional market for these products will shrink owing to increasing decentralization.

**"Old World": Simplicity, Efficiency, Profitability** What counts in the traditional business of energy retailers are maximizing margins and handling the commercial processes of the energy business with the end user completely online. Since electricity and gas discounters will not be able to recuperate their costs of customer acquisition and negative margins via higher prices, they will leave the market. Price differences and, thus, incentives to change the energy supplier will shrink.

**Future Business Will Revolve Around the Decentralized Production of Energy by End Users** For energy companies, the business model of the future will revolve primarily around the decentralized production of energy by private and commercial end users. They will cover part of their energy demand through self-consumption from solar power or CHP units, while energy companies will sell the excess electricity that is not used on site, deliver the energy for the remaining demand, and guarantee the security of supply, and with the aim of realizing the best possible economic result for the customer.

**Required Tasks and Services of Energy Companies for the "New World": Business Opportunities** For the business of the future, energy companies must be in a position to sell, install, operate, and maintain the technical systems of the end user—solar PV systems, mini CHP units, stationary battery system,s and various energy efficiency measures—and to integrate them into customer and market systems. This leads to additional tasks for the sales teams such as the sale of a product that is significantly more advanced and technically complex than the current product range, the management of additional market and regulatory risks, and the support of customers who actually demand a service from the utility, namely, that their installations at home are operating profitably.

**Already Today, the Energy Transition Requires a Dual Strategy** In the "Old World," energy retailers need to radically simplify their products, communications, customer management, processes, and IT in order to increase efficiency. Product management, which is tasked with managing margins and market shares, must become quantitative in nature. The main idea is to be simple, operate at low cost, and keep generating high margins as long as possible.

For the "New World" investments in the learning curve are the top priority. The aim is to establish the so-called factory of the future—alone or jointly with partners—and to develop products that allow for a profitable business based on decentralization and renewable energy. The current challenge for energy suppliers and municipal utilities is to free up the necessary resources and development competencies in order to establish the business of the "New World." This also includes the need to turn potential competitors into partners.

**The Energy Transition Business Field Requires a Separate Organizational Unit** Compared to the "Old World" of energy supply, the "New World" of energy transition business opportunities requires a new process and management model that rests on the three pillars of development, operations, and sales. Crucial in this regard will be the allocation of responsibility for the financial results to the development area and the proximity to the customer of the sales force.

**The Energy Transition as a Process of Innovation Management** The energy transition is a process of innovation management in an uncertain environment concerning climate and energy policies. Five years ago, nobody foresaw what today's energy market would look like, and today, nobody knows what the energy market will look like in 5 years. Regulation, technological innovation, and customer needs will witness dynamic change. Fundamental trends are becoming evident, though their exact timing and detailed shape remain uncertain. This leads to the following requirements for the sales and distribution processes: focus, flexibility, adaptation, assessment, and readjustment as needed.

# References

BDEW Bundesverband der Energie- und Wasserwirtschaft e.V. (2014). *Positionspapier—Ausgestaltung eines dezentralen Leistungsmarkts*. Accessed Nov 12, 2014, from https://www.bdew.de/internet.nsf/id/3A90CD61C49A1952C1257D0E003A0C54/$file/BDEW-Positionspapier_Ausgestaltung%20eines%20dezentralen%20Leistungsmarkts_300614_oA.pdf
BDH Bundesindustrieverband Deutschland Haus-, Energie- und Umwelttechnik e.V. (2014). *Energierevolution scheitert ohne Wärme- und Klimamarkt*. Accessed Oct 23, 2014, from http://www.bdh-koeln.de/uploads/media/PM_28012014_DWK_Keine_Energierevolution_ohne_Waerme-_und_Klimamarkt.pdf
BMUB Bundesministerium für Umwelt, Naturschutz, Bau und Reaktorsicherheit. (2013). *Konzept für die Umsetzung einer Strategischen Reserve in Deutschland—Ergebnisbericht des Fachdialogs »Strategische Reserve«*. Accessed Nov 12, 2014, from http://www.bee-ev.de/_downloads/publikationen/sonstiges/2013/20130513_Fachdialog_Strategische_Reserve.pdf
BMUB. (2014). *Nationale Klimapolitik*. Accessed Oct 22, 2014, from http://www.bmub.bund.de/themen/klima-energie/klimaschutz/nationale-klimapolitik/ (2014).

BMWi Bundesministerium für Wirtschaft und Energie. (2014a). *10-Punkte-Energie-Agenda*. p. 3. Accessed Oct 22, 2014, from http://www.bmwi.de/BMWi/Redaktion/PDF/0-9/10-punkte-energie-agenda,property=pdf,bereich=bmwi2012,sprache=de,rwb=true.pdf

BMWi. (2014b). *Zeitreihen zur Entwicklung der erneuerbaren Energien in Deutschland*, as of: Aug 2014. p. 7. Accessed Oct 27, 2014, from http://www.erneuerbare-energien.de/EE/Redaktion/DE/Downloads/zeitreihen-zur-entwicklung-der-erneuerbaren-energien-in-deutschland-1990-2013.pdf?__blob=publicationFile&v=13

BMWi. (2014c). *Energiedaten, Tab. 7: Endenergieverbrauch nach Anwendungsbereichen*. Latest update: 8 Oct 2014. Accessed Nov 12, 2014, from http://www.bmwi.de/BMWi/Redaktion/Binaer/energie-daten-gesamt,property=blob,bereich=bmwi2012,sprache=de,rwb=true.xls

BNetzA Bundesnetzagentur. (2013). *Genehmigung des Szenariorahmens für den Netzwerkentwicklungsplan 2014 (Az.: 6.00.03.05/13-08-30/Szenariorahmen 2013)*. p. 2. Accessed Oct 22, 2014, from http://www.netzausbau.de/cln_1412/DE/Bedarfsermittlung/Charlie/SzenariorahmenCharlie/SzenariorahmenCharlie-node.html

BNetzA. (2014a). *Kraftwerksliste, as of: 16 July 2014*. Accessed Oct 27, 2014, from http://www.bundesnetzagentur.de/DE/Sachgebiete/ElektrizitaetundGas/Unternehmen_Institutionen/Versorgungssicherheit/Erzeugungskapazitaeten/Kraftwerksliste/kraftwerksliste-node.html

BNetzA. (2014b). *Kraftwerksstilllegungsanzeigenliste, as of: 20 Oct 2014*. Accessed Nov 12, 2014, from http://www.bundesnetzagentur.de/SharedDocs/Downloads/DE/Sachgebiete/Energie/Unternehmen_Institutionen/Versorgungssicherheit/Erzeugungskapazitaeten/KWSAL/KWSAL_2014_10_20.pdf;jsessionid=DD4F63C3B4EF2E482C79619B9FE7D8D5?__blob=publicationFile&v=33 (2014).

BSW-Solar Bundesverband Solarwirtschaft e.V. (2014). *Statistische Zahlen der deutschen Solarstrombranche (Photovoltaik) für 2013*, as of: Mar 2014. p. 1. Accessed Oct 27, 2014, from http://www.solarwirtschaft.de/presse-mediathek/marktdaten.html

BWE Bundesverband WindEnergie e.V. (2013). *Statistiken*, as of: 31. Dec 2013. p. 2. Accessed Oct 27, 2014, from http://www.wind-energie.de/themen/statistiken

Consentec. (2012). *Versorgungssicherheit effizient gestalten—Erforderlichkeit, mögliche Ausgestaltung und Bewertung von Kapazitätsmechanismen in Deutschland. Untersuchung im Auftrag der EnBW AG*. pp. 39 following. Accessed Nov 12, 2014, from http://www.consentec.de/wp-content/uploads/2012/03/Consentec_EnBW_KapMärkte_Ber_20120207.pdf

Enervis enervis energy advisors GmbH. (2014). *Einführung eines dezentralen Leistungsmarktes in Deutschland—Modellbasierte Untersuchung im Auftrag des Verbands kommunaler Unternehmen e.V.* pp. 9 following. Accessed Nov 12, 2014, from http://www.vku.de/fileadmin/get/?28952/Einfuehrung_eines_dezentralen_Leistungsmarktes_in_Deutschland.pdf

EWI Energiewirtschaftliches Institut an der Universität zu Köln. (2012). *Untersuchungen zu einem zukunftsfähigen Elektrizitätsmarktdesign. Endbericht im Auftrag des BMWi*, pp. 55 following. Accessed Nov 12, 2014, from http://www.bmwi.de/BMWi/Redaktion/PDF/Publikationen/endbericht-untersuchungen-zu-einem-zukunftsfaehigen-elektizitätsmarktdesign.pdf

Focht, P. (2014, May 14). Grüner Strom für Gelbes Viertel. *E&M POWERNEWS*.

Öko-Institut, LBD-Beratungsgesellschaft, RAUE LLP. (2012). *Fokussierte Kapazitätsmärkte. Ein neues Marktdesign für den Übergang zu einem neuen Energiesystem. Studie für den WWF Deutschland*. Accessed Oct 22, 2014, from http://www.wwf.de/fileadmin/fm-wwf/Publikationen-PDF/Fokussierte-Kapazitaetsmaerkte.pdf

Öko-Institut. (2014). *Erneuerbare-Energien-Gesetz 3.0 (Kurzfassung). Studie im Auftrag von Agora Energierevolution*. Accessed Oct 22, 2014, from http://www.agora-energy revolution.de/fileadmin/downloads/publikationen/Impulse/EEG_30/Agora_Energy revolution_EEG_3_0_KF_web.pdf

Schlemmermeier, B. (2012, February 1). Vom Energielieferanten zum Systemdienstleister. *Energie & Management, E-World Spezial*, p. 42.

Schlemmermeier, B. (2014). *Plädoyer für einen Kapazitätsmarkt. Vortrag zum Elektrizitätsmarktdesign 2.0*. Accessed Nov 12, 2014, from http://www.lbd.de/cms/pdf-vortraege-praesentation/1410_Landesregierung-BW_Vortrag_LBD_Handout.pdf

Schlemmermeier, C. (2013, June 15). Menschen und Meinungen. Sagen Sie mal Claudia
    Schlemmermeier. *Energie & Management*, p. 32.
Schorsch, C. (2014, July 15). Eigenstrom-Service. Genug Potenzial für alle. *Energie & Manage-
    ment*, p. 10.
Schorsch, C., & Schlemmermeier, C. (2014, March 1). Die Zukunft beginnt heute. *Energie &
    Management*, p. 21.
Shell und BDH. (2013). Shell Deutschland Oil GmbH und BDH. Hauswärme-Studie. Klimaschutz
    im Wohnungssektor—Wie heizen wir morgen? p. 27. Accessed Oct 27, 2014, from http://s08.
    static-shell.com/content/dam/shell-new/local/country/deu/downloads/pdf/comms-shell-bdh-
    heating-study-2013.pdf.
TÜV Süd. (2014). *TÜV SÜD AG, Wegbereiter der Energierevolution*. Press release on 21 Oct
    2014. Accessed Oct 27, 2014, from http://www.tuev-sued.de/anlagen-bau-industrietechnik/
    aktuelles/wegbereiter-der-energierevolution

**Ben Schlemmermeier** joined LBD-Beratungsgesellschaft
mbH (LBD) in 1989. Since 1991 he has been one of three
managing partners of LBD. As a specialist in the energy
market, LBD provides advisory services for all stages of the
value chain. Among its customers are municipal utilities,
international energy suppliers, the oil and gas industry,
newcomers, state sector, service providers, politics,
associations, industry, banks, renewable energy companies,
and aggregators. The key topics of Ben Schlemmermeier
include advising on complex structures and contracts at all
stages of the energy sector's value chain, projects, mergers
and acquisitions, corporate finance, project finance, and
corporate restructuring, as well as consulting services concerning the future electricity market
design and the development of visions, targets, positioning, strategies, and sustainable business
models.

**Björn Drechsler** holds a diploma in engineering and has
been active as a management consultant at
LBD-Beratungsgesellschaft mbH (LBD) since 2006 in the
field of energy and emissions. Among his consulting topics
are analyses of the energy sector and expert opinions on
renewable and conventional energy production, emissions
trading, and the wholesale and control energy markets. He is
project manager in transactions to acquire conventional and
renewable power plants and also carries out profitability
analyses and feasibility studies for renewable energy and
combined heat and power installations, demand-side management, battery storage, and electric mobility. In recent
years he has had a consulting focus on the development
and assessment of sustainable strategies and business
models for the "New World" of the energy sector, such as self-consumption, tenant electricity,
and virtual power plants.

# Driving Renewables: Business Models for the Integration of Renewable Energy and e-Mobility in Europe

Marc Ringel

**Abstract**

Linking renewable energy sources (RES) and electric vehicles (EVs) enables various new business models and ancillary services. These services largely act to facilitate the interface between the energy and transport sectors. In this contribution we review the potential options from both energy and transport perspectives in the European context. From a transport perspective, marketing green power as fuel for EVs is the most obvious connection. From an energy systems perspective, EVs and their battery storage can offer potentially attractive options to balance grid frequency and thereby introduce more fluctuating RES into the generation mix. The so-called 'sector coupling' comprises passive and active system services to stabilize electricity networks, such as passive storage (grid to vehicle), active vehicle-to-grid recharging or balancing services. The full potential of these services is less obvious at the individual car level. It becomes evident at the level of a full fleet of EVs connected to the grid. A series of innovative business models emerges in a wider perspective: E-mobility and related transport services can be coupled to energy services. These portfolios offer the most promising deployment strategy of driving renewables in the transport sector.

**Keywords**

Sustainable transportation • Renewable energy sources • Electric vehicles • Sector coupling • Business models

M. Ringel (✉)
Faculty of Economics and Law, Energy Economics, Nuertingen Geislingen University, Parkstrasse 4, 73360 Geislingen, Germany
e-mail: marc.ringel@hfwu.de

© Springer International Publishing AG 2017
C. Herbes, C. Friege (eds.), *Marketing Renewable Energy*, Management for Professionals, DOI 10.1007/978-3-319-46427-5_12

239

# 1    Introduction

E-mobility and renewable energy sources (RES) are cornerstones of European climate change policies. The European Union (EU) has set itself the goal of reducing greenhouse gas emissions by 40 % in 2030 and by 80/95 % in 2050 compared to 1990 levels (European Council 2014). This almost total decarbonization will require a total transformation of the energy and transport systems. Both are at present largely dependent on carbon-intensive fossil fuels. On the energy side, these fuels cover at present 35.9 % of the total energy production in the EU (2015). Transport is heavily dependent on oil, with 94 % of EU transport consumption based on oil and oil products (European Commission 2015a).

RES are key drivers for substituting fossil resources in all energy uses (electricity, heating/cooling, transport). Their uptake is most pronounced in electricity generation. All EU member states have installed support mechanisms to increase the share of RES in electricity production. Underlying the support schemes is the logic to gradually lower differential costs and thus incorporate RES into the electricity market. This implies that producers are being increasingly forced to develop business models for the use of RES without any state support.

Unlike electricity generation, the potential role of RES to cover heating and cooling demand was less in focus. This changed with the introduction of a dedicated European framework, starting with the RES Directive (Directive 2009/28/EC; European Commission 2009), the Energy Performance of Buildings Directive (Directive 2010/30/EC, European Commission 2010), the Energy Efficiency Directive (Directive 2012/27/EC, European Commission 2012), and recently the European Commission's Strategy on Heating and Cooling (European Commission 2016). Owing to this regulatory framework, RES' contribution to heating and cooling has significantly increased. RES now deliver some 503.4 petajoule (PJ) or 30.5 % of Europe's gross heat generation (DG Energy 2015). A large increase in the use of electrical heat pumps has occurred, linking electricity and heating/cooling uses (so-called system coupling).

The transport sector in the EU consumes 348.5 PJ or 34 % of total final energy. With its present fuel structure, the transport sector accounts for 24 % of total EU $CO_2$ emissions (DG Climate Action 2016a, b). The heavy use of oil and oil products also raises concerns about supply security and competitiveness. The EU needs to import 90 % of transport oil products, making it subject to delivery failures and speculative price fluctuations. Both effects combined amounted to EUR 50 billion in the period 2007–2010, which is projected to grow over time (European Commission 2015a, 2011a).

To break the oil hold of the transport sector, the European Commission and member states introduced regulatory efforts. These mainly focus on expanding renewable energy use in the transport system. This can be done directly (introducing biofuels) or indirectly, following the example of the heating sector and coupling the electricity and transport sectors. In the latter case, RES electricity is used as fuel for electric vehicles (EVs). According to some estimates (Pehnt et al. 2007), RES-powered EVs could help to avoid some 600 to 800 g greenhouse gas emissions per kilowatt-hour (kWh). This in turn would be compatible with curbing transport emissions to 43 g $CO_2$/km by 2050. Stakeholders assess this figure as being compatible with the international 2° C global warming target laid down in the Paris Agreement (BMUB 2013; UNFCCC 2015).

A massive introduction of RES-powered EVs leads to a broader concept of *green mobility*. The concept rests on two pillars: (1) providing mass mobility services based on renewable energies and (2) taking advantage of the EV fleet's battery storage to develop positive synergies for the electricity system. Because these services will need to be offered in liberalized markets, system coupling will trigger new business models for RES in the transport sector.

In this connection, several questions arise: What are the key economic and technical features of EVs? Why is it only in recent times that e-mobility has emerged? What kind of regulatory framework in Europe should produce the aforementioned synergies? What are the potential interactions of RES and EVs? And most importantly: How can the system coupling of energy and transport be used to develop business models that successfully integrate RES and e-mobility?

To answer these questions, our contribution will outline the basics of e-mobility in Sect. 2. Section 3 proposes a review of the regulatory framework in the EE, both at the European and national levels. For the national level, the EV uptake strategy of Germany is presented as a case study. In Sect. 4, we will review several business models for the integration of RES and EVs. It will become clear that these models will strongly depend on the national context and can deliver benefits either from the transport or the energy angle. Section 5 summarizes our findings and draws conclusions on the successful deployment of these business models.

## 2 E-Mobility Basics

Electricity-driven engines are commonly considered a recent technical innovation. This view disregards the that the underlying technology has existed for more than 100 years. In 1891 Gustave Trouvé presented an electric tricycle in France, followed shortly by the Belgian Jenatzky's electric car. It was the first car to reach a velocity of more than 100 km/h (BMBF 2013; Ruppert 2013). The electric power train technology quickly fell behind the internal combustion engine since the latter offers considerable advantages such as greater flexibility, a simpler infrastructure for fuel supply, and a comparatively cheap operation mode—particularly in times of low-priced fossil fuels.

At present, this argument of low fuel prices still holds true to a certain extent. In the mid-term, though, all major institutions and market players project an increase in the price of fossil fuels (IEA 2016a; BP 2016; EIA 2016). Rising prices offer a strong incentive to switch to EVs. Here, the efficiency degree of primary energy use is much more advantageous, leading to lower losses of primary energy input.

As shown in Fig. 1, a standard internal combustion engine accounts for a loss of 81 % of the primary energy of the fuel employed in the operation of the power train. In comparison, a fuel cell with a loss of "only" 74 % shows much better

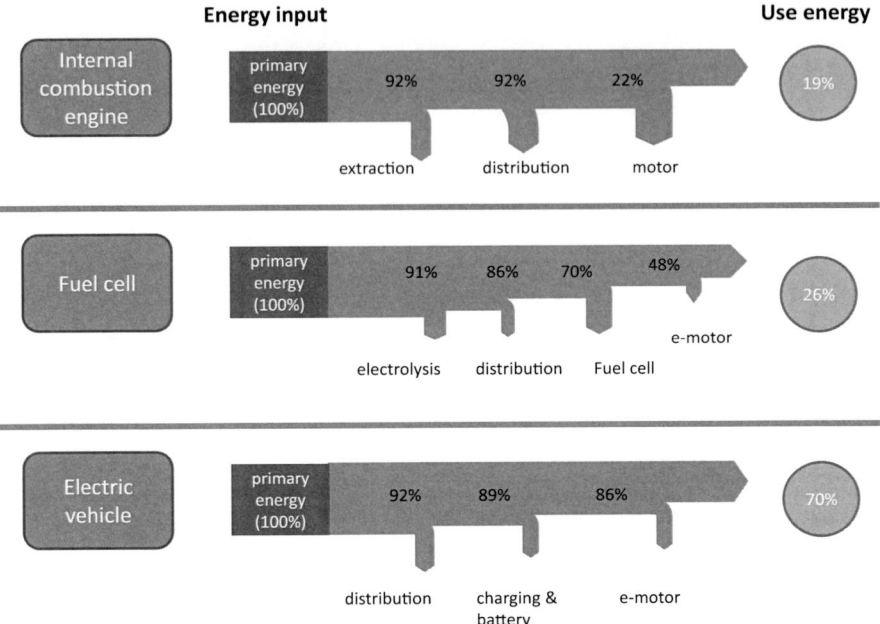

**Fig. 1** Energy flowchart of different power trains (based on BMUB 2013)

performance. EVs with a loss of some 30 % of primary energy clearly perform best in this comparison. This applies all the more once the electricity required for the power train is obtained from RES such as wind or photovoltaics with an input/ output efficiency level of 100 %.

Following this analysis, many European governments have focused their research and development (R&D) and rolling-out strategies on vehicles that are

1. powered by an electric motor and
2. supplied through the public electric grid.

Clearly the focus is on electricity as a fuel and the integration of general electricity provision and e-mobility. This comprises the following vehicle technologies:

**Battery electric vehicle (BEV):** vehicle equipped solely with an electric motor driven by a battery. The battery is either built in and loaded at charging stations via the public power network or empty batteries can be replaced at special battery-changing stations;

**Range-extended electric vehicles (REEVs):** A small combustion motor (range extender) provides supplemental power in case of a low battery level. The aim is to increase the still short/small operating distance of the batteries supplied. However, the vehicle is not directly driven by this range extender;

**Hybrid electric vehicle (HEV):** combination of a combustion motor and electric drive, that is, plug-in HEV (PHEV), that equips hybrid vehicles with direct access to the power network.

It should be noted that these power train options are not limited to cars but apply to EVs at large, including e-bikes, e-buses (so-called public e-transport), or the transport of goods (clean logistics). The larger field of applications opens up possibilities for a larger concept of e-mobility and a series of new mobility options. Approaches such as car sharing, bike sharing, or new intermodal forms of transport are increasingly being taken into consideration in EV deployment strategies. (BMWi 2014a; BMBF 2013; Canzler 2010). With these concepts, the automobile and energy sectors strongly interact and have become more and more intertwined (so-called sector blending). If in particular green electricity is us as fuel, this constitutes an important factor of success for the general acceptance of e-mobility. Also, sector coupling can allow for a multitude of innovative business models. These still strongly depend on the regulatory frameworks and the market developments in the individual regions and countries.

## 3       Regulatory and Market Framework in Europe

### 3.1       European Versus National Regulation

#### 3.1.1       Market Situation

The global stock of EVs amounted to 1.26 million in 2015. This figure is expected to almost double in 2016 (IEA 2016b). The European EV stock represents about one third of this total figure. In 2015, a McKinsey study ranked many European countries as lead markets for e-mobility. In particular, Norway, the Netherlands, and France are seen as lead markets, showing a higher market potential than the USA or China, their global competitors (McKinsey & Company 2016).

The lead market classification comprises several aspects. One is the availability of EV models by the car producers of a given country. This aspect is relevant in the case of car manufacturers in the USA, South Korea, France, Italy, or Germany that are able to compete on the global market. Another aspect is the availability of low-carbon electricity supply to allow for a scaled uptake on electric mobility. Here countries with a large share of nonfossil generation sources such as Norway and Germany are in focus. Finally, relatively dense settlement structures as in the Randstad region of the Netherlands (Regio Randstad 2014) facilitate deployment and avoid the need for range extension of EVs.

The interplay of these aspects only partially explains the diversity in EV stocks in Europe (Figs. 2 and 3).

#### 3.1.2       Regulatory Situation

Adding to the European regulatory framework on RES, the Europe-wide level provides a comprehensive regulatory framework for national policies in the transport sector. The sector contributes to the overall 2020 and 2030 sustainable energy and climate strategies (Council of the European Union 2007 and 2014). The

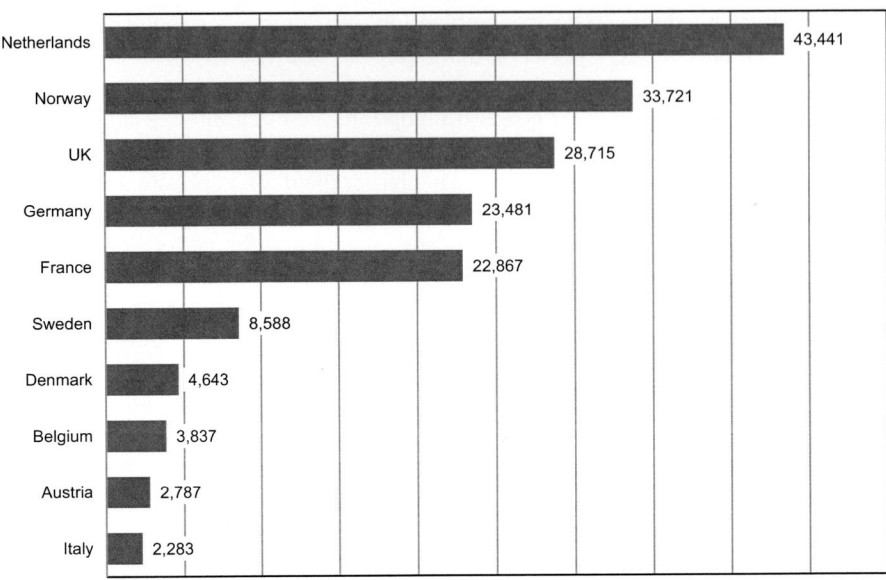

**Fig. 2**  Registration of new EVs in selected EU countries (2015) (based on Acea 2016)

European Commission's 2011 white paper on transport (European Commission 2011b) further substantiates targets for the transport sector. The paper presents the vision of a competitive and sustainable transport sector. Transport is supposed to lower its present $CO_2$ emissions by 60 % so as to achieve the overall 80/95 % decarbonization target in 2050 (European Commission 2011a, b).

The EU transport strategy summarizes and aligns existing EU legislation. The legislative acquis consists of:

- Renewable Energy Directive 2009/28: EU member states need to reach a 10 % RES share in transport fuels by 2020;
- Fuel Quality Directive 2009/30: $CO_2$ intensity of fuels sold in the EU need to be reduced by 6 % by 2020;
- EURO standards, regulating pollutant emissions of vehicles; and
- regulation of $CO_2$ emissions from vehicles: Car fleets of producers of passenger cars sold in the EU must on average remain below 130 g $CO_2$ eq/km by 2015 and 95 g/km by 2020. For light-duty vehicles the limit is fixed at 175 g/km by 2017.

In particular, the $CO_2$ emissions regulation provides strong incentives for European car makers to introduce EVs in their fleet portfolio. Vehicles with $CO_2$ emissions below 50 g/km are counted with a multiplying factor in the calculation of the fleet average, the so-called super credits (DG Climate Action 2016b). EVs run on tail-end zero $CO_2$ emissions. This is a key argument for their inclusion in car fleets (VCD 2015).

European manufacturers' strong interest in EVs is not only triggered by regulatory requirements. Meeting stringent emission standards is a condition for access to global

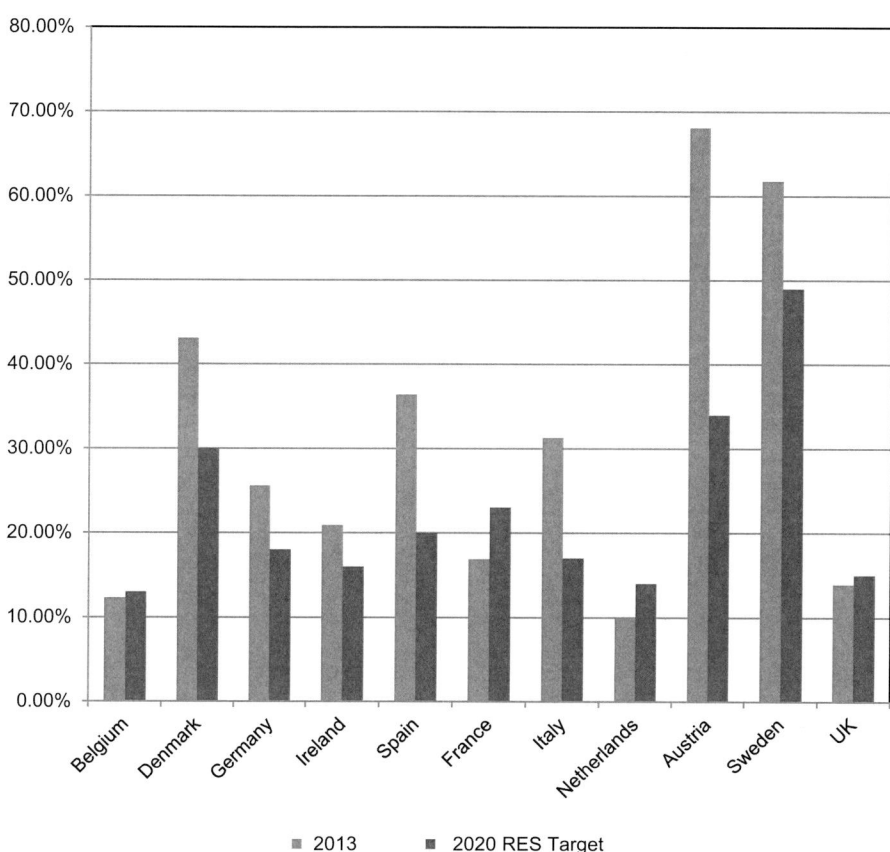

**Fig. 3** Share of RES in electricity production in selected EU countries (2015) (based on figures from DG Energy 2015)

EV markets, since by now many major economies have set up comparable standards (Fig. 4). Meeting and surpassing these standards is a key competitive advantage.

Two additional pillars are relevant for the European support of EVs. First is the increase in (energy and resource) efficiency in all transport modes by dedicated R&D support; second is the overall substitution of oil products as the main energy carriers in the transport sector. This logic asks for systematically switching to alternative fuels and related mobility technologies (European Commission 2013a).

The European Commission has explicitly spelled out its strategy on using alternative fuels in its 2013 communication "Clean Power for Transport" (European Commission 2013a) and the subsequent clean fuels directive (Directive 2014/94/EU; European Commission 2014). Both documents treat e-mobility as one puzzle piece in a more comprehensive substitution strategy (Table 1). The Commission estimates that EV mass rollout is advantageous in short- to medium-range passenger transport and short-range road transport. As a consequence, it advocates a scaled rollout of e-mobility, especially in cities and urban regions.

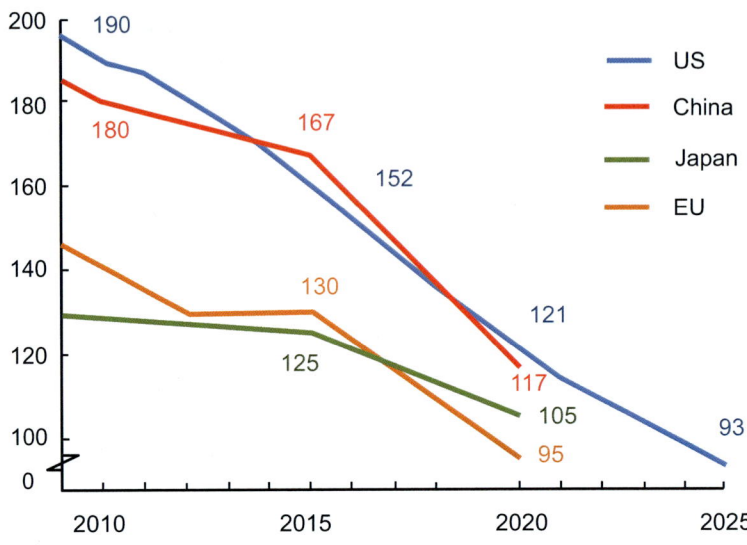

**Fig. 4** Car emission standards in selected regions (g $CO_2$/km in normalized European Drive Cycle) (based on ARF 2014)

**Table 1** European Alternative Fuels Strategy (European Commission 2013a, 2015b)

|  | Road transport (range) | | | Road freight (range) | | | Rail |
|---|---|---|---|---|---|---|---|
| Mode | Short | Medium | Long | Short | Medium | Long | |
| LPG | ■ | ■ | ■ | ■ | ■ | ■ | |
| Hydrogen | ■ | ■ | ■ | ■ | ■ | | |
| Electricity | ■ | (■) | | ■ | | | ■ |
| Biofuels | ■ | ■ | ■ | ■ | ■ | ■ | ■ |

The EU provisions clearly define the rollout of EVs as a national if not regional task. European regulations support these strategies by providing funding and reducing barriers that hinder cross-border mobility.

Analyses of the European Commission suggest that the combination of regional, national, and Europe-wide support for EVs will lead to clear economic benefits for the EU. Substitution of transport fuels is projected to save EUR 4.2 billion by 2020 and EUR 9.3 billion by 2030. Avoided price fluctuations are expected to save EUR 1 billion per annum. With the boost for regional fuels and technologies some 700,000 new jobs are anticipated, mainly due to first-mover advantages on global markets (European Commission 2013a, b).

The clean fuel directive spells out concrete actions and regulations for each fuel type. The provisions for e-mobility comprise mainly three aspects: (1) charging technology and standards, (2) charging points, and (3) consumer information:

- In terms of charging technology, the clean fuel directive harmonizes recharging plugs and standards (the so-called type 2 or Mennekes plug for slow charging; the CCS or combo standard for fast charging) while at the same time safeguarding the use of alternative plugs through 2019;
- Member states need to provide an "appropriate" number of charging points in urban/suburban and other densely populated areas. In a follow-up communication, the European Commission presented a proposal in terms of concrete numbers (Table 2). The charging points must be in place by 2020 and reported to the Commission in a national strategy document, the "National Policy Framework" (NPF), by late 2019 and every 3 years thereafter;
- Setting up appropriate consumer information on alternative fuels, including a clear and sound price comparison methodology, for example, by intelligent tagging and mapping of charging stations.

The sectoral regulations on RES and EVs are brought together in the EU's Energy Union strategy of 2015 (European Commission 2015c; Ringel 2015a). The Energy Union confirms the EU target of reaching a minimum of 27 % RES in 2030. It stipulates that the rollout strategy for alternative fuels and e-mobility needs to run in parallel in all EU member states. The Commission confirms its supporting role while underlining that the rollout is a task for the member states: "The Commission will take further action to create the right market conditions for an increased deployment of alternative fuels and to further promote procurement of clean vehicles. This will be delivered through a mix of national, regional and local measures, supported by the EU." (European Commission 2015a)

This implies that business models for the integration of RES in the transport sector will depend strongly on the national framework conditions. In many cases, these framework conditions still need to be defined. As an example, for a relatively advanced framework, we will turn to the German energy and e-mobility market.

**Table 2** Charging infrastructure: status quo 2011 and proposal 2020 (based on European Commission 2013c)

| Member State | Available infrastructure (charging points) 2011 | Proposal for publicly available charging points | Number of EVs projected for 2020[a] |
|---|---|---|---|
| Austria | 489 | 12,000 | 250,000 |
| Germany | 1937 | 150,000 | 1,000,000 |
| Denmark | 280 | 5000 | 200,000 |
| France | 1600 | 97,000 | 2,000,000 |
| Italy | 1350 | 125,000 | 130,000 |
| Netherlands | 1700 | 32,000 | 200,000 |
| Spain | 1356 | 82,000 | 2,500,000 |
| Sweden | – | 14,000 | 600,000 |
| UK | 703 | 122,000 | 1,550,000 |

[a]2015 for Italy

## 3.2    National Deployment Strategies: The Case of Germany

In 2015 Germany had a stock of 49,000 EVs in place (IEA 2016b; IEA-HEV 2015). Despite this relatively low number, the German government has committed itself to attain a target of one million EVs by 2020 (BMBF 2009). This target is to underpin a strategy of turning Germany into a leading market and a leading producer for e-mobility. Fostering e-mobility is also a key objective in terms of energy policies. On the one hand it will make it possible to introduce more RES into the energy system. On the other hand, the energy strategy of the government (*Energiewende*, or energy transition) (Ringel et al. 2016) asks for a reduction of transport's specific energy use of 10 % by 2020 and 40 % by 2050 or to 2005 consumption values (Bundesregierung 2010; Bertram and Bongard 2014).

The e-mobility strategy of 2011 (BMVI 2011) and the coalition agreement of the present government (Bundesregierung 2013) further substantiate the objectives by dedicated measures, notably in terms of R&D. As the coupling of energy and mobility markets poses both a variety of options and technical challenges, the federal government and the federal states have set up so-called model regions. In these model regions, different aspects of RES-based e-mobility are tested. The combination of lighthouse projects and regional "showcase projects" have made it possible to develop technical specifications to back up the sector convergence (BMWi 2014a). By now, a multitude of regional and local programs and measures further enhance the federal government's initiatives (e-mobil BW 2015).

After a first phase of establishing networking and coordination mechanisms, the government now aims at a mass rollout of e-mobility. This concerns both the public and private sectors. In the public sector, e-mobility is encouraged by public procurement. For private and business customers, the federal government grants a buyer's premium of EUR 4000 for BEVs and EUR 3000 for HEVs ("environmental bonus" program; BAFA 2016). The grant program started in July 2016 and covers the private procurement of vehicles with prices of up to EUR 60,000. In a first stage, the program is endowed with a total budget of EUR 1.2 million and will end automatically once this sum is spent. The German government and EV manufacturers share the subsidy volume on even terms. In addition to exemption from car tax, nonmonetary benefits, such as the use of bus lanes and free parking lots, are included in the government's support package (Bundesregierung 2016; Ringel 2015b).

Despite regulatory and monetary support for the development of e-mobility, the government's declared objective is to establish the German suppliers as competitive producers for "green mobility". Several studies (Bozem et al. 2013; Fazel 2014) underline that customers see a clear value added and additional buying argument for EV once they are coupled with RES. Along this line the government considers its support as initial aid for the market actors. These actors are strongly encouraged to develop business models combining RES and EV which are competitive on global scale.

# 4     Options and Business Models for the Integration of Renewable Energy Sources

The politically fostered integration of the transport and energy sectors paves the way for several options and business models to combine RES and EVs. Providing green power as fuel for electric mobiles is the closest and most obvious link between both fields. The use of RES electricity (in times of oversupply comparatively cheap or even zero/negative prices) increases the ecological and economic attractiveness of EVs. From this perspective, RES development serves facilitates the introduction of e-mobility.

Additional options and business models appear with an increased installation of fluctuating RES on the energy side. In this case the interaction of RES installations and EVs can provide at least three distinct ancillary services to the power grid:

- EVs as passive load shedding option in times of RES production above grid capacity,
- EVs as active storage facilities,
- EVs providing balancing services to stabilize grid frequency.

In addition, other options for business model integration are conceivable. At the individual level, combining in-home RES (mainly photovoltaics) and EVs can offer so-called energy self-reliance. This in turn would stabilize and mitigate decentralized feed-ins into the general power grids. It can be expected that business models on self-sufficient electricity generation, including individual charging of EVs at home, will expand from a niche market to a larger customer base at some point in time. Contrary to such specialized business models, traditional suppliers can be expected to focus on core and ancillary services of the EV/RES combination provided to the general electricity and transport systems.

## 4.1     RES as Supplier of "Green Power"

RES-produced electricity highlights the full panoply of ecological benefits of e-mobility. $CO_2$ and local air pollutant emissions are avoided at both the input and tail-end levels. This option is physically only possible once charging points are directly coupled to RES installations, such as wind parks or PV systems. Provided that the production capacity is sufficient to cover charging needs, connected EVs will drive fully on RES. This "green mobility" service offers 100 % carbon-free mobility.

In Europe several regions exist where the physical link between RES installations and EV fleets has been established. On the research side, several EU- or state-funded lighthouse projects test the opportunities and challenges of this direct linking. On the commercial side, many car manufacturers have by now invested in RES installations, such as wind or PV parks (VWK 2013). The key aim is to deliver the selling argument "100 % RES" for their e-mobiles in a given area.

The potential for such direct linking of RES installations and EVs used to be judged as relatively small. Still the option becomes an increasingly realistic one thanks to a growing decentralization of energy production and larger shares of RES self-production by home-owned PV installations. For the time being, though, a physical direct coupling of RES and EVs is economically inefficient on larger scales (Pehnt et al. 2007). In addition, synchronizing the production of regionally and seasonally fluctuating RES with e-mobility charging patterns remains challenging (Linsen et al. 2013).

It follows that the general electricity grid infrastructure will provide the power for the large majority of EVs. This in turn implies that the overall generation park close to charging points will determine the "greenness" of the traction current. Because it is physically impossible to attribute a RES generation technology to a dedicated customer—Pehnt et al. (2007) refer to this theoretical option as the "renewable electron"—the EV fleets of a country will be charged on average using the prevailing pool of generation technologies. Depending on the primary energy production mix, the ecological balance of EVs can vary significantly. Figure 5 shows a comparison of the electricity generation mix of selected European countries.

As the feed-in from the various generation sources varies constantly over time, a stronger link between EVs and RES can be established in physical terms by charging EV batteries at times when the share of RES in the generation pool is high. To match times of high RES feed-ins with EV charging, a suitable information and communication technology (ICT) infrastructure needs to be in place. This infrastructure would send the "charging" signal to a grid-connected EV to start the loading process. Even with ICT use, synchronizing feed-in peaks and charging processes remains challenging because it could also affect user behavior (Linsen et al. 2013). Some associations (BEE 2010) plead for the integration of smart meters into EVs to facilitate information exchange between grid conditions and charging needs.

Some authors suggest that the ICT solutions could enable a dialer concept like that in telecommunications between EVs and the power grid. A customer would have the possibility to book the charging process with a dedicated producer of green power via a chip card. The RES producer would then increase its feeds into the grid on demand.

The preceding discussion shows that a synchronized physical delivery of RES for EV charging will remain challenging with existing technologies. The use of commercial relationships could serve as a back-up. Charging volumes could be covered by the additional purchase of green certificates or guarantees of origin certifying that an equivalent amount of power was produced by RES and fed into the grid.

With the EU RES Directive and subsequent national registers in place, a more transparent tracking of production and trading of these so-called European Energy Certifications (EECs) is possible (Seebach and Mohrbach 2013). Already by now

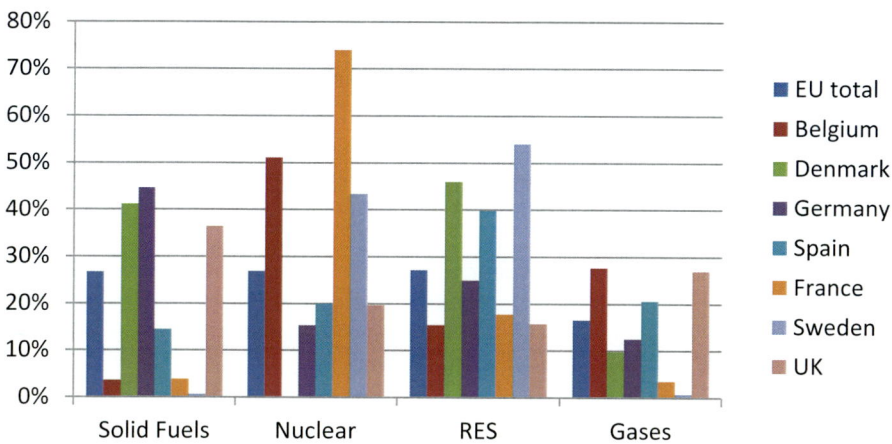

**Fig. 5** Electricity generation mix of selected European countries (DG Energy 2015)

ecolabels and EECs constitute the basis for direct green power commercialization in mainstream electricity markets. Their use in providing commercial backup for green operating power for EVs seems realistic in the absence of more enhanced ICT solutions.

The additional market volume for RES, which is connected with EV charging, is judged to be relatively small by several stakeholders (BEE 2010; Pehnt et al. 2007). In the German case, the projected one million EVs are expected to consume some additional 2–3 terawatt-hours (TWh). Compared to the overall 629 TWh electricity production (BMWi 2014b), this additional consumption seems subordinate to other uses. It is likely to be covered by a moderate increase in RES facilities. This in turn implies that "green charging power" is a limited business model on its own, mainly in connection with marketing. However, the concept can develop a larger scope if combined with additional commercialization opportunities.

## 5    System Services as Power Consumer and Storage Facility

Grid capacities and the stabilization of grid frequency are challenges in almost all countries developing RES on a larger scale. A significant increase in fluctuating RES like wind energy or photovoltaics have led in many EU countries to situations with electricity oversupply. With dispatch obligations in place this can lead to situations with negative market prices at the power exchanges. Conversely, insufficient supply requires increased use of so-called backup or balancing energy. Both situations are inherently destabilizing for the overall electricity networks.

In these situations, ICT and smart metering can prove an important enabler of system service offerings. Two-way communication channels installed at charging points can log connected EVs into the electricity network. Power grid operators could then have access to the connected EVs to use them as passive storage facility for RES oversupply. In undersupply situations, the system operator can send signals to charging points to stop or slow down EV charging if battery capacity is still sufficient (grid to vehicle, G2V). In both cases, detailed information about charging points and connected EVs is necessary and would include charge status, place of charging, and allocation to grid balancing area.

The business model in the case of RES over- or undersupply would link charging processes or charging slowdowns/stops to dedicated flexible tariffs, the so-called demand response. Consumers willing to flexibly adapt their charging behavior to grid system availability would get preferential charging rates. These rates would be cross-financed by the grid operators. Having access to additional power storage and load shedding potential enables operators to forego expensive system services to stabilize network frequency in connection with other actors.

It needs to be underlined that several conditions need to be met for a full application of this model. Dedicated tariffs must be on offer and attractive enough for consumers. The EV fleet offering the storage capacity would logically need to be associated with the balancing area where the electricity oversupply is occurring.

In Europe, the case of Germany, which has a one million EV target, might be able to provide the greatest storage potential. If we assume a storage capacity of 12 kWh/vehicle, this in turn provides the country with an additional storage capacity of some 12 GWh. Opposed to this theoretical storage potential is an estimated oversupply of some 66,000 GWh projected for 2050 (Nitsch 2007; Wietschel 2006), half of which is already in place at present.

This implies that using EVs for storage can again only be one component in a more comprehensive strategy to increase the use of RES by means of e-mobility. It strongly depends on (1) the development of ICT solutions to allow for two-way communication and (2) innovations in storage technology allowing larger power storage volumes for EVs.

## 6    Green Balancing Power

Grid-connected EVs in an intelligent, ICT-based grid infrastructure can enable the business model of actively supporting grid stability. In this case, the connected EVs would actively feed back stored electricity into the grid. These feeds can be marketed as balancing services to stabilize grid frequency in times of undersupply. Such a vehicle-to-grid (V2G) business model was originally tested in the USA. It has been taken up in several research projects in Europe (Garcia-Valle and Lopes 2013). Using EVs as a power source or power sink can offer a much broader business model as the passive storage function discussed in Sect. 5. Balancing power in the EU is provided across the European network area as a whole. This reduces the need for large volumes of balancing power since a multitude of plants can cover the balance by comparatively small contributions to so-called secondary and tertiary control.

The attractiveness of the so-called balancing power business model is more pronounced in a scaled perspective and when combined with the green charging model discussed in Sect. 4.1: Rather than offering this service at the individual EV level, several EVs can be coupled into a virtual balancing power plant. Estimates suggest that one million EVs could cover a balancing power of some 3 gigawatt. Especially for countries with limited options for RES balancing services like pump storage, V2G balancing might be attractive. It would enable and facilitate a larger RES share in the system. For the case of Germany, grid operator figures (BNetzA 2014) list an installed pump storage capacity of some 9.3 GW installed and a further 4.5 GW planned. Including the V2G business model in this framework would imply covering two thirds of the additional capacity needs by e-mobility balancing.

Offering balancing services via e-mobility might prove especially attractive for regional and local suppliers. Often these suppliers run their own e-mobility fleets. As a result, they can easily integrate balancing service options in their planning of charging infrastructure. Second, regional providers often run low- and medium-voltage grid systems. They are directly concerned with fluctuating feed-ins from RES that occur at this grid level. In-grid balancing can help them to stabilize their operation systems. Using a two-way coupling of the power grid and the e-mobility feed-in and storage, an active balancing can be achieved. G2V and V2G combined enable a broader and more stable increase of fluctuating RES in the generation mix. Oversupply of RES electricity is fed into the EV fleet and released into the grid as green balancing energy.

As with the other business models, the necessary conditions for these services are smart meters and smart grids. ICT solutions need to act as the logistical backbone for operating the system and enabling a so-called green balancing power business model. This ICT intervention involves finding answers to various legal problems such as data ownership or data protection (Mayer and Klein 2013).

## 7    Discussion: Opportunities and Limits of Green Mobility Business Models

The foregoing analysis shows that RES deployment and e-mobility concepts can be bundled into various synergetic business models. This synergy works directly via the provision of RES power supplies for EVs. Indirectly, e-mobility concepts and EVs support the integration of RES via storage and balancing systems, making G2V and V2G a mutually beneficial combination.

Our analysis also shows that business model options are highly correlated with a further development of ICT and key system components. At the demand-side level, battery storage and overall production costs of EVs will remain key issues in the coming years (Williamson 2013). On the supply side, smart grids/smart infrastructure and ICT tools need to be developed. Innovative business models depend strongly on an optimal synchronization of supply and demand at large, but also feed-ins and feed-outs of renewable-energy-sourced power at the individual level.

Tariff regulation and tariff structures are an additional issue for developing innovative business models. Wrong regulation or commercial tariff setting might significantly hinder the development of green mobility. This would be the case once the high procurement costs for EVs are surpassed by comparatively high variable fuel costs by the mark-up of a green power tariff. Here the high CAPEX costs of EVs cannot be recovered by lower maintenance costs through the use of RES in low-price periods.

It is worthwhile to note that the opportunities to replace traditional internal combustion engines with RES-powered EVs seem to be limited at the individual level. Still they become more attractive once the whole EV fleet of a larger area or even a country is considered, which would require a scaling of EV deployment.

Shifting to a broader perspective on e-mobility changes this analysis significantly. Here EVs are not approached in terms of replacing cars one by one. They are part of integrated and intermodal transport concepts. These services will be offered mostly in cities and urban areas (Brand and Schmidt 2014; Institute for Mobility Research 2013). Here, a number of green e-mobility services are conceivable that relate to business segment vehicles, power, infrastructure, and supporting services (Muratori et al. 2013). Experts anticipate that the clear-cut differentiation between services offered by the automotive industry and the energy sector will blend. Energy actors are expected to offer mobility services that are close to their core business. In turn, mobility providers will consider bundling strategies to include RES fuel in their portfolio.

Whereas it is still too early to clearly identify the emerging trends of this blending, three examples can serve as illustrations of such strategies. First, a wind farm operator might be interested in including "green power for mobility" concepts in its portfolio. Such an operator could provide neighbors with physically connected green power for their EVs, free of charge, with the aim of raising acceptance for local wind power installations. This model mainly relates to marketing. Still, the positive support effect for RES could potentially be strong in densely populated areas.

Second, dedicated green charging stations delivering 100 % RES power as "premium fuel" might enter the market. With the planned increase in the number of charging points in Europe, such a specialization of charging points might be conceivable. This model could be implemented either physically or commercially using intermediaries like ecolabels or green certificates. Like the present compensation of $CO_2$ emitted on business trips by paying into compensation schemes, such a segment might be interesting for companies both at business-to-consumer (B2C) and business-to-business (B2B) levels.

Finally, business models for RES uptake in the e-mobility sector could be used as an add-on to the core business of established third-party service providers. Parking lots or supermarkets could offer their clients "recharge as you park"/ "recharge as you shop" services. In these business models, EVs would be recharged during parking. Such services might even be offered for free in RES surplus conditions in the grid. EVs in charge mode could be used to offer balancing services and thus compensate for the charging costs for the parking lot operators or cross subsidize green power tariffs.

# 8 Conclusion

The presented review of European markets for RES and e-mobility shows that regulatory framework conditions favor an uptake of business models for RES in the transport sector. Whereas the European framework largely supports installing appropriate infrastructure, it is up to the member states to develop and support dedicated schemes to increase the use of EVs and e-mobility at large.

Combing RES and EVs (sector coupling) could trigger synergies for both sectors. Enabling clean mobility not only by avoiding end-of-pipe emissions but also on the fuel input side is a strong selling argument for EVs. As could be shown, business models involving the coupling of RES with EVs are limited at the individual level but become more attractive once scaled to the full vehicle fleet. Likewise, a portfolio of various business models bundled around the core operations of an electricity or transport provider can generate a clear value added.

The most promising and innovative business models occur with e-mobility and dedicated transport services. Here EV deployment and RES development are clearly mutually beneficial. The different applications are still limited by battery storage capacities and dedicated ICT solutions to enable a two-way communication between the power grid and EVs. Still, this barrier should only be a temporary one. The underlying technologies—especially as concerns ICT—already exist and require no disruptive innovations.

Once the relevant technologies are applied, the synchronization of RES and e-mobility will allow a multitude of innovative business models to emerge, from the perspective of both transport and energy sector operators. Besides a clear blending of these two sectors, new actors will emerge that will further foster the uptake of RES in Europe's transport sector.

## References

ACEA (European Automobile Manufacturers Association). (2016). *Electric and alternative vehicle registrations*. Accessed July 2, 2016 from http://www.acea.be/statistics/tag/category/electric-and-alternative-vehicle-registrations

ARF—Amsterdam Roundtable Foundation and McKinsey & Company The Netherlands. (2014). *EVolution. Electric vehicles in Europe: gearing up for a new phase*. Accessed 22 June 2016.

BAFA. (2016). *Elektromobilität (Umweltbonus)*. Accessed July 2, 2016, from http://www.bafa.de/bafa/de/wirtschaftsfoerderung/elektromobilitaet/

BEE—Bundesverband Erneuerbare Energie e.V. (2010). *Elektromobilität und Erneuerbare Energien: BEE-Position*. Accessed May 15, 2016, from http://www.bee-ev.de/_downloads/publikationen/sonstiges/2010/1005_BEE-Position_Erneuerbare-Elektromobilitaet.pdf

Bertram, M., & Bongard, S. (2014). *Elektromobilität im motorisierten Individualverkehr: Grundlagen, Einflussfaktoren und Wirtschaftlichkeitsvergleich*. Wiesbaden: Springer.

BMBF—Bundesministerium für Bildung und Forschung. (2009). *Nationaler Entwicklungsplan Mobilität der Bundesregierung*. Bonn: BMBF.

BMBF—Bundesministerium für Bildung und Forschung. (2013). *Elektromobilität—das Auto neu denken*. Bonn: BMBF.

BMUB—Bundesministerium für Umwelt, Naturschutz, Bau und Reaktorsicherheit. (2013). *Erneuerbar mobil: Marktfähige Lösungen für eine klimafreundliche Elektromobilität*. Berlin: BMUB.

BMVI—Bundesministerium für Verkehr und digitale Infrastruktur. (2011). *Elektromobilität— Deutschland als Leitmarkt und Leitanbieter*. Berlin: BMVI.

BMWi—Bundesministerium für Wirtschaft und Energie. (2014a). *Smart energy made in Germany smart energy made in Germany: Erkenntnisse zum Aufbau und zur Nutzung intelligenter Energiesysteme im Rahmen der Energiewende*. Berlin: BMWi.

BMWi—Bundesministerium für Wirtschaft und Energie. (2014b). *Energiedaten— Gesamtausgabe*. http://bmwi.de/DE/Themen/Energie/Energiedaten-und-analysen/ Energiedaten/gesamtausgabe,did=476134.html. Accessed 1 July 2014.

BNetzA—Bundenetzagentur. (2014). *Zweiter Monitoring-Bericht "Energie der Zukunft"*. Bonn: BMWi.

Bozem, K., et al. (2013). *Elektromobilität: Kundensicht, Strategien, Geschäftsmodelle: Ergebnisse der repräsentativen Marktstudie Future Mobility*. Berlin: Springer.

BP. (2016). *Energy outlook*. London: BP.

Brand, M., & Schmidt, A. (2014). Herausforderung Elektromobilität: Lehren aus den Entwicklungen des Strom-, Bahn- und Mobilfunkmarktes. *Energiewirtschaftliche Tagesfragen, 64*(3), 105–107.

Bundesregierung. (2010). *Energiekonzept für eine umweltschonende, bezahlbare und sichere Energieversorgung*. Accessed June 23, 2016, from http://www.bundesregierung.de/ ContentArchiv/DE/Archiv17/_Anlagen/2012/02/energiekonzept-final.pdf.

Bundesregierung. (2013). *Deutschlands Zukunft gestalten. Koalitionsvertrag zwischen CDU, CSU und SPD. 18. Legislaturperiode*. Accessed June 23, 2016, from http://www.bundesregierung. de/Content/DE/_Anlagen/2013/2013-12-17-koalitionsvertrag.pdf?__blob=publicationFile

Bundesregierung. (2016). *Elektromobilität: Einigung auf Kaufprämie für E-Autos*. Accessed July 2, 2016, from https://www.bundesregierung.de/Content/DE/Artikel/2016/04/2016-04-27- foerderung-fuer-elektroautos-beschlossen.html

Canzler, W. (2010). Mobilitätskonzepte der Zukunft und Elektromoblität. In R. Hüttl, B. Pischetsrieder, & D. Spath (Eds.), *Elektromobilität* (pp. 39–61, Acatech Diskutiert). Berlin: Springer.

Council of the European Union. (2007). *Presidency conclusions of 2 May 2007. Doc. 7224/1/07 REV 1*. Brussels: EU Council.

Council of the European Union. (2014). *European Council (23 and 24 October 2014). Conclusions. Doc. CO EUR 13 CONCL 5*. Brüssel: European Council.

DG Climate Action. (2016a). *Reducing emissions from transport*. Accessed June 23, 2016, from http://ec.europa.eu/clima/policies/transport/index_en.htm

DG Climate Action. (2016b). *Reducing emissions from passenger cars*. Accessed June 23, 2016, from http://ec.europa.eu/clima/policies/transport/vehicles/cars/index_en.htm

DG Energy. (2015). *EU energy in figures. Statistical pocketbook 2015*. Brussels: European Commission.

EIA—U.S. Energy Information Administration. (2016). *International energy outlook*. Report number DOE/EIA-0484(2016). Accessed July 2, 2016, from http://www.eia.gov/forecasts/ ieo/world.cfm

e-mobil BW. (2015). *Strukturstudie BWe-mobil*. Stuttgart: Landesregierung Baden-Württemberg.

European Commission. (2009). *Directive 2009/28/EC of the European Parliament and of the Council of 23 April 2009 on the promotion of the use of energy from renewable sources and amending and subsequently repealing Directives 2001/77/EC and 2003/30/EC (L 140, 5.6.2009)*. Brussels: European Commission.

European Commission. (2010). Directive 2010/31/EU of the European Parliament and of the Council of 19 May 2010 on the energy performance of buildings (L 153, 18.6.2010).

European Commission. (2011a). *Commission Communication "A Roadmap for moving to a competitive low carbon economy in 2050", COM (2011)112*. Brussels: European Commission.

European Commission. (2011b). *White Paper. Roadmap to a Single European Transport Area—towards a competitive and resource efficient transport system: COM(2011) 144 final.* Brussels: European Commission.

European Commission. (2012). Directive 2012/27/EU of the European Parliament and of the Council of 25 October 2012 on energy efficiency, amending Directives 2009/125/EC and 2010/30/EU and repealing Directives 2004/8/EC and 2006/32/EC.

European Commission. (2013a). *Communication clean power for transport: European alternative fuels strategy. COM(2013)17 final.* Brussels: European Commission.

European Commission. (2013b). *Impact Assessment accompanying the document proposal for a directive on the deployment of alternative fuels infrastructure. SWD(2013)5 final.* Brussels: European Commission.

European Commission. (2013c). *Press release: EU launches clean fuel strategy. IP 13/40.* Brussels: European Commission.

European Commission. (2014). Directive 2014/94/EU of the European Parliament and of the Council of 22 October 2014 on the deployment of alternative fuels infrastructure.

European Commission. (2015a). *Communication. A framework strategy for a Resilient Energy Union with a Forward-Looking Climate Change Policy.* COM/2015/080 final. Brussels: European Commission.

European Commission. (2015b). *A European strategy for alternative fuels. Policy perspective.* Presentation of 03 December 2015 to the Automotive intergroup meeting, Brussels.

European Commission. (2015c). *Communication a framework strategy for a Resilient Energy Union with a Forward-Looking Climate Change Policy.* COM(2015) 80 final. Brüssel: European Commission.

European Commission. (2016). *An EU Strategy on Heating and Cooling. COM(2016) 51 final.* Brussels: European Commission.

Fazel, L. (2014). *Akzeptanz von Elektromobilität: Entwicklung und Validierung eines Modells unter Berücksichtigung der Nutzungsform des Carsharing (Schriften zum europäischen Management).* Wiesbaden: Springer.

Garcia-Valle, R., & Lopes, J. (2013). *Electric vehicle integration into modern power networks (Power electronics and power systems).* New York: Springer.

IA-HEV—IEA Technology Collaboration Programme on Hybrid and Electric Vehicle Technologies. (2015). *Hybrid and electric vehicles: The electric drive delivers.* Paris: IEA/OECD.

IEA—International Energy Agency. (2016a). *Mid-term oil market report.* Paris: IEA/OECD.

IEA—International Energy Agency. (2016b). *Global EV Outlook 2016.* Paris: IEA/OECD.

Institute for Mobility Research (ifmo). (2013). *Megacities mobility: How cities move on in a diverse world.* Berlin: Springer.

Linsen, J., et al. (2013). Netzintegration von Elektrofahrzeugen und deren Auswirkungen auf die Energieversorgung. *Energiewirtschaftliche Tagesfragen, 63*(1/2), 121–124.

Mayer, C., & Klein, C. (2013). Ladeinfrastruktur für Elektrofahrzeuge—rechtliche Fragestellungen und Herausforderungen. *Energiewirtschaftliche Tagesfragen, 63*(8), 73–75.

McKinsey&Company. (2016). *Automotive revolution—perspectives towards 2030: How the convergence of disruptive technology-driven trends could transform the auto industry.* Accessed June 22, 2016.

Muratori, M., Marano, V., Rizzo, G., & Rizzoni, G. (2013). Electric mobility: From fossil fuels to renewable energy, opportunities and challenges. *IFAC Proceedings Volumes, 46*(21), 812–817. doi:10.3182/20130904-4-JP-2042.00161.

Nitsch, J. (2007). *Leitstudie 2007: "Ausbaustrategie Erneuerbare Energien": Studie im Auftrag des Bundesumweltministeriums.* Stuttgart.

Pehnt, M., Hoepfner, U., & Merten, F. (2007). *Elektromobilitaet und erneuerbare Energien (Arbeitspapier Nr. 5).*

Regio Randstad. (2014). *Randstad monitor.* Accessed July 1, 2016, from http://www.randstadregion.eu/uploads/2014/09/randstadmonitor-voorkant.pdf

Ringel, M. (2015a). Die europäische Energieunion—Ein Paradigmenwechsel der europäischen Energiepolitik? *Wirtschaftswissenschaftliches Studium—WiSt, 44*(8), 456–460.

Ringel, M. (2015b). Elektromobilität als Absatzmarkt für Strom aus Erneuerbaren Energien: Möglichkeiten und Grenzen des Geschäftsmodells "Grüne Mobilität". In C. Herbes & C. Friege (Ed.), *Marketing Erneuerbarer Energien. Grundlagen, Geschäftsmodelle, Fallbeispiele* (pp. 299–316).

Ringel, M., Schlomann, B., Krail, M., & Rohde, C. (2016). Towards a green economy in Germany? The role of energy efficiency policies. *Applied Energy.* doi:10.1016/j.apenergy.2016.03.063.

Ruppert, W. (2013). Herrschaft über Raum und Zeit—Zur Kulturgeschichte des Automobils. In M. Keichel (Ed.), *Das Elektroauto: Mobilität im Umbruch* (pp. 9–44). Wiesbaden: Springer.

Seebach, D., & Mohrbach, E. (2013). Wie können Herkunftsnachweise zur Differenzierung des Ökostrommarktes in Deutschland beitragen? *Energiewirtschaftliche Tagesfragen, 63*(3), 62–64.

UNFCCC—United Nations Framework Convention on Climate Change. (2015). *Adoption of the Paris Agreement.* Document FCCC/CP/2015/L.9/Rev.1. Bonn: UNFCCC.

VCD—Verkehrsklub Deutschland. (2015). *CO_2-Grenzwert—Neue Vorgaben für Pkw ab 2020.* Accessed July 1, 2016, from https://www.vcd.org/themen/auto-umwelt/co2-grenzwert/

VWK—Volkswagen Kraftwerks GmbH. Lagebericht 2013. Accessed July 1, 2016, from http://www.volkswagenag.com/content/vwcorp/content/de/investor_relations/annual_general_meeting/Aktionaersversammlung_2014/Ordentliche_Hauptversammlung/Top7.bin.html/contentparsys/superteaser/tabs/tab_0/teaserlists/teaserlist_2/teasers/talksandpresentation

Wietschel, M., et al. (2006). Ein Vergleich unterschiedlicher Speichermedien für überschüssigen Windstrom. *Zeitschrift für Energiewirtschaft, 30*, 103–114.

Williamson, S. (2013). *Energy management strategies for electric and plug-in hybrid electric vehicles.* New York, NY: Springer.

**Marc Ringel** reads energy policy, energy efficiency and environmental economics as professor at Nuertingen Geislingen University, Germany. As a former official with the Directorate-General for Energy of the European Commission and with the German Federal Ministry of Economics and Energy, he continues to be deeply involved in the development and economic assessment of national and European energy policy frameworks. He supports policy-oriented research for several national energy action plans and key strategic EU policy actions. Marc Ringel holds a master's degree in economics from Mainz University, Germany, and Université d'Angers, France. In his PhD thesis, he analyzed and developed energy policy measures to combat climate change.

# Smart Battery Systems Driving Renewable Energy Markets

Benjamin Schott and Oliver Koch

**Abstract**

Many countries across the globe are on the verge of experiencing a second wave of renewable energy adoption driven by the advent of intelligent energy storage systems. In this chapter we will explore the drivers of this new phase of a sustainable energy revolution and the role played by technology and business models on the basis of smart energy systems.

**Keywords**

Solar • Energy Storage • Peer-to-Peer • Energy Market • Renewables

## 1    Storage as Enabler for Continuous Growth of Renewables

Significant research has gone into exploring the reasons for the different rates of adoption of renewable energy by countries across the globe in the last two decades. While in some countries photovoltaic (PV) installations are already less than what they were during their peak several years ago (e.g. Germany, Australia), others are just in the middle of their first boom in renewable energy (USA). However, the policy instruments and economic frameworks are very different in each market, leading to unique nuances in the development of each country. While some countries used feed-in tariffs to boost PV deployment (Germany, France, UK, Japan, Australia), others used tax credits to achieve the same goal (USA). Policies like net metering (Canada, USA, Italy) or tax depreciation (Italy) were also instruments as well as direct financial support (Croatia) and standards for a renewable energy generation portfolio (Australia, China, Republic of Korea, USA). The

B. Schott (✉) • O. Koch
sonnen GmbH, Am Riedbach 1, 87499 Wildpoldsried, Germany
e-mail: b.schott@sonnenbatterie.de

© Springer International Publishing AG 2017

C. Herbes, C. Friege (eds.), *Marketing Renewable Energy*, Management for Professionals, DOI 10.1007/978-3-319-46427-5_13

For what reasons have you decided to buy a SB?

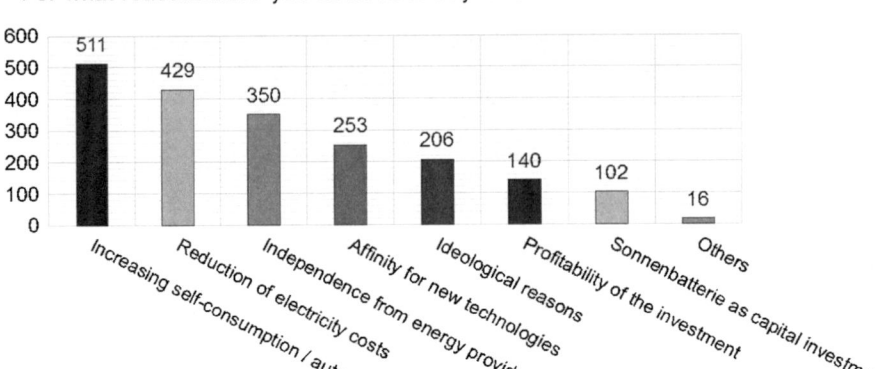

**Fig. 1** Results of customer survey of existing sonnenBatterie owners in 2014/2015 (internal analysis by sonnen)

timing and choice of policy instruments have led to an environment in which significant local differences exist between countries when it comes to renewable energy deployment.

One of the early adopters of a comprehensive renewable energy development plan was Germany. After a feed-in tariff induced a boom in PV installations from 2005 to 2013, new installations quickly fell to the current level of just over 1000 MW of added capacity per year as the feed-in tariff was reduced to the current level of approx. EUR 0.12/kWh. While the reduction in incentives proved to have devastating effects on the German PV market, it also allowed new technologies and business models to develop. A prime example of such a development was the rapid introduction and advance of decentralized small-scale energy storage systems.

Grid-connected storage started to make an appearance in the German market from 2010. At that time, the feed-in tariff was significantly higher than the average retail rate of electricity, creating an environment where an investment into storage did not make financial sense. Instead, the early adopters of this technology focused more on emotional factors: freedom from future price increases, independence from the utility, and backup in case of power failure were all reasons contributing to the purchase decisions of early adopters (Fig. 1). However, like the situation with PV deployment, the market for storage products showed equally diverse characteristics across the globe. While some countries had a mix of financial support and incentives (Germany), others offered financial incentives, even though there was no financial case for storage (California). Simplifying legislation and permitting processes were another driving factor in some countries when it came to the deployment of distributed small-scale residential storage. Depending on the policy framework deployed, storage started to move from an early-adopters phase into the mainstream market at different speeds, with Japan and Germany at the forefront of this development.

As feed-in tariffs started to drop in Germany to their current level and retail prices for electricity continued to climb to about EUR 0.29/kWh, an additional component was added to the factors influencing the purchasing decision: return on investment (ROI). Optimizing self-consumption was no longer just an emotional decision; it started to make financial sense. With this, the early-adopter phase in Germany ended and storage started to enter the mainstream. Rather than being an accessory to a PV installation, storage started to become the driver behind more renewable energy installations with ca. 50 % of all residential PV systems in Germany being installed with storage in 2015 (Kairies et al. 2016). With a combination of financial and emotional factors, storage developed into an enabler of future growth in residential renewable electricity production.

The technical requirements for those early-generation systems were relatively straightforward: Optimize self-consumption during the day and make sure that the batteries are fully charged in the early evening. Relatively simple algorithms were used in those early systems to control the charge and discharge behaviour of the storage system to make sure batteries were sufficiently charged to last the individual consumer through the night. This was what we now characterize as the first wave of residential energy storage deployment.

Two factors, however, led to a change in the system design of storage systems towards a more intelligent approach to small-scale energy storage. First, with increased internationalization of small-scale, decentralized storage, the functionality requirements increased as well. Additional countries meant different additional business models, which storage systems needed to cater for. In addition to maximizing self-consumption, companies now developed a software platform that caters for backup power, time-of-use shifting, or peak shaving. The regulatory frameworks in each country brought with them additional requirements for the intelligence of the system.

The second factor driving the development of a more sophisticated approach to the controls surrounding energy storage was the move from a purely emotional buy towards a more rational, ROI-focused purchasing decision. While ever decreasing hardware cost was naturally conducive to a better ROI for the customer, smarter storage application had the power to further drive down payback periods and increase the ROI. Examples of this development was the deployment of daily weather forecast data to optimize charging strategies or the introduction of smart home ready storage systems by companies like sonnen to increase the level of autarky. End customers are able to turn on electrical appliances such as a dishwasher either using their sonnen smartphone app or letting the system automatically determine when to turn on those devices in order to maximize self-consumption. This was the second phase of residential energy storage deployment.

While decentralized storage got smarter over time, optimization happened only on the individual system level. Factors such as grid friendliness were a by-product rather than the declared goal of any system intelligence developed during this phase of the introduction of decentralized small-scale storage. What was lacking at this point were concepts on how to increase the individual and collective utility one could derive from storage systems by starting to utilize network effects once decentralized storage had reached a critical mass.

With this we see residential storage enter a new, the third, phase: virtually connecting decentralized storage systems to create benefits partially outside the location where the system is being installed.

While the first and second phases have already contributed to the growth of renewable generation capacity in early-adopter markets, we believe that it will be the value created by the third phase that will have the power to dramatically change the way people produce, distribute and consume energy. With this, distributed energy storage could become the catalyst for a worldwide transformation towards a carbon-neutral energy economy.

## 2    Storage as the Missing Piece for Sustainable Renewable Energy Markets

Connecting assets like energy storage, especially decentralized small-scale systems installed in private households, and renewable generation in a virtual power plant as the next phase of technology development will further drive the growth of renewable generation in many countries and will be the solution for sustainable renewable energy markets.

One can observe that markets for energy storage are evolving in those countries where already a high penetration of renewables in the electricity mix has been achieved and the energy market is facing the challenge of integrating fluctuating and volatile sources of energy in the existing energy market design and infrastructures. In particular, switching from a support programme based on fixed feed-in tariffs to more price-based mechanisms, for example direct marketing or the recently introduced new auction mechanisms in Germany, will drive the demand for flexible management to maximize the value of the renewable feed-in tariff compared to the investment security of a fixed tariff, which does not incentivize flexible management. Most types of renewable generation are intermittent, relying on weather, like solar and wind, and thus cannot be managed in a flexible way. Today only biogas plants and, with some restrictions, hydropower plants provide flexible loads to systems. The biggest portion of power generation in future will be based on solar and wind, which will require flexible management and especially energy storage as the missing piece to stabilize systems, as many energy studies in Germany have shown (Buttler 2015). Maximizing or optimizing the value of solar and wind generation will thus be the main challenge and interest for plant operators and owners. In Germany, for example, today already almost 20 % of electricity generation is covered by wind and solar (BMWi 2016).

The reduction or complete phase-out of fixed feed-in tariff schemes can be seen as a major tipping point in many countries when flexible management will be introduced by stakeholders in the energy market and is the main reason why new ways of marketing renewables will be designed and launched. Recent years have already witnessed a trend in Germany and other countries, like the UK or the

Netherlands, towards new business models trying to find a way to optimize the marketing of renewable energy.

These new business models can be clustered into the following use cases and will be described subsequently:

1. Grid-friendly renewable energy usage on site,
2. Optimized renewable energy trading,
3. Renewable energy retail.

**Grid-friendly renewable energy usage on site** This use case is the typical so-called prosumer case, as described in Sect. 13.1 of this chapter. Usually the load profile of a typical household in Germany will allow for using 20–30 % of generated solar energy directly (Weniger et al. 2015). For commercial customers this figure can be significantly higher as the load profile shows high consumption during the day and thus matches with solar generation directly. Adding a storage system to the solar system will increase this value for a private household to 60–80 % depending on the size of the solar system and the size of the storage system as well as the consumption pattern (Weniger et al. 2015). In combination with electric heating, for example, solar self-consumption can reach up to almost 100 %. In many cases the household will still consume 25–30 % from the grid, while in some cases the combination of storage with micro CHP on site could allow customers to minimize grid consumption. A storage system allows end users to reduce the electricity bill by using more of their own generated solar energy. As several analyses show, the shrinking costs of solar and batteries will drive this use case to reach the mass market very soon (Franz 2016; Farid et al. 2016). Figure 2 shows the typical profile of a prosumer site in Germany.

An important building block of this use case is intelligent energy management, which allows one to schedule the discharging/charging of batteries taking into account solar and load forecasts as well as other generation assets and external requirements. For example, in Germany one regulation says that solar systems not equipped with a decentralized relay controlled by the distribution system operator (DSO) must limit solar feed-in to 70 % of the nominal power of the system, and in combination with storage—if the storage system is funded by the KfW subsidy program (KfW 2016)—even a 50 % limit is required. This is done to reduce the solar peak to relieve the grid but will lead to a loss of produced solar energy of up to 10 % per year as a result of the curtailment (Weniger et al. 2016). A storage system equipped with an intelligent energy manager as well as a load and production forecaster is able to avoid curtailment through delayed charging (Figs. 2 and 3) (Weniger et al. 2016).

This approach to intelligent peak shaving optimizes the use of renewable energy for end users and helps them maximize the value of their solar usage on site while at the same time helping to relieve the burden places on the grid. Several studies have shown the positive impact of smart battery systems for the integration of renewables. For example, an analysis by HTW Berlin determined that the installed capacity of solar in Germany could be doubled with no impact on grid infrastructure

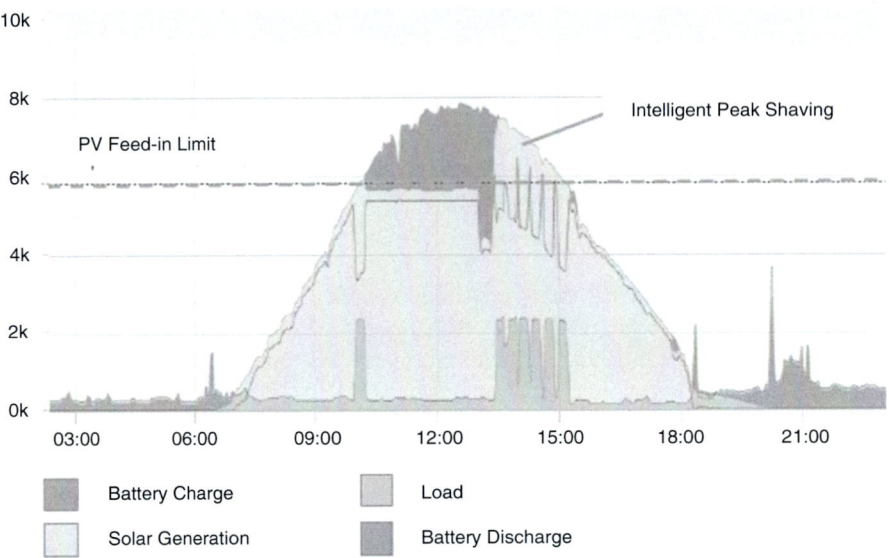

**Fig. 2** Data monitoring of typical solar + storage system in private household with electric heating and intelligent peak shaving (sonnen GmbH)

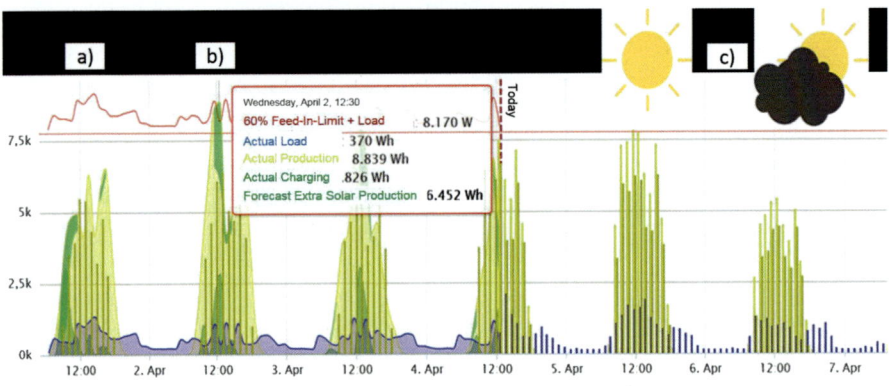

**Fig. 3** Monitoring of intelligent energy management of sonnenBatterie including forecast. (**a**) Normal charging—production below limit (*red line*); (**b**) delayed charging—shaving production peak around noon; (**c**) forecast next 3 days (sonnen GmbH)

owing to peak shaving (Weniger et al. 2015). Furthermore, a study by Prognos showed that peak shaving- and forecast-based control of batteries would make it possible to avoid EUR 120 million in grid upgrades and investments in Southern Germany (Krampe and Peter 2015).

**Optimized renewable energy trading** Efficient and profitable trading of renewable energy generation will be one of the major challenges facing plant operators and energy trading companies in the future energy system. Franz (2015) showed that decentralized energy trading will create tremendous value in Germany's energy markets. Solar and wind are intermittent weather-dependent resources. Although excellent forecasting helps to minimize trading risks, optimizing profits is only possible under a regime of flexible management that makes it possible to hedge low and high price periods. One option to bring flexibility to the management of solar and wind assets is the installation of an energy storage system on site. This would make it possible to store energy when spot market prices are low and sell it on a delayed basis to the market when prices are high again to maximize profitability. One example of this is the solar park installed by Belectric in Alt-Daber, Germany, a 2 MW lead acid storage system that is the first project of its kind in Europe (Petersen 2014). Typically real-time trading of renewable energy is only attractive to traders for generation assets >100 kW or even >500 kW [cf. data of direct marketing companies in Franz (2015)]. This is due to the fact that the cost efficiency for trading small-scale assets is very poor for most trading companies because of the high operational costs, transaction costs and additional hardware costs per kilowatt. Thus, today this business model has clear limitations in terms of the size of assets, and the typical installation of storage systems at renewable sites for the primary use of optimizing energy trading is >100 kW to the megawatt scale. Minimizing these integration costs for small-scale assets is a major challenge. The example of the sonnenCommunity[1] in Germany shows that the combination of small-scale solar generation with smart storage systems on site, installed for the purpose of solar self-consumption, enables an automated and standardized low-cost process for integration in direct marketing.

**Renewable energy retailing** Another innovative way to market renewables is by creating new energy tariffs for energy retailing. This means that a specific segment of energy consumers, whether private households or commercial customers, is interested in purchasing green electricity from a region or from known suppliers based on transparency of origin. Energy retail companies offer certificates of origin for the power they purchase or use platforms displaying the producing power plants to create transparency for their customers. Examples include the sonnenCommunity by sonnen GmbH in Germany, Vandebron[2] in the Netherlands, and Open Utility[3] in the UK. Consumers are willing to pay a premium for this transparency for several reasons. Green electricity supply of eligible sources at any given time, though, is only possible with a massive oversupply of renewables in the supplier's portfolio, or storage can be used to manage the demand side using the battery as a flexible buffer for excess renewable electricity. The combination of storage-optimized demand-side management with storage installed on the site where the renewable energy is generated to optimize the supply side makes a virtual power plant. This

---

[1]www.sonnenbatterie.de/sonnencommunity (Accessed 25 June 2016).

[2]www.vandebron.nl (Accessed 25 June 2016).

[3]www.openutility.uk (Accessed 25 June 2016).

**Fig. 4** sonnenCommunity as example of a new peer-to-peer energy marketplace (sonnen GmbH). The image shows an example of an end customer live platform with the display of energy providers within the peer-to-peer network, which shows the participants with photos, descriptions and their location on a map

makes storage the game changer that allows for managing and balancing supply and demand to optimize the whole portfolio. This has already been started by sonnen through the set-up of the sonnenCommunity, were a portfolio of solar and other renewable assets like biogas and wind, combined with the use of flexible storage units, is managed so as to unsure the demand of Community members is met at any time. This fleet forms the basis of the Community and is combined with a suitable portfolio of generation assets managed through virtual power plant software. The sonnenCommunity thus only manages assets owned by private homeowners or others. This model can be compared to the idea of Uber, AirBnB and others where the platform manages assets without owning them. New models like an "AirBnB of Energy" or a peer-to-peer energy platform can change dramatically the way energy is managed in the future. Prosumers can now be energy suppliers feeding their solar exports into the sonnenCommunity and provide it to others peer to peer (Fig. 4).

## 3      Outlook: Energy Storage Combining Multiple Use Cases Will Drive New Energy Markets Based on Renewables

In all the different models described in this chapter as new ways of marketing renewables, the combination of renewable energy generation with (aggregated) storage can be seen as the driver for new business models. Looking at the market for energy storage today there are examples of each use case described earlier.

Although in most cases a primary use case as the main driver for renewable or storage investment can be identified, the profitability that will make it possible to bring these concepts to the mass market will be driven by a combination and optimization of different use cases.

Although already today several gigawatt-scale utility storage projects are in operation, under construction, or being planned, most of them are aimed at the primary use case to help stabilize the grid through frequency and voltage control; only a few are on the site of renewable power plants. According to the US Department of Energy's (DOE) Energy Storage Database, 1.35 GW has been installed since 2012, most of it thermal storage (DOE 2016). In Germany alone 150 MW of storage systems will be installed by the end of 2016 to be offered mainly in the primary frequency control market (Franz 2016). As the market for primary frequency control is under pressure as a result of these new capacities, prices are already expected to decrease (ibid.). Thus storage operators like WEMAG in Germany are already looking into combining use cases to maximize value and stabilize the investment case (WEMAG 2016).

On the other hand, solar self-consumption is a mega trend in many countries at the residential and small commercial levels (Farid et al. 2016). Thousands of small-scale storage systems have been deployed in recent years and are expected to boom in the coming decades, as described in Sect. 13.1. Already today aggregators are trying to pool storage assets to build up a virtual power plant with several megawatts of capacity for the energy and grid service markets. This combination of use cases makes it possible to achieve a maximum utilization rate and value maximization through a second dispatch. On the one hand this is the combination of the primary use case of prosumers with energy trading and energy retailing. The combination of renewable energy marketing and storage as a demand response resource for grid services will allow companies to maximize value for investors. Especially behind-the-meter, prosumer storage allows for a combination of a broad variety of applications. Being producer, consumer and storage operator at the same time makes it possible to compound several different levels of value for investors. At the same time, aggregated behind-the-meter storage at the prosumer level can also be used for energy trading to hedge or to provide any needed grid support services for the grid.

For example, the Rocky Mountain Institute has assessed how combinations of different use cases can stack this value (Farid et al. 2016). As shown by Fitzgerald et al. (2015) and by Sterner et al. (2015), for the German market, energy storage can be used for a variety of grid services. Storage units can, for example, be dispatched at times when they are not being used by prosumers for frequency control, with the owner receiving a capacity and utilization payment, or they can be used for intraday trading for arbitrage, much like in the case of pumped hydro storage. The results of Fitzgerald et al. (2015) especially show that behind-the-meter storage systems give the broadest variety of use cases and offer the greatest opportunities for stacking value through multiple applications. Stacking value will be important for bringing storage to the mass market. This variety of possible use cases gives prosumers with storage assets a specific role in new approaches to marketing renewable generation.

The primary use case is strong enough to attract investors, while additional applications increase the value and accelerate the time to market of new models. Thus prosumers are already investing in the required infrastructure to create new markets for renewables at a much lower interest rate. The flexibility to combine or optimize all use cases for customers is the key for models like the sonnenCommunity as the regulatory environment in many countries is still under development and can change in many directions. Thus also today many regulatory barriers exist to full deployment of multi-use cases, for example due to system fees that must be paid for behind-the-meter storage or a lack of smart meter technology roll-out.

# References

BMWi. (2016). *Erneuerbare Energien auf einen Blick*. German Federal Ministry for Economic Affairs and Energy. Accessed June 25, 2016, from http://www.bmwi.de/DE/Themen/Energie/Erneuerbare-Energien/erneuerbare-energien-auf-einen-blick.html

Buttler, A. (2015, Januar). *Kampf der Studien*. VDI Vortragsreihe Kraftwerkstechnik.

DOE. (2016). *DOE Global Energy Storage Database*. Department of Energy. Accessed June 25, 2016, from http://www.energystorageexchange.org/projects/data_visualization

Farid, A., Anderson, N., Rosser, J., & Armstrong, C. (2016, February). *Battery adoption at the tipping point*. Berenberg Thematics.

Fitzgerald, G., Mandel, J., Morris J., & Touati, H. (2015, September). *The economics of battery energy storage: How multi-use, customer-sited batteries deliver the most services and value to customers and the grid*. Rocky Mountain Institute. Accessed June 25, 2016, from http://www.rmi.org/electricity_battery_value

Franz, S. (2015). Fit für die nächste Phase der Energiewende: Durch Speicher und Digitalisierung erfolgreiche Geschäftsmodelle finden. *Energiewirtschaftliche Tagesfragen, 65*(11), 43–45.

Franz, S. (2016, Januar). *Neue Geschäftsmodelle in dezentralen und digitalisierten Strommärkten*. Berlin: Studie Energy Business Lab des des Büro F.

Kairies, K., et al. (2016). *Wissenschaftliches Mess- und Evaluierungsprogramm Solarstromspeicher. Jahresbericht 2016*. Institut für Stromrichtertechnik und Elektrische Antriebe RWTH Aachen. www.speichermonitoring.de, Aachen 2016.

KfW. (2016). *KfW-Programm Erneuerbare Energien "Speicher"—275 Kredit*. Kreditanstalt für Wiederaufbau. Accessed June 25, 2016, from https://www.kfw.de/inlandsfoerderung/Unternehmen/Energie-Umwelt/F%C3%B6rderprodukte/Erneuerbare-Energien-%E2%80%93-Speicher-%28275%29/

Krampe, L., & Peter, F. (2015). *Auswirkungen von Batteriespeichern auf das Stromsystem in Süddeutschland. Studie von prognos im Auftrag der sonnen GmbH*, Berlin. Accessed June 25, 2016, from https://www.sonnenbatterie.de/de/neue-prognos-kurzstudie-nutzen-von-heimspeichern-fuer-geringeren-netzausbau-bislang-vollkommen

Petersen N. H. (2014). *Belectric startet Stromspeicher am Solarpark Alt Daber*. Accessed June 25, 2016, from http://www.photovoltaik.eu/Archiv/Meldungsarchiv/Belectric-startet-Stromspeicher-am-Solarpark-Alt-Daber,QUlEPTYyMDk4OCZNSUQ9MTEwOTQ5JlBBR0U9MQ.html

Sterner, M., Eckert, F., Thema, M., & Bauer, F. (2015). *Der positive Beitrag dezentraler Batteriespeicher für eine stabile Stromversorgung*. Forschungsstelle Energienetze und Energiespeicher (FENES) OTH Regensburg, Kurzstudie im Auftrag von BEE e.V. und Hannover Messe, Regensburg, Berlin, Hannover.

WEMAG. (2016). *Eine Batterie für alle Fälle: WEMAG-Speicher zeigt Schwarzstartfähigkeit*. WEMAG Press release. Accessed June 25, 2016, from https://www.wemag.com/ueber_die_wemag/presse/pressemeldungen/2016/20160127_Batteriespeicher_Schwarzstartfaehigeit

Weniger, J., Bergner, J., Tjaden, T., & Quaschning, V. (2015). *Dezentrale Solarstromspeicher für die Energiewende*. Hochschule für Technik und Wirtschaft Berlin, BWV Berliner Wissenschafts-Verlag GmbH, Berlin. Accessed June 25, 2016, from http://pvspeicher.htw-berlin.de

Weniger, J., Bergner, J., Tjaden, T., & Quaschning, V. (2016). *Effekte der 50%-Einspeisebegrenzung des KfW-Förderprogramms für Photovoltaik-Speichersysteme*. Hochschule für Technik und Wirtschaft Berlin, Berlin. Accessed June 25, 2016, from http://pvspeicher.htw-berlin.de

**Benjamin Schott** graduated in business and chemistry at the University of Ulm in 2007. He obtained his Ph.D. in Economics in 2016 based on an analysis of policy-driven market entry strategies of electric vehicles in Germany. During his professional career he held several positions in business development, consulting and research, for example as new business development manager in the chemical industry or as project manager at the battery research and consulting institute Zentrum für Sonnenenergie- und Wasserstoff-Forschung in Ulm. Since 2014 he has been leading the Business Innovation Team at sonnen GmbH, one of the pioneers in the battery storage market with a broad product range.

**Oliver Koch** is Managing Director and Chief Operating Officer (COO) of sonnen GmbH. He joined the company in 2014. Before joining sonnen, he was COO at the US financial services and solar company Paramount Equity which was successfully sold to SolarCity in late 2013. Prior to his engagement in the solar industry, he held various management positions in international subsidiaries of the Bertelsmann Group in Asia, North America and Australia. Oliver Koch holds a Master's Degree in Sustainability Leadership from the University of Cambridge. In addition, he is an alumnus of Harvard Business School's AMP program and holds a Master of Business Administration degree from the University of Münster.

# Part III

# Marketing of Renewables in Regional Markets

# Exploiting the Economic Opportunities of the Energy Transition

Claudia Kemfert

**Abstract**

This paper provides a concise view on the state of the German Energiewende (energy transition), converting energy generation from fossil fuels to renewable generation. It discusses various aspects of the regulatory framework, including the phasing out of coal-fired power plants, emission trading, the new tender process for investments in subsidized renewable power generation, and other aspects. It also discusses employment effects and other impacts on German society at large.

**Keywords**

German Energiewende (energy transition) • Nuclear phase-out • Energy efficiency • Regulatory framework • Tender process

## 1 Introduction

In times of falling oil prices, fears of deflation, and disputes over European Union (EU) austerity policies, a major project, Germany's so-called energy transition (*Energiewende*), is being completely overlooked. The goal of the energy transition is to increase the share of electricity produced from renewable energy sources from almost 30 % today to 80 % by 2050. The nuclear power plants that are still in operation, mainly in Southern Germany, will be decommissioned by 2022. The energy transition also aims to improve energy efficiency in the construction sector and to make mobility more sustainable. As a result, the energy transition should result in a permanently sustainable energy supply.

C. Kemfert (✉)
German Institute for Economic Research (DIW), Berlin, Germany

Hertie School of Governance, Berlin, Germany
e-mail: sekretariat-evu@diw.de

In 2010, the German government initiated a new Energy Concept for a substantial transition of energy use to reduce carbon emissions in all these sectors simultaneously, which adjusts previous strategies and climate policy packages. An additional policy push came after the nuclear catastrophe in Fukushima. A societal and political consensus in Germany emerged, which determined that the risks of nuclear energy and the burden of final storage of nuclear waste were too high. Although in 2010 nuclear power accounted for more than 22 % of Germany's electricity, in July 2011 (3 months after the Fukushima disaster) the German government decided to completely phase out nuclear power generation in Germany within 10 years. Since that time the nuclear phase-out has been an integral part of the German energy transition – a transition to carbon neutral energy supply.

The concept aims at ensuring "a reliable, economically viable and environmentally sound energy supply"—the so-called energy policy triangle—and is connected to the following targets (see also Table 1):

- Complete nuclear phase-out by 2022;
- Significant increase in energy efficiency in all sectors, resulting in significant energy savings;
- Substantial increase in the share of renewable energies in satisfying final energy consumption;
- Reduction of $CO_2$ emissions by 40 % by 2020 compared to 1990.

The power generation structures will undergo substantial transformation, moving toward more decentralized energy supply structures in which renewable energy sources, combined heat and power systems, and smart grids and storage solutions will all be interconnected. This will also require effective load management capable of efficiently coordinating supply and demand. All these developments will generate numerous innovations, opening up future markets through investment.

The task of the energy transition is therefore to restructure the current power system with a view to more decentralization, flexibility, and dynamism, and this will include smart grids, optimum control of supply and demand, and more storage in the medium term. There is currently a huge excess in power supply capacity from old coal and nuclear power plants and, at times, from renewable energy sources. These excess capacities push prices downward on the electricity stock exchange market, meaning Germany is exporting cheaper electricity. This oversupply, in combination with low power market prices, has consequences: the profitability of conventional power plants is diminished. For this reason and because $CO_2$ prices are at historically low levels, lignite-fired power plants are still economically viable. Consequently, the use of lignite-fired power plants has risen and, at the same time, so too have greenhouse gas emissions. As a result, the business model and the innovative and flexible gas power plants and pumped storage power plants required for the energy transition are not viable owing to the massive excess power supply and associated low electricity prices on the energy exchange market.

If the German government is serious about achieving its climate targets, building energy, mobility, and the power sector in particular must make a substantial

**Table 1** Ambitious targets versus status quo (BMWi 2015; Löschel et al. 2015)

| | 2013 | 2020 | 2030 | 2040 | 2050 |
|---|---|---|---|---|---|
| Greenhouse gas emissions (compared with 1990) | −22.6 % | −40 % | −55 % | −70 % | −80 %– −95 % |
| Renewable energy | | | | | |
| Share in gross final energy consumption | 12.0 % | 18 % | 30 % | 45 % | 60 % |
| Share in gross electricity consumption | 25.3 % | 35 % | 50 % | 65 % | 80 % |
| Energy efficiency | | | | | |
| Primary energy consumption (compared with 2008) | −4.0 % | −20 % | – | – | −50 % |
| Energy productivity (final energy consumption) | 0.26 % p.a. (average 2008–2013) | 2.1 % p.a. | | | |
| Gross electricity consumption (compared with 2008) | −3.3 % | −10 % | – | – | −25 % |
| Thermal refurbishment of residential buildings | ~1 % p.a. (2012 value) | 2 % p.a. | | | |
| Final energy consumption of transport sector (compared with 2005) | 1.0 % | −10 % | – | – | −40 % |

contribution to emission reductions. This can happen only if old and inefficient coal-fired power plants are replaced with renewable energy sources, combined heat and power systems, and gas power plants. Outdated, inefficient coal-fired power plants not only produce an enormous power supply surplus but also generate too many greenhouse gases. Moreover, they are too inflexible in combination with renewable energy sources. Despite claims to the contrary by proponents of coal-fired power plants, they are not suitable as a bridging technology for a sustainable energy transition. Gas power plants are much better suited for that because they are more flexible than coal-fired power plants and generate fewer harmful greenhouse gases. However, these new gas power plants, which are efficient and therefore important for the energy transition, are increasingly standing idle because they are not profitable.

Instead of paying new subsidies for fossil-based energy sources, the electricity market should be restructured. Only a better market clearance will achieve the scarcity prices needed on the electricity market to improve the situation again. If the oldest, most inefficient coal-fired power stations were to disappear from the market, this could pay a double dividend: the market would be transformed, electricity prices on the energy exchange market and the profitability of the remaining power plants would rise again, and climate targets could be met.

The current EU Emissions Trading System (ETS) is a complete failure of effective climate policy: emissions trading still suffers from too high allocations of emission allowances in the early years and from the economic slump, static emission reduction targets, and an increase in additional certificates from abroad.

Even if the EU ETS were to be repaired as envisaged by the EU and, fortunately, supported by Germany, the $CO_2$ price would not rise to a sufficiently high level for the coal surplus on the German electricity market to disappear. These measures are likely to have little impact on the already very low $CO_2$ price, and as a result, it is expected to remain low. Instead of 7 euros per ton of $CO_2$, between 40 and 60 euros per ton of $CO_2$ would be required to provide sufficient financial incentives for the use of gas power plants instead of coal-fired plants. Consequently, there is a lack of suitable market signals for an increase in the price of $CO_2$.

## 2 Renewable Energy Sources—Problem Solver and Economic Factor

Renewables are the building block of a sustainable energy supply: they are climate friendly, ensure supply security as a domestic energy producer, and can also increase competitiveness as a stimulus for growth and jobs. Renewable energy sources are of interest for all energy sectors: for power and heat generation and as alternative fuels in the transport sector. The renewables industry promotes expansion and innovation and has become a growth industry like no other sector in recent years (Blazejczak et al. 2013; OECD 2010). The majority of employees work in the wind energy and biomass power generation fields, followed by the solar and geothermal industries. In the course of the political process of phasing out nuclear energy and reducing the high $CO_2$ emissions from coal-fired power plants, renewable energy sources can also make an outstanding contribution to combating climate change and to improving the security of supply by reducing our dependence on imports. Because renewables are usually used for decentralized energy supply by means of wind turbines, biomass plants, and combined heat and power, the use of renewable energy sources increases the security of supply. There are plenty of opportunities for expanding renewables. Depending on how global demand for renewable energy sources develops, the export potential can be enhanced considerably.

## 3 New Market Integration of Renewables—More Problems Than Solutions

A key objective of the German government is to develop renewable energy sources at the lowest cost. In addition, the pace at which renewable energy sources grow should be kept within the expansion corridor. To date, renewables have been supported by fixed feed-in tariffs. Since the costs of renewable energy sources have plummeted, the intention is now to switch to a tender system—also in accordance with EU regulations. Tenders are to be introduced for both onshore and offshore wind turbines and for solar plants with an output of at least 1 MW, thereby making it possible to calculate and specify tariffs and a sliding market premium. This approach should allow 80 % of all renewable energy systems to be determined through competitive tenders in the future.

The main objectives of the tender process are therefore reduced costs through lower subsidy rates, better compliance with the expansion corridor, limited influence of interest groups, and greater acceptance of the expansion of renewable energy generation and, consequently, the energy transition. In theory, tenders provide various options for combining competitive pricing with elements of control. They would, for example, allow system compliancy, that is, how renewables can relieve pressure on the energy system, or regional aspects, to be included in the award criteria.

As tempting as it might sound to use the tender process to automatically achieve low costs and compliance with the expansion corridor, its successful implementation entails difficulties. The devil is in the details. Essentially, determining subsidies competitively through tendering can work only if there is adequate competition. There must be enough investors compared to the amount of land available to be used for renewable energy sources. To encourage enough bidders, the process must be comprehensible and the preliminary work and securities that bidders are expected to provide must be affordable. The need to encourage the highest possible number of bidders to participate forces regulators to make the tender regulations attractive to bidders.

Against this background, it is worth taking a look at the second key objective of the tender: better compliance with expansion objectives. This is to be achieved by putting potential renewable energy expansion out to tender in several processes each year. Experience from other countries shows that low requirements with regard to preparatory work and securities and low financial penalties for noncompliance lead to a high proportion of bidders failing to make use of the permission they have been granted to build plants (Grau 2014; GIZ and Ecofys 2013). This would mean that the expansion targets will not be met. Securities and preparatory work are thus urgently required in order for development objectives to actually be achieved (Deutscher Bundestag 2016). They do, however, increase the cost of participation and inevitably lead to a low number of participants. Little competition in the tenders also means higher funding rates, that is, higher costs. In addition, if potential investors are willing or able to bear risks associated with the tenders to varying degrees, this will not lead to an efficient allocation of licenses because efficient bidders may decline to participate from the outset.

Consequently, there is a clear conflict of interest between the two key objectives of introducing a tender process: automatically generating low subsidy rates, and thus lower costs, and meeting expansion targets. Hence, it is by no means certain that costs will fall and that expansion objectives will be achieved more efficiently. Financial risks are presumably factored in, and low participation due to high barriers to entry could cause prices to rise. Lowering qualification requirements and penalties, on the other hand, involves the risk that not all approved plants will actually be constructed and planned expansion corridors will be missed by a considerable margin.

The additional objective of addressing quality aspects such as demand-driven electricity generation, regional criteria, or maintaining the variety of participants would seem very difficult to implement in this context. In particular, small investors and citizen energy cooperatives that could have stimulated the market to date could lose out in the new system. This is confirmed by the first experiences with tender schemes of photovoltaic open area tenders: first tender rounds have shown that most often big bidders and almost no cooperatives participate. Although quite high competition and low prices might have been reached, the actual realization rate remains unknown. First experiences with tendering schemes in Germany do not allow for any conclusion to be drawn but provide first insights into the potential challenges (Klessmann et al. 2015).

Although tenders may increase transparency and identify the most attractive provider among participating bidders, there are, nevertheless, many disadvantages: there is a great deal of red tape involved, there are conflicts between the objectives to be met by the tenders, and strategic behavior in combination with low participation can be a problem and push up bid prices. Bidder transaction costs will rise and are probably not affordable for all stakeholders. There are also bidding risks such as time and effort spent carrying out preparatory work without any certainty of ultimately being awarded the contract.

Tenders do not automatically lead to lower costs, and there is a high risk of not meeting renewable energy expansion targets (Held et al. 2014). In addition, there is a danger that the variety of participants, and therefore acceptance of the energy transition, will diminish. Tenders can bring benefits. However, many challenges remain in their specific design.

## 4     An Energy Transition "Made in Germany"

The "made in Germany" energy transition is an important contribution to global climate policy. However, Germany is not at the top of the energy transition rankings but is only mid-tier behind countries such as Sweden, Brazil, and Italy (Fig. 1).[1]

However, the German energy transition has both upsides and downsides. On the one hand, Germany has managed to considerably reduce emissions in the power sector through its expansion of renewables. On the other hand, Germany still uses too much coal for power generation (Kemfert 2013). This means that Germany's self-imposed climate target of a 40 % reduction by 2020 will most likely not be achievable. There are no binding climate policy measures, and emissions trading in

---

[1]See Fraunhofer ISE, *The Energy Transformation Index* https://www.ise.fraunhofer.de/de/downloads/pdf-files/aktuelles/ise-ises-eti.pdf (2015), and IEA (2014). The Climate Change Performance Index also sees Germany only in the middle tier on issues such as emissions development, expansion of renewables, improving energy efficiency, and policy measures (Burck et al. 2016). In contrast, other studies like the Global Green Economy Index (GGEI) see Germany alongside Sweden in first place on global green markets (Tamanini et al. 2014). A study by the Handelsblatt Research Institute sees Germany in eighth place out of 24 countries.

## ETI Ranking of Selected Countries
### ETI Growth between 1990 and 2013

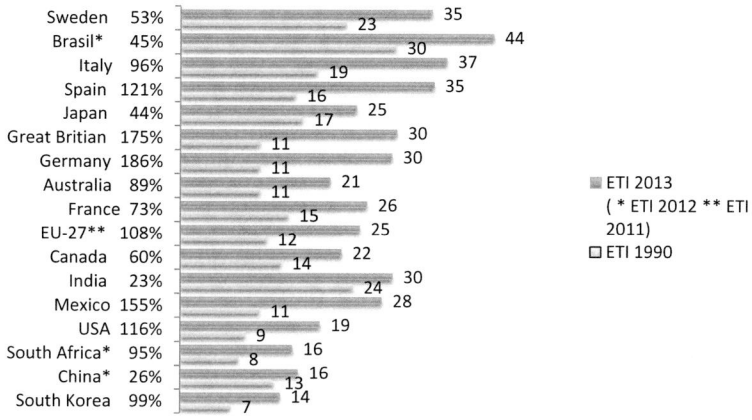

**Fig. 1** Energy Transformation Index (ETI) ranking of selected countries: ETI growth between 1990 and 2013 (Fraunhofer ISE 2014). **The Fraunhofer Institute for Solar Energy Systems (ISE)** developed the **Energy Transformation Index (ETI) in 2013.** It measures the progress of countries toward an energy transition: https://www.ise.fraunhofer.de/de/presse-und-medien/presseinformationen/presseinformationen-2013/energy-transformation-index-eti

its current form is ineffective because the $CO_2$ price is far too low. Therefore, accompanying measures are required, such as a carbon tax and a structured fossil-fuel phase-out. In addition, Germany has not done enough in the area of sustainable mobility; the recent VW diesel scandal is a stark example of environmental protection that is made in Germany. If Germany is to be celebrated as a model for climate policy and Chancellor Merkel wants to be known as the "climate chancellor," the carbon problem must be solved, and we also need to do more to save energy and restore our credibility, particularly in the area of sustainable mobility. Consequently, policymakers are now being called upon to implement measures that combat climate change more effectively but that run counter to economic interests. Nevertheless, the German energy transition policy has initiated a global upheaval, and perhaps Germany can learn more from other countries when it comes to issues like the fossil-fuel phase-out or properly measuring exhaust emissions.

The German energy transition is nevertheless setting an important precedent: thanks to investments from Germany, the rising demand and associated economies of scale have caused the costs of renewable energy sources to fall massively worldwide. For the first time, more global investment is being made in renewables than in fossil fuels.

More and more countries will follow the German example and prefer to invest in renewable energy sources than fossil fuels or atomic energy: more opportunities, fewer risks.

## 5    Conclusion

The energy system must become more flexible, more intelligent, and more holistic. This will require smart grids and, in the medium term, more storage than fossil fuels and outdated structures. Germany has made a substantial contribution to the success of the global energy transition, with more investment now being made in renewables than fossil fuels. By promoting renewable energy sources, Germany has contributed to a situation in which the costs of renewable energy sources continue to fall and in which renewables are thus becoming more competitive. Germany should not jeopardize these successes. Putting the promotion of renewable energy sources out to tender comes with huge risks and threatens to stifle the energy transition.

The energy transition provides enormous economic opportunities. Today five times as many people work in the renewables sector in Germany as in the coal industry (Fig. 2) (Kemfert et al. 2015). If we add energy efficiency in general to the field of renewable energy sources, the number of employees rises even more. It is of the utmost importance to continue to implement the structural changes needed to convert the energy supply to renewables and greater energy efficiency and to monitor them in the coming decades. This is the only way Germany can continue to be the role model for a sustainable global energy transition in the future.

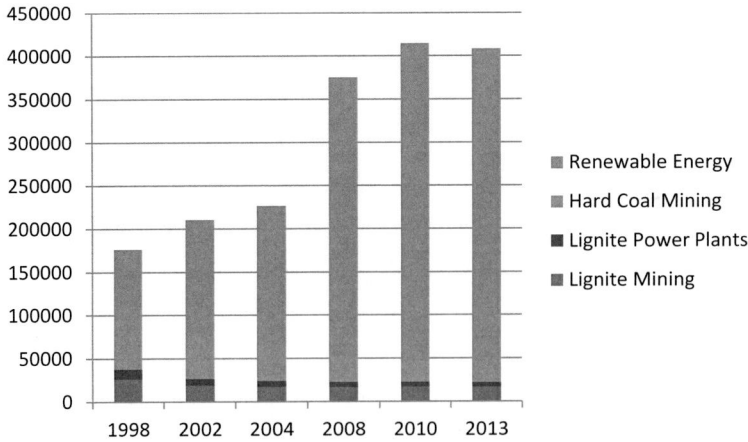

**Fig. 2** Employment in coal and renewable energy sector (Ulrich and Lehr 2014; Statista 2016; Kohlenstatistik 2015)

# References

Blazejczak, J., Diekmann, J., Edler, D., Kemfert, C., Neuhoff, K., & Schill, W.-P. (2013). Energy transition calls for high investment. *DIW Economic Bulletin*, 9, p. 11.

BMWi. (2015). *Vierter Monitoring-Bericht "Energie der Zukunft"*.

Burck, J., Marten, F., & Bals, C. *Climate Change Performance Index 2016*. Accessed July 17, 2016, from https://germanwatch.org/en/download/13626.pdf

Deutscher Bundestag. (2016). *Ausschreibungsbericht nach § 99 des Erneuerbare-Energien-Gesetzes*. Drucksache 18/7287. Accessed July 17, 2016, from http://dip21.bundestag.de/dip21/btd/18/072/1807287.pdf

Fraunhofer ISE. (2014). *Energy Transformation Index 2013 Freiburg 2014*. Accessed July 17, 2016, from https://www.ise.fraunhofer.de/de/downloads/pdf-files/aktuelles/ise-ises-eti.pdf

GIZ and Ecofys. (2013). *Lessons for the tendering system for renewable electricity in South Africa from international experience in Brazil*. Morocco and Peru, Berlin. Accessed July 17, 2016, from http://www.ecofys.com/files/files/ecofys-giz-2013-international-experience-res-tendering.pdf

Grau, T. (2014). *Comparison of feed-in tariffs and tenders to remunerate solar power generation*. DIW Discussion paper 1363. Accessed July 17, 2016, from https://www.diw.de/documents/publikationen/73/diw_01.c.437464.de/dp1363.pdf

Held, A., Ragwitz, M., Gephart, M., De Visser, E., & Klessmann, C. (2014). *Design features of support schemes for renewable electricity*, Utrecht.

IEA. (2014). *International Energy Agency (2014): Key World Energy Statistics 2014, country ranking, energy consumption, $CO_2$-emissions: Paris*.

Kemfert, C. (2013). *Battle about electricity, myths, power and monopolies*. Hamburg: Murmann.

Kemfert, C., Opitz, P., Traber, T., & Handrich, L. (2015). *Deep decarbonization in Germany a macro-analysis of economic and political challenges of the 'Energiewende' (Energy Transition)*. DIW Berlin Politikberatung Kompakt 93/2015.

Klessmann, C., Tiedemann, S., Wigand, F., Tobias Kelm, T., Winkler, J., Ragwitz, M., et al. (2015). *Ausschreibungen für PV-Freiflächenanlagen Auswertung der ersten zwei Runden*, Berlin. Accessed July 17, 2016, from https://www.erneuerbare-energien.de/EE/Redaktion/DE/Downloads/ausschreibungen-pv-freiflaechanlagen-auswertung-erste-zwei-runden.pdf?__blob=publicationFile&v=3

Kohlenstatistik. (2015). *Statistik der Kohlenwirtschaft e.V. (2015) Datenübersichten zu Steinkohle und Braunkohle in Deutschland 2014*. Accessed July 17, 2016, from http://www.kohlenstatistik.de/3-0-Uebersichten.html

Löschel, A., Erdmann, G., Staiß, F., & Ziesing, H.-J. (2015). *Expertenkommission zum Monitoring-Prozess "Energie der Zukunft". Stellungnahme zum vierten Monitoring-Bericht der Bundesregierung für das Berichtsjahr 2014*. Accessed July 17, 2016, from https://www.bmwi.de/BMWi/Redaktion/PDF/M-O/monitoringbericht-energie-der-zukunft-stellungnahme-2014,property=pdf,bereich=bmwi2012,sprache=de,rwb=true.pdf

OECD. (2010). *Transition to a low-carbon economy: Public goals and corporate practices*. OECD. doi:10.1787/9789264090231-en. Accessed 17 July 2016.

Statista. (2016). *Erneuerbare Energie seit 1998*. Accessed July 17, 2016, from http://de.statista.com/statistik/daten/studie/156645/umfrage/anzahl-der-beschaeftigten-in-der-erneuerbare-energien-branche-seit-1998/

Tamanini, J., Bassi, A., Hoffman, C., & Valenciano, J. (2014). *Green Economy Index GGEI 2014 Strategic Review 2014*, Washington, New York. Accessed July 17, 2016, from http://dualcitizeninc.com/GGEI-Report2014.pdf

Ulrich, P., & Lehr, U. (2014). *Erneuerbar beschäftigt in den Bundesländern: Bericht zuraktualisierten Abschätzung der Bruttobeschäftigung 2013 in den Bundesländern*. Osnabrück: GWS mbH.

**Claudia Kemfert** has been professor of energy economics and sustainability at Berlin's Hertie School of Governance since 2009 and head of the Department of Energy, Transportation, Environment at the German Institute of Economic Research (DIW Berlin) since April 2004. Her research activities concentrate on the evaluation of climate and energy policy strategies. In 2016 Claudia Kemfert was appointed by the Federal Ministry for the Environment, Nature Conservation, Building and Nuclear Safety as a member of the German Advisory Council on the Environment. She advised European Union president José Manuel Barroso in a High-Level Group on Energy and Climate. In 2006, Claudia Kemfert was named a top German scientist by the Helmholtz and Leibniz Association, a German research foundation. In 2011 she was awarded the Urania Medaille as well as the B.A.U.M. environmental award for best science. In 2013, she published her book *The Battle about Electricity* in which she describes the myths of the energy debate.

# Building a Renewables-Driven Power System. Successes and Challenges in Germany

Patrick Graichen, Christian Redl, and Markus Steigenberger

**Abstract**

The German energy transition (*Energiewende*) is a long-term energy and climate strategy aimed at a low-carbon-energy system based on renewable energy and energy efficiency. The focus of the Energiewende so far has been on the power sector, especially the deployment of renewables. Wind energy and solar photovoltaics (PVs) form the backbone of the German Energiewende. Due to a dramatic price decrease in recent years, they are now mature technologies and cost-competitive with conventional energy sources for new investments. As wind power and solar PV systems are variable sources, flexibility is the new paradigm of the German power system. Baseload capacities are no longer needed: power markets and power systems are now built around variable renewable energy sources.

This chapter provides a fact-based overview of the German Energiewende. It explains the current status of the energy transition in Germany and outlines the challenges ahead.

**Keywords**

Energy transition • Decarbonisation • Market design • Renewables deployment

## 1   Introduction

When it comes to Germany's energy transition, it seems that there are only two opinions: it is either a curse or a blessing. Some praise it for creating jobs, producing clean and distributed energy, reducing import dependency, and

P. Graichen (✉) • C. Redl • M. Steigenberger
Agora Energiewende, Berlin, Germany
e-mail: patrick.graichen@agora-energiewende.de; christian.redl@agora-energiewende.de;
markus.steigenberger@agora-energiewende.de

© Springer International Publishing AG 2017
C. Herbes, C. Friege (eds.), *Marketing Renewable Energy*, Management for Professionals, DOI 10.1007/978-3-319-46427-5_15

minimising climate and nuclear risks. Others blame it for escalating costs and increased grid instability.

Although energy systems around the world are changing, the famous German Energiewende is attracting explicit attention—and for good reasons. With an overwhelming three-fourths majority, the German Parliament took steps in 2011 that can no longer be seen as temporary political phenomena but as a clear commitment to transform the energy system from fossil/nuclear to predominantly renewables based. Citizens firmly support the energy transition, as surveys regularly prove. The German Parliament's decisions are quite radical: Germany has decided to decarbonise its energy sector in favour of renewable energy and energy efficiency. No other decarbonisation technology, including nuclear or carbon capture and sequestration (CCS), is considered feasible in the domestic context. While many countries within the European Union (EU) share similar goals, the scope of the German transition is unique: Germany is the world's fifth largest economy and has a strong industrial basis, so no blueprint exists. Accordingly, the energy transition is a stepwise exploration of new territory for Germany. Obviously, decision-makers will encounter pitfalls, difficulties, and challenges, and, again unsurprisingly, a transition as fundamental as the one Germany is undertaking—from a fossil–nuclear to an almost entirely renewable energy supply—is creating uncertainty. At times, market participants, consumers, and decision-makers feel threatened by the dimensions of this renewable energy transition.

Starting from this observation, we aim to provide insights on what Germany's energy transition is all about, describing the state of affairs, trends, and challenges.[1] The chapter focuses on the power sector, which many studies have shown will be crucial in this transition. To reach the EU's goal of reducing greenhouse gases by 80 to 95 % below 1990 levels by 2050, the EU's power system will have to be completely carbon-free by 2050 (e.g. ECF 2010; European Commission 2011). Thus, in this chapter we first describe the most important developments and the current state of affairs. Then we identify the key characteristics of the transition in the power sector resulting from high proportions of variable renewable sources in Germany's future energy mix. We conclude by briefly discussing the key challenges and taking a look at upcoming developments.

## 2    Targets and State of Affairs of the Energy Transition

With the energy concept of 2010 and the legislative package of 2011, Germany has decided on a set of concrete goals in different energy sectors. The track record of the energy transition can best be judged against these targets. Although officially no hierarchy exists, two main goals can be identified[2]:

---

[1]This chapter is based on Agora Energiewende (2015a, 2016) and Graichen and Steigenberger (2016).

[2]This classification into goals of first and second order was first developed by the independent expert group for monitoring of the energy transition; see http://bmwi.de/DE/Themen/Energie/Energiewende/monitoring-prozess.html.

- Reducing greenhouse gas emissions by 40 % by 2020 and 80–95 % by 2050 (vs. 1990 levels) and
- Phasing out nuclear energy by the end of 2022.

To achieve these two main goals, a set of additional targets, policies, and instruments has been established over the years. Among the targets, two stand out:

- Increasing the share of renewable energy and
- Using energy more efficiently.

Table 1 summarises the main targets.

Progress towards achieving these targets has been made in recent years in Germany, albeit to varying extents. By the end of 2014, greenhouse gas emissions were reduced by roughly 28 % over 1990 levels but increased slightly from 2014 to 2015. More than a third of the emission reductions occurred between 1990 and 1995 and can thus be attributed both to the industrial breakdown in eastern Germany after 1990 and the modernisation of the inefficient power plants of the former German Democratic Republic.

Regarding the nuclear phase-out, since 2000, 11 GW of nuclear capacity has been shut down. According to the phase-out law passed by a 90 % majority in

**Table 1** German energy and climate policy targets and status quo

|  | Status quo | 2020 | 2025 | 2030 | 2035 | 2040 | 2050 |
|---|---|---|---|---|---|---|---|
| Greenhouse gas emissions (vs. 1990) | −27.2 % (2015) | −40 % | – | −55 % | – | −70 % | −80 to −95 % |
| Nuclear phase-out | 11 power plants shut down since 2000 | Stepwise phase-out of remaining 8 power plants by end of 2022 | | | | | |
| Overall renewable energy (share in consumption) | 13.8% (2014) | 18 % | – | 30 % | – | 45 % | ≥60 % |
| Electric renewable energy (share in electricity consumption) | 31.5 % (2015) | – | 40–45 % | – | 55–60 % | – | ≥80 % |
| Primary energy efficiency (primary energy use, vs. 2008) | −8.9 % (2014) | −20 % | – | – | – | – | −50 % |
| Electric energy efficiency (electricity demand, vs. 2008) | −3.4 % (2015) | −10 % | – | – | – | – | −25 % |

Targets according to energy concept of 2010, except for nuclear target (taken from Nuclear Phase Out Act 2011) and share of renewable energy in electricity (taken from Renewable Energy Act 2014)

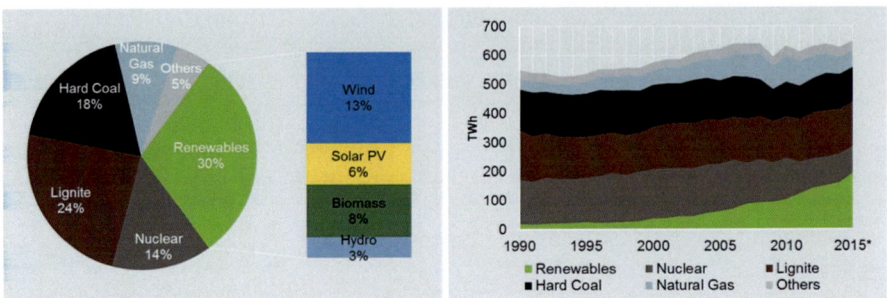

**Fig. 1** Share in gross electricity generation by fuel 2015 (*left*) and gross electricity generation by fuel 1990–2015 (*right*). 2015 figures are preliminary (author's illustration based on AGEB 2016)

Parliament in 2011, the remaining roughly 11 GW nuclear power plants will be shut down in a stepwise approach (2017, 2019, 2021) by the end of 2022 at the latest.

In the field of energy efficiency, the target translates into an annual increase in energy productivity of 2.1 %. Between 1990 and 2013, the overall efficiency of the German economy increased by 1.7 % per year, thus falling short of the target. Moreover, progress has slowed in recent years. While overall efficiency increased by 2.2 % between 1990 and 2000, the rate dropped to 1.3 % between 2000 and 2013 (Blazejczak et al. 2014; AG Energiebilanzen 2013).

Finally, renewable energy has grown significantly. In 1990, renewables accounted for approximately 3 % of Germany's electricity consumption and 2 % of overall energy consumption. Today, gross consumption from renewable sources across all sectors is at 14 % (2015), with the electricity sector reaching 31.5 % in 2015 (AG Energiebilanzen 2015). Renewables development changed the power market structure tremendously: Triggered by the main support instruments—feed-in tariffs and feed-in premia—almost 50 % of today's renewable generation is owned by citizens—a fact that is considered one of the reasons why the energy transition remains highly popular in Germany (Trend Research 2014). Figure 1 shows that in 2015 renewables were the most important source in the electricity system, followed by lignite and hard coal.

## 3    Second Phase of German Energy Transition: Towards 50 % Renewables in Germany's Power System

Breaking the 30 % threshold of renewable electricity in final power consumption in 2015 symbolically represented entering into the second phase of the energy transition. While the first phase was characterised by the innovation and development of a variety of renewable technologies, the current debate is about wind and photovoltaics (PV) becoming the dominant generation source. This is bringing about fundamental changes and challenges to the power system. We will describe them in the following sections in more detail.

## 3.1    It Is All About Wind Power and Solar PV

Wind energy and solar PV form the backbone of the German Energiewende. Owing to a dramatic price decrease in recent years, they are now mature technologies and cost-competitive with conventional energy sources for new investments. In 2015, generation costs in Germany ranged between EUR 0.06 and 0.09/kWh for wind energy[3] and EUR 0.08 and 0.09/kWh for solar PV. Costs have decreased further also in 2016. For example, the recent tender for ground-mounted PV in Germany (December 2016) has cleared at an average remuneration of EUR 0.069/kWh. All other renewable technologies are either significantly more expensive or have limited potential in Central Europe. Figure 2 shows the development of gross electricity generation from renewables from 2000 to 2035. Development till 2035 is based on the main scenario according to the German regulatory authority.

The technological development of wind power and solar PV systems has been rapid in recent years (Fig. 3). Today's wind turbines are 20 times more powerful than those 20 years ago (average 3 MW instead of 170 kW), and costs for solar PV have fallen by up to 90 % in the last 25 years. Furthermore, the end of the learning curve has not yet been reached.[4] The potential for both technologies is significant. Even in densely populated Germany, 1200 GW of wind power onshore could theoretically be installed (UBA 2013). The technical and ecologically sound potential of PV is estimated at about 275 GW (UBA 2011).

In contrast, other renewables will not be able to increase their share in the power mix significantly. Biomass today already accounts for 8 % of Germany's electricity generation. However, projections expect it to remain below 10 % in the long run. The reason is resource constraints—acreage is limited, and the use of wood and

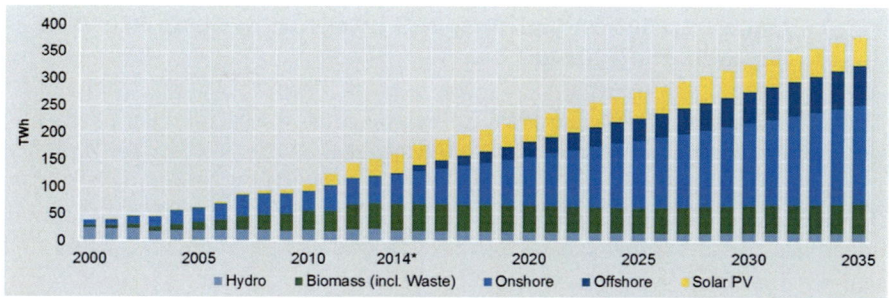

**Fig. 2** Gross electricity generation of renewable energies, 2000–2035 (author's illustration based on 2000–2014: AGEB 2015; 2015–2035: author's calculation based on BNetzA 2015)

---

[3]The dominant technology is wind onshore. While the installed capacity of wind offshore was less than 1 GW in 2013, the government is planning to increase this figure to 6.5 GW by 2020 and 15 GW by 2030.

[4]For projections of future cost developments of renewable technologies see, for example, Fraunhofer ISE (2013) and Hirschhausen et al. (2013).

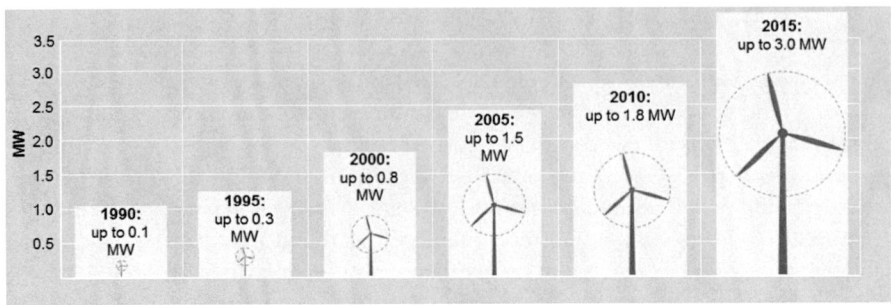

**Fig. 3** Size development of wind turbines 1990–2015 (author's illustration based on IEA 2014)

energy crops directly competes with other needs, especially raising food crops or raw materials for industry—in Germany and other countries. The quantity of low-cost biomass is more or less exhausted, and remuneration levels for biomass plants in the German support scheme are rising rather than falling.[5] Hydropower currently contributes approximately 4 % to Germany's electricity mix and will remain on this level because the potential for expanding hydropower capacity is very limited in Germany. The situation with geothermal is similar; limited potential and high costs suggest that this technology will not increase its share significantly. Other technologies, such as wave power or osmosis, are still in the development stage. Whether they will ever be able to play a significant role is unclear.

### 3.2 Wind Onshore and Solar PV Are the Cheapest Decarbonisation Options, and Integration Costs Are Well Defined and Rather Low

Today, wind onshore and solar PV are cost-competitive with all other newly built conventional energy sources (in terms of levelised costs of electricity generation, or LCOE), with generation costs in Germany ranging in 2015 between EUR 0.06 and 0.09/kWh for wind and EUR 0.08 and 0.09/kWh for solar PV (Fig. 4). Furthermore, additional cost decreases can be expected, especially for solar PV, with LCOE ranging from EUR 0.055 to 0.08/kWh by 2025.[6]

From a system perspective that considers the costs of integrating variable wind and PV technologies into the power system, the picture does not change substantially. Integration costs of adding wind onshore or solar PV into the German system, even at high penetration rates, may range around EUR 0.005 to 0.02/kWh (see Fig. 5).

---

[5]In 2002, basic remuneration for small biomass plants was EUR 0.101/kWh, while in 2012 it was EUR 0.143/kWh.

[6]Fraunhofer ISE (2015).

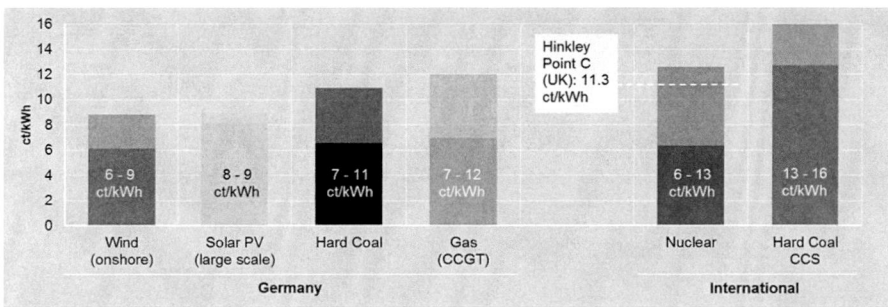

**Fig. 4** Range of levelised cost of electricity (LCOE) 2015. The range is based on varying utilisation, $CO_2$ price, and investment cost (Agora Energiewende 2015a, b)

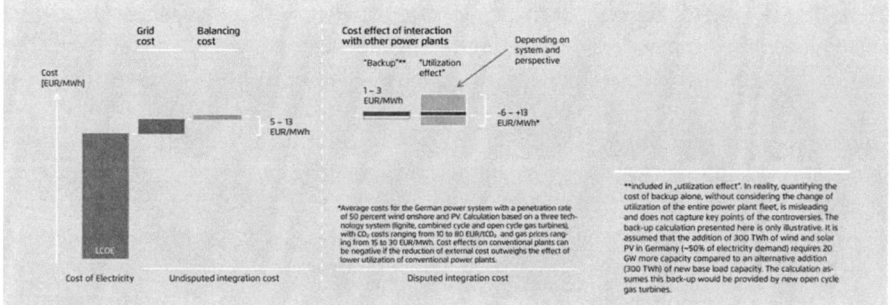

**Fig. 5** Components of integration costs (Agora Energiewende 2015a, b)

Three components are typically discussed under the term *integration costs* of wind and solar energy: grid costs, balancing costs, and cost effects on conventional power plants (so-called utilisation effect).[7] The calculation of these costs varies widely depending on the specific power system and methodologies applied. Moreover, opinions diverge concerning how to attribute certain costs and benefits, not only to wind and solar energy but to the system as a whole.

Integration costs for grids and balancing are well defined and rather low. Certain costs for building electricity grids and balancing can be clearly classified without much discussion as costs that arise from the addition of new renewable energy. In the literature, these costs are often estimated at EUR 5 to 13/MWh, even with high shares of renewables.

However, experts disagree on whether the utilisation effect can (and should) be considered as integration costs, as it is difficult to quantify and new plants always modify the utilisation rate of existing plants. When new solar and wind plants are

---

[7]For further details see IEA (2014).

added to a power system, they reduce the utilisation of the existing power plants and, thus, their revenues.

Thus, in most cases, the cost for "backup" power increases. Calculations of these effects range between EUR $-6$ and $+13$/MWh in the case of Germany at a penetration of 50 % wind and PV, depending especially on the $CO_2$ cost. Despite the debate about integration costs, the comparison of the total power system costs of different scenarios is a more appropriate approach to analysing the question: What are the implications of choosing path A or path B?[8]

## 3.3    Flexibility Is the New Paradigm of Germany's Power System

The characteristics of PV and wind are radically different from those of fossil fuel power plants. Wind energy and solar PV have variable output and provide electricity only when the wind blows and the sun shines. Given their short-term variability, they cannot be turned based on the demand for electricity. Furthermore, they are characterised by high capital costs and virtually no operating costs. Once installed, wind and solar power plants produce electricity at almost zero marginal cost. Therefore, they change the utilisation patterns of the conventional generation fleet, encouraging less baseload operation and more middle and peak operation.

These features fundamentally alter power systems and power markets, which must cope with highly fluctuating power generation.[9] This new power system is characterised by enhanced flexibility to respond quickly to changes in variable generation and changes in the load. Baseload capacities will no longer be needed, but relatively more mid-merit and peak load capacities that quickly adjust their production will be. Fossil power plants will need to become very responsive; in essence, they will have to ramp up and down more frequently, operate often at partial loads, and be turned on and off with greater regularity. Figure 6 illustrates this need for flexibility for three sample weeks in 2023 for situations with both high and low shares of renewables generation.

A geographically widespread expansion of wind and solar PV will help to reduce the burden of increasing flexibility (Fig. 7). Wind and solar PV complement each other as their generation patterns are different. While solar radiation is strongest in summer and most sunshine occurs during mid-day, the wind can blow at any time, and it usually blows stronger in the winter in Europe.[10]

---

[8]A greenfield power system, for example, consisting of 50 % newly built wind and solar combined with 50 % newly built gas-fired power plants would yield total power generation costs of around EUR 70 to 80/MWh (including integration costs). These costs are 21 % lower than a system with the same emission performance but consisting of 50 % nuclear generation and 50 % gas-fired generation (Prognos 2014).

[9]At the latest, by 2030 renewables will provide some 50 % of total German electricity demand. At that time, wind power and solar PV will have a share of some 35 % in the generation mix.

[10]For further details, see Fraunhofer IWES (2015).

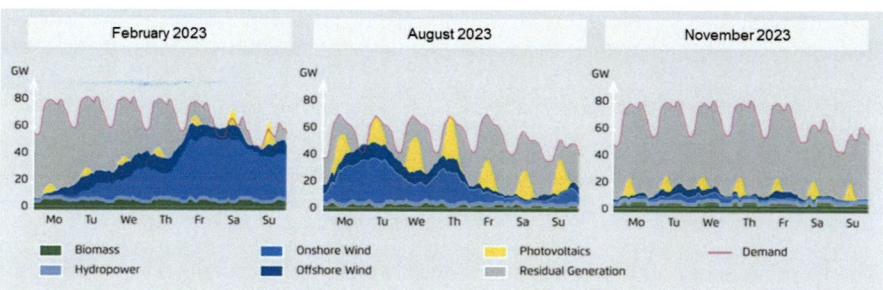

**Fig. 6** Electricity generation and consumption in three sample weeks, 2023. The modelling is based on 2011 weather and load data (Agora Energiewende 2013 based on Fraunhofer IWES)

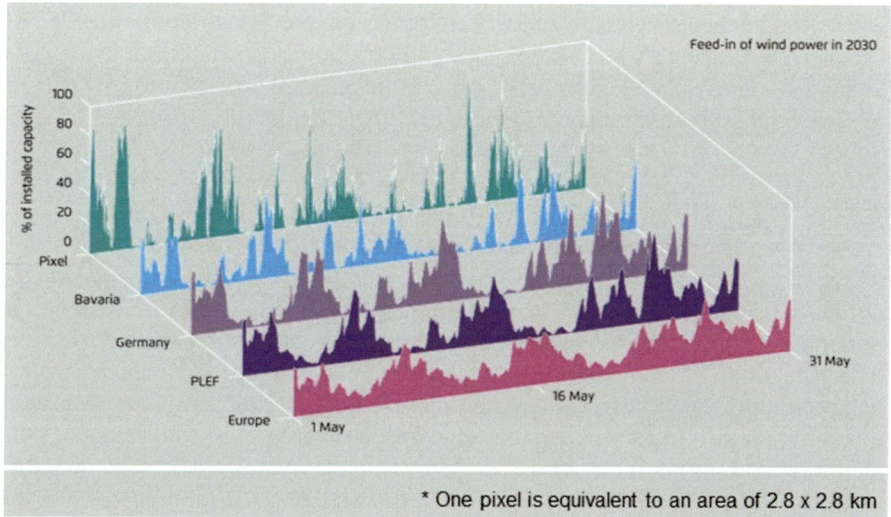

**Fig. 7** Time series of onshore wind power generation in a simulation for May 2030 at different levels of aggregation (as a percentage of installed capacity at specific aggregation level). Note that one pixel is equivalent to an area of 2.8 × 2.8 km. The Pentalateral Energy Forum (PLEF) region comprises Austria, Belgium, France, Germany, Luxembourg, the Netherlands, and Switzerland (Fraunhofer IWES 2015)

## 3.4    Power Systems Possess a Broad Range of Flexibility Options

In addition to flexible fossil power plants, several other flexibility options exist to incorporate variable energy sources in the power system. These include demand-side management, the expansion of (smart) grid infrastructure, bioenergy power plants, the temporary curtailment of wind and PV energy, new storage technologies, and new electricity demands from other sectors such as power-to-heat and electric cars.[11]

---

[11] For further details, see Agora Energiewende (2016).

Currently, the German power system offers abundant technical potential for flexibility (much higher than the actual demand for flexibility). Nevertheless, efficient market incentives need to be designed to translate the flexibility needs into market prices and leverage this technical potential in the most cost-efficient way.

## 4 Forthcoming Challenges

The German energy transition is a long-term industrial and societal transformation. Given the transformational nature of the project, the stakeholders of this energy transition (policymakers, consumers, utilities, industries) will face many new challenges and opportunities in the coming years and decades.

We will focus on the issues of grid infrastructure, reducing $CO_2$ emissions, improving energy efficiency, and scaling up regional cooperation in this chapter. Chapter "A Holistic Power Market Design Framework and Implications for Marketing Renewables" will then deal with the issue of power market design, financing of renewables, and their integration into the power markets.

### 4.1 Upgrading the Grid from North to South Is Crucial for the German and European Electricity Market

Grid expansion—both nationally and with neighbouring countries—constitutes an important flexibility option to balance volatile influx from wind and PV over long distances (Fig. 7). In geographically larger areas, supply and demand can be balanced more easily, especially when complementary production and consumption profiles exist. In addition, a well-developed grid enables the utilisation of other flexibility options across regions.

In Germany, the major challenge lies in accelerating the reinforcement of the grid on the South–North axis (Fig. 8). This is also important for European market integration, for example to avoid loop flows, especially in Central-Eastern Europe. The four German transmission system operators publish an annual national network development plan that contains the expansion and reinforcement measures required for the next 10 years to ensure stable and reliable operation of the grid. The grid development plan estimates a need for 3500 km of new transmission lines, of which 2000 km are allocated to the erection of four HVDC North–South corridors. This expansion need arises predominantly from the fact that the additional wind power in Germany will be built in the North close to the coast, while major industrial power consumers are situated in Southern Germany. Surplus power produced in the North can lead to unplanned power flows through the grids of Germany's neighbours to reach Southern Germany. Over the last few years, these unplanned flows have led to a decrease in commercial power exchange between Germany and Poland. This loop flow issue has been provisionally solved by establishing a virtual phase-shifter (binational re-dispatch mechanism), which will be followed in the next year by the introduction of physical phase-shifters on the German–Polish and later on the

**Fig. 8** Necessary expansion and restructuring of German electricity grid up to 2024, according to German network development plan NEP 2024 (Agora Energiewende 2015)

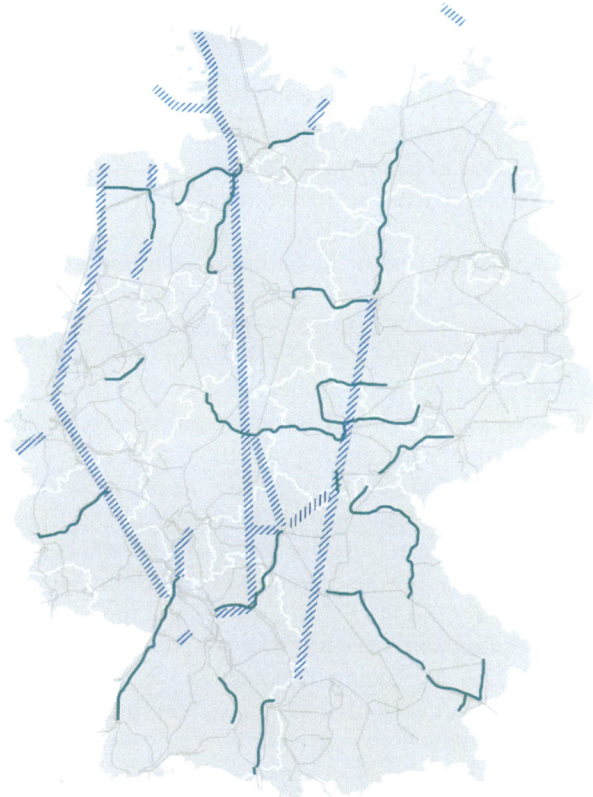

German–Czech borders—the first having been taken online in June 2016. Unsolved loop flow issues can slow down the process of integrating Europe's power markets.

Furthermore, expansion and reinforcement measures in the distribution network are absolutely necessary because a large part of wind energy onshore and PV are directly connected at this level. Grid developments often face acceptance problems from local populations, leading to delays in the deployment of this infrastructure. Building consensus at the local level through enhanced dialogue with a variety of stakeholders will be key for widening the necessary public acceptance for these projects.

## 4.2     Meeting Climate Targets

$CO_2$ emissions from the power sector fell sharply in 2014 (5 % reduction compared to 2013 levels) owing to favourable developments in renewable energy and energy efficiency, together with a mild winter and a decrease in power produced using hard coal (to its second lowest level since 1990). The emissions of the power sector are expected to decline further in a business-as-usual scenario by about 37 MtCO$_2$ by 2020. However, this declining trend in the power sector is not sufficient to meet 2020 reduction targets, making further reduction efforts necessary. A set of complementary policy measures has been discussed in Germany in order to fill the gap, resulting in policies to enable the early retirement of 2.7 GW of Germany's oldest lignite power plants by October 2019.

German coal power plants, especially lignite-based ones, are currently extremely competitive. This is the result of two factors. First, the European Emissions Trading System (ETS), the main European instrument for internalising the costs of CO$_2$ emissions in the power sector, is weak (as a consequence of the vast over-allocations of CO$_2$ certificates). Unless remedied, this will lead to persistent low CO$_2$ prices. Second, coal prices are currently very low on the world market (a side effect of the US shale gas revolution). This has led to an increasingly wide spread between cheap coal and expensive gas in Europe. As a result, German coal power plants produce at very high levels, contributing to historically high export levels and a crowding-out of gas power plants both in Germany and in neighbouring countries. This trend has led to an increase in CO$_2$ emissions between 2011 and 2013 in Germany, despite a rise in renewable energy during the same period (a counterintuitive development known as the "Energiewende paradox").

Germany and the EU as a whole need to fix their carbon policies, to reverse the coal resurrection witnessed recently all over Europe. But also strengthened domestic policies are required to reduce the large stock of emission-intensive coal capacities in the German power system. Indeed, Germany can only reach its 2020 emission reduction targets if it drastically reduces electricity generation from lignite and hard coal. An analysis from Agora Energiewende has shown that the shares of lignite and coal need to drop from 45 % in 2014 to at most 28 % in 2030 to meet the 2030 climate target. By 2040, a complete phase-out of coal would be required (Fig. 9).

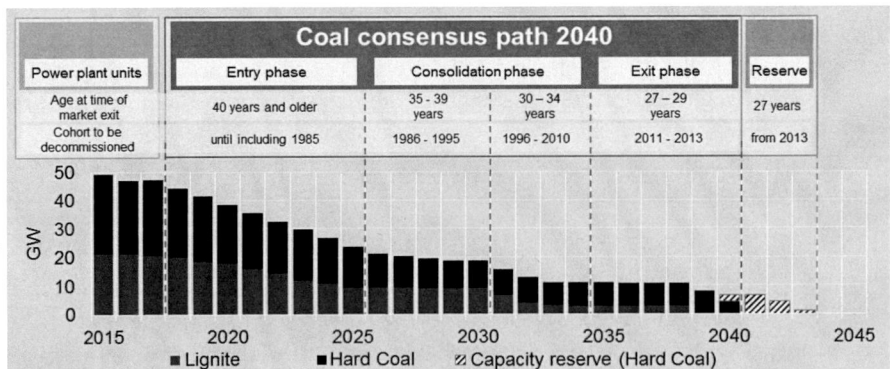

**Fig. 9** Installed lignite and hard coal capacities in a proposed Coal Consensus Path 2040 (Agora Energiewende 2016)

## 4.3    A Coherent Strategy and New Business Models Must Be Developed to Leverage the Potential of Energy Efficiency Measures

In addition to renewable energy targets, the Energiewende sets ambitious goals in terms of energy efficiency: Primary energy consumption is to be reduced by 20 % by 2020 and 50 % by 2050 (compared to 2008 levels). This requires an increase of energy productivity of 2.1 % per annum. In the electricity sector, consumption is to decrease by 10 % by 2020 and 25 % by 2050.

Although Germany has managed to decouple economic growth from energy consumption, further policies and measures are needed to consolidate recent trends in electrical efficiency and to speed up the decline in primary energy consumption.

There is already a broad mix of energy efficiency instruments and programmes in Germany, but more effort is needed. Market-based solutions for triggering investments in energy efficiency do not appear sufficient since business models that focus on energy efficiency are usually associated with high initial costs and long payback periods. A coherent strategy and additional instruments need to be implemented to support investments. One possibility—as offered by the EU Energy Efficiency Directive—is to require energy suppliers or network operators to implement efficiency obligation schemes (to comply with binding targets). In Germany there is still no majority to implement such obligations. Therefore, in its National Action Plan on Energy Efficiency (NAPE), the German government aimed instead at alternative measures, including funding and support programmes, regulatory instruments for setting standards, information and advice programmes, and financial incentives. The NAPE adopted in December 2014, together with the Climate Action Programme 2020, aims at setting forth an Energy Efficiency Strategy for the current (18th) legislative term to meet the national and European energy efficiency targets.

## 4.4    Enhancing Cooperation Between Neighbouring Countries and Deepening European Power Market Integration

Cooperation among European neighbours makes the energy transition easier and less costly. Significant gains could be expected if the optimisation of the power market design were realised on the Pan-European level. Studies have estimated annual net benefits of a Pan-European power system integration in the range of EUR 12.5 to 42.6 billion (Booz and Co. 2013; ECF 2010).

As a first step, it is likely that a regional approach will continue to evolve, starting on the level of a few already well-integrated countries, and consider scaling up to the European level at a later stage. From a German perspective, closer cooperation with the Nordic countries as well as the Alpine region could be interesting as the hydro capacity in these countries could match quite well with increasing variable renewable generation in Germany. A good example of regional cooperation and integration is the so-called Pentalateral Energy Forum. This forum was established to integrate energy markets in France, Germany, the Netherlands, Belgium, Luxembourg, Austria, and Switzerland in the context of the market coupling process. The governments of the Pentalateral Energy Forum countries decided to take the next steps and address joint system adequacy issues.

Sharing resources and developing joint regulatory frameworks could help achieve system adequacy at lower costs and balance variable power generation across Europe. National energy policy instruments are, however, still very fragmented (for example, on renewable energy support schemes and adequacy measures, including capacity mechanisms), which can lead to inefficiencies and distortive effects. For all EU member states it is thus key to work closely with their neighbours, both bilaterally and within regional initiatives.

## 5    A Holistic Power Market Design Framework and Implications for Marketing Renewables

Power sector decarbonisation rests on continuous investments in renewables and flexibility options. Usually it is expected that the energy market will deliver these investments, in combination with the ETS. However, we argue that this rather theoretical view to power market design is not the way forward. Instead, a more pragmatic approach is needed that takes into account the complex practical, political, and economic challenges of the transition to a carbon-free power system.

A pragmatic market design approach consists of five elements:

- More flexible energy and balancing markets: A well-functioning, broad, liquid and—through further grid expansion and enhanced market coupling—increasingly EU-wide integrated wholesale market with low barriers to entry on both the demand and supply side in order to manage the flexibility challenge;
- A European ETS that provides a long-term, declining cap on power sector emissions;
- Complementary EU-level measures enabling member states with a high share of coal-fired power plants to chart a pathway out of coal (smart and managed retirement);
- Market-based instruments for revenue stabilisation for renewable energy source (RES) investments paid through premiums on market prices or long-term contracts to unlock the needed investments in new renewable energy capacities;
- Clear rules on government interventions to safeguard system adequacy consistent with the need for increasingly flexible power systems and long-term decarbonisation.

Together, these elements form a so-called Power Market Pentagon (Fig. 10); all of them are required for a functioning market design. Their interplay ensures that, despite legacy investments in high-carbon and inflexible technologies, fundamental

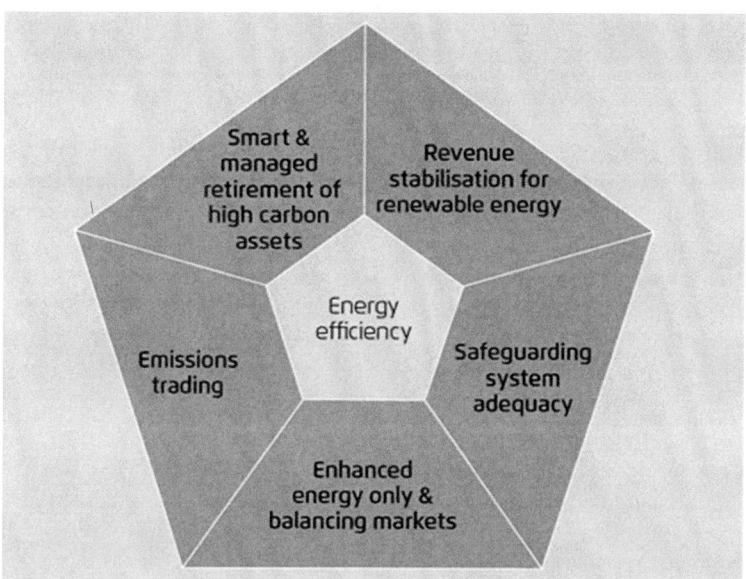

**Fig. 10** Power Market Pentagon (Agora Energiewende 2016)

uncertainties about market dynamics, and $CO_2$ prices well below the social cost of carbon, the transition to a reliable, decarbonised power system can occur in a cost-efficient manner. In what follows, we will focus on Element 4 of the Power Market Pentagon—providing for stable market revenues for new investments in renewables (RES-E).[12]

## 5.1    Providing for Stable Market Revenues for New RES-E Investments

There is an ongoing and important academic debate concerning the electricity market prices achieved by RES installations during the hours they produce when the power system has a high share of variable renewables.[13] Beyond theory-based arguments, there is some evidence that a higher share of variable renewables (vRES) is associated with falling market revenues for each kilowatt-hour of vRES electricity produced. Some questions remain, however: Does this reduction in market revenues decline slower or faster than the still falling LCOE of newly

---

[12]For further details on all elements of the Power Market Pentagon, see Agora Energiewende (2016).

[13]See for example Agora Energiewende (2015) and references therein, Hirth (2013), and Hartner et al. (2015).

built RES capacity? Furthermore, does an increase in flexibility options in the power system result in a bottoming out of the market price? In other words, does the market value of wind and PV decline as a function of the speed of their deployment? Does their market value increase in relation to the speed with which the overall power system becomes more flexible? If the market revenues achieved by wind and PV were to fall faster than LCOEs, this would support the argument that at higher shares of wind and PV, new investments in these two technologies typically cannot be fully financed from wholesale market revenues.

Furthermore, at high shares of RES-E, the marginal price in the wholesale market will, during an increasing number of hours, be set by RES-E and nuclear, not by fossil fuel fired plants falling under the European ETS. During those hours, the ETS will thus not add to the market price obtained by RES-E producers. The moment the last fossil fired power plant is not dispatching, the market price could drop to the marginal cost of nuclear and/or the marginal cost of RES-E installations—i.e. zero for wind and PV. By 2030 at the latest, RES-E investors would anticipate such developments and not invest in new RES-E capacities unless there is some mechanism for generating stable market revenues, even in presence of large shares of zero-carbon capacity.

Thus, the framework of laws, policies and measures relevant to renewable energy is important for providing investors with confidence and stability to invest into RES-E capacities. Stable and reliable conditions are seen as lower risk, which translates into lower costs for project developers and lower rates of return needed to make an investment profitable. Lower rates of return means that—at the same power price—low risk projects will need no or less help in closing possible revenue gaps.

Important elements for fit-for-purpose renewables support policies are as follows:

- Revenue stabilisation mechanisms to close gaps between projected revenues from electricity sold in the market and needed returns on investment. Various mechanisms are available to this end (feed-in tariffs, market premiums). Such interventions could occur at a national level (as is currently the case) or be coordinated at a regional level. The latter would create more opportunities for consistent planning of RES deployment and grid infrastructure development;
- Competitive bidding for the construction of new capacities to identify where and when project developers and investors regard market conditions as sufficiently stable to allow for project development without additional revenue stabilisation. Small-scale projects will still benefit from administratively set feed-in tariffs;
- Maintaining the principle of priority grid access and priority dispatch;
- Technology-specific support measures, improved provisions for siting of new RES capacity, and development of cross-border participation in national renewable support schemes.

## 5.2     Marketing RES-E on Power Markets

Deeper reflection is also likely to take place in the years to come on how to finance low-carbon assets in the most effective way while at the same time promoting the marketing of these technologies on the power markets. With the share of renewables rising to some 50 % by 2030, periods where renewables will meet all electricity demand will increase substantially. With their marginal costs close to zero, PV and wind energy will therefore deeply influence price formation on the energy markets. New revenue stabilisation mechanisms thus need to be able not only to close the cost recovery gap (see preceding discussion) but also to minimise overall system costs and wholesale market price distortions. Some proposals, including one published by Agora Energiewende, suggest a move towards technology-specific capacity payments for renewable producers (to complement the revenues made by selling electricity on the market). This would distort as little as possible the electricity price signals of the energy market for power plant dispatch and produce additional incentives for renewables plant design that is compatible with the flexibility needs of the future electricity system.[14]

---

## 6     Conclusion

As we have shown, wind and solar PV will provide the backbone of Germany's future zero-carbon power system. The reason is straightforward: Even including the integration cost, they are cheaper than any other new zero-carbon technology. To integrate progressively higher shares of volatile renewable electricity in a cost-effective way, the power system must react flexibly on the supply and demand side to the more variable patterns of electricity generation from wind and solar PVs. Consequently, the flexibility challenge has been identified as the main paradigm shift in Germany's energy policy: smart grids, renewable assets, fossil power plants, electricity demand, and storage facilities will have to be highly flexible. Many technological solutions are already available to solve this issue at comparable low cost, so that a key task to be solved by German energy politics is to develop a market design to ensure system adequacy, the necessary build-up of renewables, and a fair competition between the different flexibility options.

A pragmatic and solution-oriented power market design approach would maximise the value of the energy-only market and the established ETS, but expand it by three elements:

- Smart retirement of inflexible high carbon capacity from the system,
- Continued measures to provide stable market revenues for new RES-E investments, and
- Measures for safeguarding system adequacy.

---

[14]Agora Energiewende (2014).

The German energy transition is embedded in a wider European policy framework designed to bring greater sustainability, energy security, and competitiveness to Europe. According to the decision of the European council in October 2014, Europe will continue its ambitious energy and climate policies—decreasing $CO_2$ emissions, developing renewables, and increasing energy efficiency. The EU's climate change objectives demand a full decarbonisation of Europe's power sector by 2050 at the latest. Importantly, the successful climate summit in Paris in December 2015 demonstrates that Europe is not alone in this endeavour.

Accordingly, many other European member states have equally ambitious short- and long-term targets. Thus, the challenges faced by Germany are a snapshot of what is likely to occur in several countries in the medium to long term. The transition of electricity systems towards renewable energy is taking place not only in Europe but also on the global scale: For the third year running, worldwide investment in new renewable capacity exceeded investment in fossil-fuel power.

## References

Agora Energiewende. (2013). *Cost optimal expansion of renewables in Germany. Analysis by Consentec and Fraunhofer IWES.* Accessed June 30, 2016, from https://www.agora-energiewende.de/en/

Agora Energiewende. (2014). *Erneuerbare-Energien-Gesetz 3.0: Konzept einer strukturellen EEG-Reform auf dem Weg zu einem neuen Strommarktdesign.* Analysis by Öko-Institut. Accessed June 30, 2016, from https://www.agora-energiewende.de/en/

Agora Energiewende. (2015a). *Understanding the Energiewende. FAQ on the ongoing transition of the German power system.* Accessed 30 June 2016 from https://www.agora-energiewende.de/en/

Agora Energiewende. (2015b). *Integration costs of wind and solar power.* Concept paper. Berlin, Agora Energiewende. Accessed June 30, 2016, from https://www.agora-energiewende.de/en/

Agora Energiewende. (2016). *The power market Pentagon: A pragmatic power market design for Europe's Energy Transition.* Accessed June 30, 2016, from https://www.agora-energiewende.de/en/

AG Energiebilanzen. (2013). *Ausgewählte Effizienzindikatoren zur Energiebilanz Deutschland. Daten für die Jahre von 1990 bis 2012.* Accessed June 30, 2016, from http://www.ag-energiebilanzen.de/4-1-Home.html

AG Energiebilanzen. (2015). *Stromerzeugung nach Energieträgern, 1990–2014.* Accessed June 30, 2016, from http://www.ag-energiebilanzen.de/4-1-Home.html

Blazejczak, J., Edler, D., & Schill, W. P. (2014). Improved energy efficiency: Vital for energy transition and stimulus for economic growth. *DIW Economic Bulletin, 4*(2014), 3–15.

BNetzA. (2015). *Genehmigter Szenariorahmen für einen Netzentwicklungsplan Strom 2015.* Accessed June 30, 2016, from http://www.bundesnetzagentur.de/cln_1432/DE/Home/home_node.html

Booz & Co. (2013). *Benefits of an integrated European Energy Market.* Final Report Prepared for Directorate-General Energy European Commission. Accessed June 30, 2016, from https://ec.europa.eu/energy/sites/ener/files/documents/20130902_energy_integration_benefits.pdf

European Commission. (2011). *Energy roadmap 2050.* Accessed June 30, 2016, from https://ec.europa.eu/energy/sites/ener/files/documents/2012_energy_roadmap_2050_en_0.pdf

ECF. (2010). *Energy roadmap 2050—a practical guide to a prosperous, low carbon Europe*. Accessed June 30, 2016, from http://www.roadmap2050.eu/attachments/files/Volume1_fullreport_PressPack.pdf

Fraunhofer ISE. (2013). *Levelised costs of electricity—renewable energy technologies*. Accessed June 30, 2016, from https://www.ise.fraunhofer.de/en/publications/veroeffentlichungen-pdf-dateien-en/studien-und-konzeptpapiere/study-levelized-cost-of-electricity-renewable-energies.pdf

Fraunhofer ISE. (2015). *Current and future costs of photovoltaics*. Study on Behalf of Agora Energiewende. Accessed June 30, 2016, from https://www.agora-energiewende.de/en/

Fraunhofer IWES. (2015). *The European power system in 2030: Flexibility challenges and integration benefits*. Study on Behalf of Agora Energiewende. Accessed June 30, 2016, from https://www.agora-energiewende.de/en/

Graichen, P., & Steigenberger, M. (2016). The renewable energy transition – Insights from Germany's energiewende. A New Dynamic 2: Effective systems in a circular economy.

Hirschhausen, C., Kemfert, C., Kunz, F., & Mendelevitch, R. (2013). European electricity generation post-2020: Renewable energy not to be underestimated. *DIW Economic Bulletin, 9*(2013), 16–25.

Hirth, L. (2013). The market value of variable renewables. *Energy Economics, 38*, 218–236.

IEA. (2014). *The power of transformation—wind, sun and the economics of flexible power systems*. Accessed June 30, 2016, from https://www.iea.org/

Prognos. (2014). *Comparing the costs of low carbon technologies. What is the cheapest option?* Study on behalf of Agora Energiewende. Accessed June 30, 2016, from https://www.agora-energiewende.de/en/

Trend Research. (2014). *Definition und Marktanalyse von Bürgerenergie in Deutschland*. Accessed June 30, 2016, from https://www.buendnis-buergerenergie.de/fileadmin/user_upload/downloads/Studien/Studie_Definition_und_Marktanalyse_von_Buergerenergie_in_Deutschland_BBEn.pdf

UBA. (2011). *Stromerzeugung aus erneuerbaren Energien*. Accessed June 30, 2016, from https://www.umweltbundesamt.de/

UBA. (2013). *Potenzial der Windenergie an Land. Studie zur Ermittlung des bundesweiten Flächen und Landnutzungspotenzials der Windenergienutzung an Land*. Accessed June 30, 2016, from https://www.umweltbundesamt.de/

**Patrick Graichen** is Director of Agora Energiewende since January 2014, having served as its Deputy Director in 2012/2013. From 2001 to 2012, he worked at the Federal Ministry for the Environment. From 2004 to 2006 he served as personal assistant to the secretary of state in the ministry, and from 2007 he was Head of Unit for Energy and Climate Change Policy. During this time, he oversaw negotiations on the design of the economic instruments of the Kyoto Protocol, the Integrated Energy and Climate Programme of the Federal Government (2007), the European Union's Climate and Energy Package (2008), and the legislative procedures in the area of energy legislation. The Federal Environment Ministry has granted Mr. Graichen leave for his work at Agora Energiewende. He studied economics and political science, earning a PhD with a thesis on the topic of municipal energy policy at the Interdisciplinary Institute for Environmental Economics, University of Heidelberg.

**Christian Redl** is part of the European Energy Cooperation team of Agora Energiewende, where he focuses on power market design and its link to decarbonising power systems. He also works on options for cross-border cooperation to reap the benefits of integrated power systems and coordinated policies for energy transition targets. Previously, he worked for an energy consulting firm in the Netherlands and as a researcher at Vienna University of Technology focusing on power market design and system planning, renewables integration, and decarbonisation scenarios. Christian holds a PhD in energy economics.

**Markus Steigenberger** is Deputy Executive Director of Agora Energiewende. He oversees Agora's international activities, heading a team of senior project managers. Before working at Agora, Markus Steigenberger led the Germany programme of the European Climate Foundation (ECF). In this role, he developed ECF's strategy for Germany and led its grant-making activities. Previously, Markus held various positions in German and international civil society organisations like Friends of the Earth (BUND) or as a board member of the European Environmental Bureau. Markus completed his MBA as well as a master of arts degree in economic history, political science, and law.

# Marketing Renewable Energy in France

Michel Cruciani

**Abstract**

A law adopted in 2015 marked a change in French policy in favour of renewable energy. A number of ministerial decrees and orders were issued starting in early 2016 introducing significant changes to the financial support of renewable electricity. In various sectors and for several years France has launched tender procedures as a means to fix the guaranteed purchase price when feed-in tariffs apply (e.g. photovoltaic, offshore wind, wood). Beyond certain capacity floors, the country is now applying the principle of direct selling on the market with additional compensation. The promotion of heat from renewable sources has undergone fewer changes since the previous regulatory framework proved to be quite effective, especially for wood and heat pumps. Only the biofuel sector faces uncertain future demand. France holds great potential for many renewable sources; alongside the most common sectors (biomass, hydro, wind, photovoltaics) the country can also develop less widespread sources, ocean energy or high-temperature geothermal energy, for example. France is also striving to stimulate research and innovation in this area.

**Keywords**

France • Marketing renewables • Financial support

M. Cruciani (✉)
CGEMP, Université Paris-Dauphine, Place du Maréchal De Lattre de Tassigny, 75016 Paris, France
e-mail: michel.cruciani@dauphine.fr

# 1    Introduction

France possesses considerable resources for most renewable energy sources: wind, solar, bioenergy, and so forth. To date, the country has exploited most of its potential for hydropower and ranks among the first European biofuel producers, but the untapped potential available for renewable heat and power is still large. Since 2005 all French governments have gradually adapted legislation to stimulate the development of these two sectors, but with frequent adjustments, to reflect changes in public opinion, the lessons learned from the first experiences and new developments in the European regulatory framework.

In particular, national policymakers keep trying to limit the cost of the financial support for renewable sources, reflecting a widespread concern among economic players anxious to keep the rather low price of electricity that the French nuclear fleet has procured them until recently. This desire led to multiply tender procedures in several sectors (e.g. biomass, offshore wind, photovoltaic farms), which in return accustomed French actors to a practice now recommended in the guidelines adopted by the European Commission in 2014.

The year 2016 has emerged as a major step in this process. After the adoption of an ambitious law in favour of the energy transition in 2015, the government is now working to implement its content by issuing ministerial decrees and orders that give substance to the law. The first months of 2016 saw the arrival of a series of regulations that could radically overhaul the French renewable energy sector. This chapter intends to situate these recent developments in historical perspective and, for each energy source, highlight those new provisions that are already known as of mid-2016.

# 2    Starting Point

In 2014, renewable sources totalled 22.8 million tonnes of oil equivalent (Mtoe), of which 12.5 Mtoe (55 %) is in the form of heat. Figure 1 provides details.

The top four sources account for 80.5 % of the total: wood (42.2 %), hydropower (23.8 %), heat pumps (8 %) and wind power (6.5 %).

France had a target of 23.3 % of renewables in final energy consumption in 2020 (Directive 2009/28/EC). In its National Action Plan submitted to the European Commission, France presented plans to achieve that level by bringing renewable sources to 33 % in the heat sector, to 27 % in the electricity sector and to 9.6 % in the transport sector. The situation observed in 2014 is depicted in Fig. 2. It shows that considerable efforts are still required.

Most of the expected contribution will come from already confirmed technological sectors, offering off-the-shelf products. However, France also intends to spur innovation. To this end, the Agency for the Environment and Energy Management (*Agence de l'Environnement et de la Maîtrise de l'Energie*) (ADEME) regularly issues calls for projects on emerging energy sources, for example, hydrokinetic

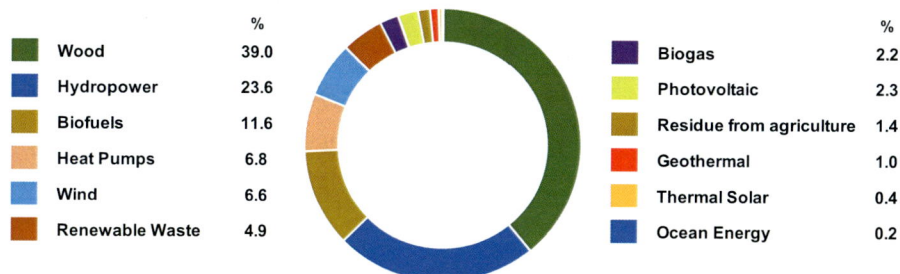

| | % | | | % |
|---|---|---|---|---|
| Wood | 39.0 | Biogas | | 2.2 |
| Hydropower | 23.6 | Photovoltaic | | 2.3 |
| Biofuels | 11.6 | Residue from agriculture | | 1.4 |
| Heat Pumps | 6.8 | Geothermal | | 1.0 |
| Wind | 6.6 | Thermal Solar | | 0.4 |
| Renewable Waste | 4.9 | Ocean Energy | | 0.2 |

**Fig. 1** Production from renewable sources in 2014 (all of France depending on authors; statistics concern mainland France, mainland France + Corsica, or all of France (mainland France + Corsica + Overseas Departments), Primary Energy (CGDD 2015)

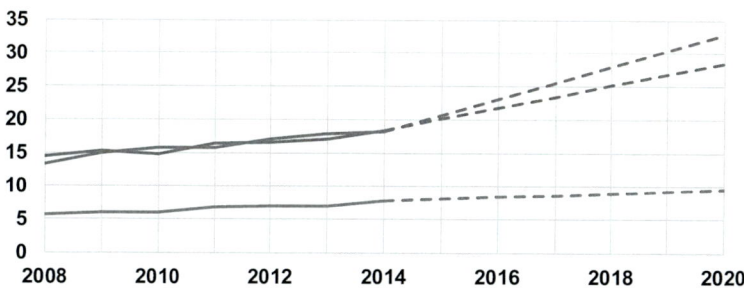

**Fig. 2** Situation in 2014 and targets for 2020. These results are calculated according to EU methodology, which differs from that applied in general in France. French statistics show climate-adjusted data (all of France) (CGDD 2015 and PAN 2012)

turbines, floating wind turbines and advanced biofuels. Some calls for tenders also include specific technological requirements.

# 3 Electricity

## 3.1 Overview

France has the largest hydroelectric generating capacity in Europe. It also has the largest installed nuclear capacity, and opposition to nuclear energy remains moderate. With this fleet, in 2000 France generated more than 90 % of its electricity without emitting greenhouse gases and without dependency on fossil fuel prices. This explains why France has not felt the need to quickly develop renewable sources in the power sector at a time when other countries were undertaking proactive policies in their favour, such as Denmark, Germany and Spain. Thus, although the law introduced as early as 2000 the principle of feed-in tariffs for

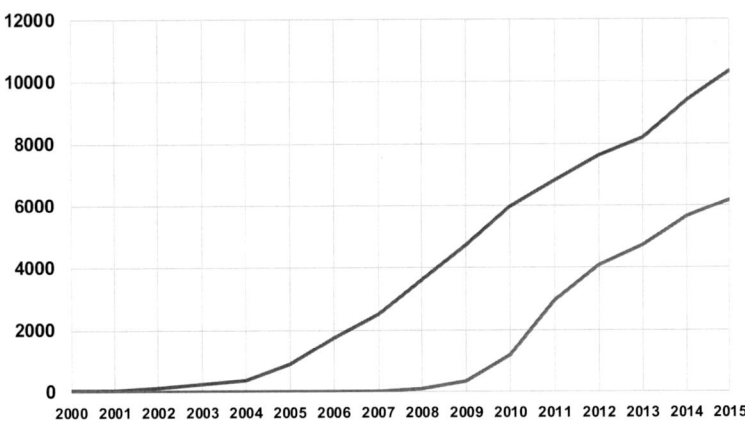

**Fig. 3** Capacity of wind (blue) and photovoltaic (red) sources 2000–2015 (all of France) (SOeS 2015; RTE 2016)

renewable electricity, prices were intentionally set at a low level, enabling very few projects.

A new law, adopted in 2005, required the government to set a target for each source and authorized the minister to take decisions favouring their achievement. The targets were set in 2009. However, before they were passed, the government had raised the level of feed-in tariffs in 2006 for photovoltaics (PVs) and 2008 for wind energy. The capacity then substantially increased (Fig. 3).

The government elected in 2012 launched a national debate that ended with the adoption of the Law for the Energy Transition and Green Growth, 17 August 2015. This law enacts a reduction of the share of nuclear energy by 2025 and higher ambitions for renewable energy, so that they account 32 % of energy consumed and 40 % of electricity produced in 2030. Under the terms of the law, the government adjusts the target set for every energy source every 5 years.

Figure 4 shows the old 2020 objectives, adopted in 2009, the situation observed in 2015 and the goals adopted 24 April 2016, including an intermediate step in 2018 and two possible levels for 2023, low and high. Developments reflect the difference between the paths drawn in 2009 and the actual results achieved in 2015. As an example, PV production progressed much faster than had been envisaged in 2009, which justifies a much higher target. Conversely, maritime wind projects have fallen behind, causing a postponement of targets to 2023 and beyond. In addition, two new objectives have been introduced, for wood and biogas. These targets cover only mainland France; specific targets will be set later for Corsica and the Overseas Departments.

To accelerate achievements, the decree of 3 April 2016 sets a maximum period of 2 months for connecting to the grid for facilities up to 3 kW and 18 months for facilities above 3 kW. The cost of connection depends on a Regional Scheme (*Schéma Régional de Raccordement au Réseau des Energies Renouvelables*) (S3R EnR), whose formulation was simplified by another decree, this one from 13 April 2016.

| CAPACITY (MW) | Former target 2020 | Achievement 2015 | New target 2018 | New target 2023 Low | High |
|---|---|---|---|---|---|
| Photovoltaïcs | 5,400 | 6,186 | 10,200 | 18,200 | 20,200 |
| Wind onshore | 19,000 | 10,269 | 15,000 | 21,800 | 26,000 |
| Wind offshore (fixed foundation) | 6,000 | 0 | 500 | 3,000 | |
| Ocean energy | | 0 | | 100 | |
| Geothermal | | 2 | 8 | 53 | |
| Wood | | 365 | 540 | 790 | 1,040 |
| Biogas | | 358 | 137 | 237 | 300 |
| Renewable urban waste | | 2,382 | | | |
| Hydropower | 28,300 | 25,203 | 25,300 | 25,800 | 26,050 |
| Total | 58,700 | 44,765 | 51,685 | 69,980 | 76,743 |

**Fig. 4** Former targets, achievements, new targets for power from renewable energy (mainland France) (Sources: Former target 2020: Arrêtés du 15 décembre 2009 relatifs à la programmation pluriannuelle des investissements de production d'électricité; Achievement: CGDD Tableau de bord PV 2015T4, CGDD Tableau de bord Eolien 2015T4, Observ'ER Baromètre Electrique Intégral France 2015, CGDD Tableau de bord Biogaz 2015T4, Observ'ER-Baromètre Electrique Chap-04-Hydraulique 2015; Target 2018 and 2023: Arrêté du 24 Avril 2016 relatif aux objectifs de développement des énergies renouvelables)

A third decree, published on 29 April 2016, reduces the burden for obtaining authorization to operate an electrical generating facility.

## 3.2 Financial Support Mechanism

Following adoption of the law of 17 August 2015, two support mechanisms arose for renewable electricity:

1. Feed-in tariff (FiT) at a guaranteed price, with two variants, a price set by ministerial order (the so-called open window principle with no ceiling volume for generation) or a price determined by a tender procedure (with limited quantities);
2. Direct sales on the market, with additional compensation, a feed-in premium (FiP). In most cases this compensation will be determined by a tender procedure (with limited quantities).

The second mechanism was enacted by three ministerial decrees, published on 27 and 28 May 2016. They concern new facilities and indicate that additional compensation (*complément de rémunération*) obeys the equation

$$\text{Additional compensation} = \text{Energy premium} + \text{Management Premium} - \text{Capacity Payment}$$

- The energy premium fills the gap between the average market price and a reference price, which will usually be determined by a tender procedure. It is therefore a floating premium.
- The management premium was supposed to be constant and determined by ministerial order. In fact, in all tenders launched in 2016, the management premium was internalized in the energy premium.
- The capacity payment will be established upon entry into force of the capacity market, as of 1 January 2017 (if the European Commission gives the green light).

Three features deserve comment:

1. Several ministerial orders must still clarify this support mechanism, which is entirely new in France. However, the law changes nothing in connection with the additional cost of renewable electricity compared to the market price. For the end consumer, this cost is still included in a specific charge, called the contribution to the public electricity service (*contribution au service public de l'electricité*) (CSPE). This charge was worth EUR 22.5/MWh on 1 January 2016, of which EUR 15.1/MWh was due to renewable energy (CRE 2015).
2. The reference price may be affected by a decreasing coefficient during the period in which the contract is in force. Moreover, if the average market price exceeds the reference price, the energy premium becomes negative; this mechanism is thus akin to the "contract for difference" that exists in the UK. The energy premium is zero during periods when the market price becomes negative. The average market price is to be determined every month by the Energy Regulatory Commission (CRE).
3. Contracts signed within the framework of FiTs usually include an escalation clause, which links the level of the guaranteed price to indices such as the hourly cost of labour, the price of industry production or the price of steel products. The decrees of 27 and 28 May 2016 do not specify whether these provisions will be renewed within the framework of FiPs.

Facilities subject to the obligation of direct selling in the market may entrust this function to an aggregator. The aggregation business began to develop in France a few years ago. The pioneers include Hydronext, an aggregator that gathers the production of small installations, including hydropower plants. Their adjustable production is a major asset in terms of balancing, which commits generators to fulfil the volumes placed on the day-ahead market for each time slot. Aggregators try to combine various types of facilities in various regions; they can incorporate into their basket industrial customers that are able to adjust their consumption.

Holders of FiTs whose contracts expire (after 15 or 20 years) will also engage in direct selling of their production. These producers will seek to garner additional revenues by selling every guarantee of origin (GO) they can get for each megawatt-hour from renewable sources. Until 2014, the French market remained extremely narrow: only 0.5 % of residential consumers and 0.7 % of non-residential consumers subscribed to a so-called green offer, amounting to 2.5 TWh against a total consumption of 465.3 TWh (CRE 2015). Two reasons may explain this poor result:

- Most providers of green offers generally cover the corresponding sales by purchasing GOs from producers of electricity from renewable sources. In France, 93 % of GOs were issued by two producers operating old dams that had been paid for long ago and were not eligible for FiTs, namely EDF and Engie directly or through its subsidiary Compagnie Nationale du Rhône (CNR). Consumer organizations know this and discourage their members from bringing additional revenues to the two national giants by subscribing to a green offer.[1]
- France has retained regulated tariffs (Tarifs Réglementés de Vente, TRV) for households, set by the government at a deliberately low level to preserve the purchasing power of citizens. Only EDF and local distribution companies are entitled to offer them. Their competitors have experienced difficulties with offering tariffs as low as the regulated tariffs, which aroused the distrust of consumers in connection with any kind of market offer.

The situation may change in the coming years. The Paris Conference on Climate Change (COP 21) held in December 2015 has raised awareness of the climate benefits of renewable energy. Several suppliers now provide green offers, including a significant proportion of electricity generated from sources such as wind and PVs, or agree to pay a fraction of their income to funds that finance new installations (e.g. Enalp, Enercoop, Proxelia). Lower prices on wholesale markets also enable alternative suppliers to purchase the power they need at a lower cost than that of the French nuclear fleet, which largely determines the level of regulated tariffs. On 1 July 2016, several green offers already led to an annual bill that is lower than the bill calculated with regulated tariffs (e.g. Alterna Idea Vert, Direct Energie 100 % Pur Jus, Lampiris or Planète Oui). Finally, since 2015, the French market has been open to GOs from other EU member states, which expands the options for buyers.

## 3.3    Photovoltaics

Between 2002 and 2010, support for PV power was based entirely on FiTs at a price fixed by ministerial order with 20-year contracts. In 2006, this price benefited from a sharp rise. When solar panel prices dropped in 2008, the installation of PV

---

[1]Under French law, renewable electricity producers who have signed a FiT contract are obliged to sell their production to EDF or to a local distribution company (*entreprise locale de distribution*, ELD). EDFs and ELDs receive compensation equivalent to the difference between the guaranteed purchase price and the market price. Until recently, the compensation received for 1 MWh from a new renewable source (e.g. wind, photovoltaic, biomass) was much higher than the price of a GO, which the CRE assesses at a level below EUR 4/MWh. Hence it was not in the interest of EDFs to issue GOs from these new sources. While 102.5 TWh of renewable electricity was generated in France in 2013, only 20.3 TWh of GOs was released (approximately 20 % of generation).

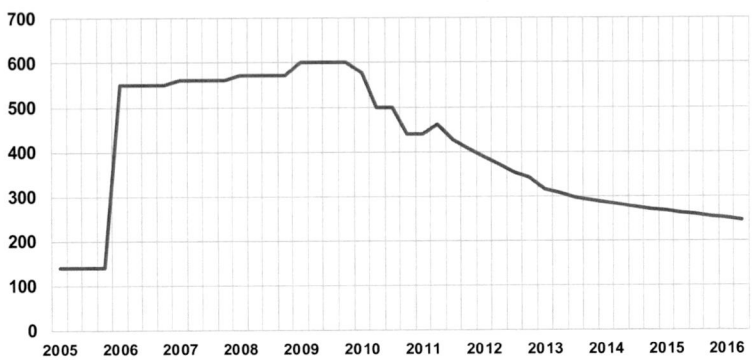

**Fig. 5** FiT for built-in roof panels with power between 0 and 9 kWp (PV Info 2016; CRE 2015)

systems became extremely lucrative, causing a rush for projects. Given the risk of an explosive increase of the CSPE, the government decided to reduce every quarter the guaranteed purchase price and keep it at a low level for capacity above 100 kWp. For such capacity, the low level of the tariff was meant as an incentive for project developers to respond to calls of tender launched from 2011 on.[2] The decrease of the price set for FiTs appears in Fig. 5.

As regards tenders, two different procedures are implemented:

1. Fast-track tenders for power from 100 to 250 kWp per generation unit: All bids are made over the Internet. Selection in 2013 was based on two criteria, with a total score of 30 points:
   - Required purchase price, from 20 points (= EUR 80/MWh) to 0 (>EUR 180/MWh; bids with a score of 0 were discarded).
   - Carbon footprint of panels ($kgeqCO_2/kWp$) from 10 points (=295) to 0 (>2118).
2. Detailed tenders for power above 250 kWp per generation unit: Specifications differ according to the site (ground, rooftop). Specifications may promote new technologies (e.g. sun trackers, concentrated PVs, combined heat and power solar) as well as good integration into the grid (e.g. management of reactive power, power forecast, storage). Specifications may also value PV farms located on poor quality grounds (e.g. contaminated sites, brownfields, fallow).

Available results displayed in Fig. 6 show that, so far, the tendering procedure for PVs in France seems satisfactory. Competition works: there are a sufficient number of bidders and more bidders than required to meet the tendered volume, risks are manageable, no strategic bidding appears and purchase price is declining.

---

[2]For facilities outside tendering of a capacity above 100 kWp, the FiT price was EUR 58/MWh on 27 May 2016.

| | | Applications Number | | Capacity MW | | | Purchase Price |
|---|---|---|---|---|---|---|---|
| | | Total | Selected | Target | Proposed | Selected | €/MWh |
| C>250 | 2012 | 425 | 105 | 450 | 1891 | 520 | 210 |
| | 2013 | 396 | 121 | 400 | 1721 | 380 | 142 |

| | | | | | | | |
|---|---|---|---|---|---|---|---|
| 100<C<250 | 2014-Q1 | 594 | 177 | 40 | 124 | 40 | 168 |
| | 2014-Q2 | 706 | 193 | 40 | 144 | 41 | 165 |
| | 2014-Q3 | 932 | 217 | 40 | 189 | 41 | 153 |

**Fig. 6** Results of tender procedures for PVs in 2013 and 2014 (Sources: CRE: Cahier des charges des appels d'offres 2012, 2013 et 2014; PV > 250 kW: CRE: Avis sur les appels d'offres; PV 100 à 250 kW: Communiqué de presse du Ministère de l'Environnement, de l'Energie et de la Mer du 17 Novembre 2014)

Projects are also realized after the tender, and thus the political targets for the roll-out of PVs are met.

Discussion is still open on the following issues:

- Prequalification criteria;
- Penalties in case of default;
- Ability to transfer or trade obligations on a secondary market;
- Ensuring a variety of actors, including local and civic participation; consideration may be given to separate auctions for small projects, small tender sizes and easy access (a one-page application form), defined share of tender reserved for small or local bidders. But who defines small local actors and civic participation?

According to information available in June 2016, support mechanisms will be as follows:

- Rooftop PVs with a capacity less than or equal to 100 kWp: FiT with guaranteed price set by ministerial order (open window);
- PV farms of a capacity higher than 100 kWp: FiT with price determined by tender (limited volume).

Several tenders remained open as of early 2016, and on 28 June 2016 the ministry announced new calls for tenders in the coming months, with a volume of 1000 MW per year for 6 years, as well as an upcoming tender in 2016 for PV systems with self-consumption of electricity generated.

## 3.4    Onshore Wind

According to a study by the European Environment Agency in 2009, France has one of the four best wind potentials in Europe (EEA 2009). Yet by late 2014 France was ranked 15th for installed capacity in relation to population, with 145 W/capita (vs. 862 in Denmark, 471 in Portugal and 246 in Austria, for example) (Eurobserv'ER 2015). This poor result reflects the reluctance of a significant part of the population to adopt this source of energy. The causes are many: for example, devotion to the landscape, presence of many historical and natural protected sites, importance of tourism in economic activity, scattered housing with owners fearing a devaluation of property assets. This distrust is reflected in the frequently changing rules by Parliament, which has put in place heavy and cumbersome procedures that lengthen construction times by up to 5 years on average per project. Despite the strict rules, about 35 % of projects are the subject of legal action on the part of local residents.

Onshore wind energy is nevertheless encouraged by public authorities. The support mechanism has until now rested entirely on the FiT with a guaranteed price for 15 years. Unlike the PV industry, the price has not changed since 17 November 2008. However, one of the most powerful French anti-wind associations, Wind of Anger, sued over the ministerial order introducing this price before the Court of Justice of the EU. After a long procedure, the European Commission authorized the French government to maintain this price, which had been set by a ministerial order of 17 June 2014. The case is not yet over because Wind of Anger is now requiring that wind farms that benefited from the tariff before its legalization in 2014 pay compensation to the state.

The guaranteed purchase price that prevails in 2016 has a special feature: it is designed with a view towards the equality of regions. Since 2008, the purchase price will be applied for 15 years, in two different periods:

- During the first 10 years, the purchase price is set at EUR 82/MWh;
- This price will apply again for the next 5 years in low wind areas (load factor below 27 %). In other areas, the price drops to a level depending on the load factor (Fig. 7).

The aim of the legislation was that at EUR 82/MWh for 15 years, even areas poorly endowed with wind may host wind farms. Figure 8 shows the results of this policy: the regions hosting the largest generation capacity are not always the windiest. From an economic point of view, the CSPE transfers to the electricity consumer a charge that aims to support renewable energy, not at the lowest possible cost, but at a level increased by the cost of a policy of national territorial development.

The decision of the European Commission would extend the wind FiT until 2019, but the French government announced that a FiP mechanism would apply to wind power as soon as 2017.

## 3.5    Offshore Wind

By launching very early onshore wind support mechanisms, several European countries have fostered a powerful national wind industry, Germany, Denmark and Spain in particular. France, which got off to a late start, lost most of the market share: by late 2014, French manufacturers had provided only 4 % of the installed

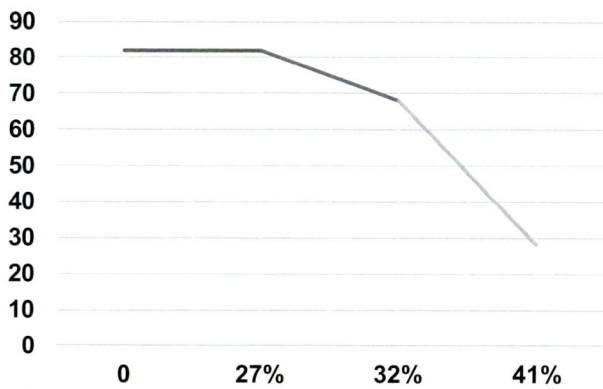

**Fig. 7** Purchase price for wind energy for last 5 years of FiT contract (Order 11/17/2008). (In this example, a load factor of 36 % on average during years 1–10 would imply a guaranteed purchase price below EUR 50/MWh in years 11–15)

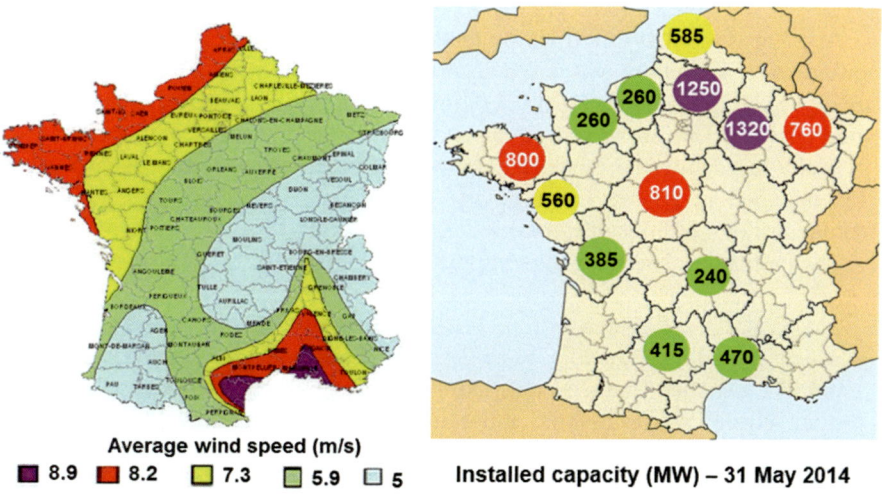

**Fig. 8** Comparison of windiest regions and locations of wind capacity (Sources: Carte des vents dominants en France: http://www.meteo10.com/carte-des-vents.php; Maps: http://commons.wikimedia.org/wiki/File:Carte_France_geo_dep3.png. Accessed 30 June 2016; Capacity: Le journal de l'éolien, Hors-série n°15, June 2014)

wind capacity in the country. After 2009, successive governments have deemed that it was still possible to develop a national industry for offshore wind. A support mechanism aimed at this objective was thus designed. The mechanism being currently implemented relies exclusively on FiTs at a guaranteed price determined by tender.[3]

---

[3]Until 2014, the regulation left the possibility of establishing offshore wind farms benefiting from a FiT with a price determined by ministerial decree. This price was identical to that of wind power on land (EUR 82/MWh in 2008), which remained far too low to bring about any real results.

A first call of tender was launched in July 2011 for 3000 MW in five locations. The specifications made it possible to classify the candidates according to three criteria, summarized in what follows:

| | |
|---|---|
| 1. Required purchase price | (40 points) |
| 2. Industrial quality | |
|    Industrial reliability | (14 points) |
|    Impact of industrial facilities | (2 points) |
|    Mastery of technical and financial risks | (22 points) |
|    Action beneficial to R& D | (2 points) |
| 3. Environmental and social quality | |
|    Minimizing the number of installed units | (10 points) |
|    Reducing impact on existing activities | (4 points) |
|    Compensation for environmental impacts | (4 points) |
|    Environmental monitoring system | (1 point) |
|    Decommissioning plan | (1 point) |
| Total score from 0 to 100 points | |

As regards the first criterion, the tender did not provide for any floor or ceiling for the price. The rating was based according to the location, on a curve defined by area, to take into account local conditions (e.g. distance to shore, depth of water, wind quality). This curve is reproduced in Fig. 9.

While respecting the European competition rules, the second group of criteria was intended to confer an advantage on companies that planned to establish manufacturing facilities in France.

The first tender had the following results:

• The consortium Eolien Maritime France was chosen on three of the four sites for which it had submitted a bid (Fécamp, Courseulles and Saint-Nazaire). It brought together EDF-EN and Dong Energy; two other partners will play a smaller role: WPD Offshore and Nass & Wind. The turbines will be built by Alstom (which had been acquired by General Electric in the meantime).
• The consortium Ailes Marines received one site (Saint-Brieuc). It was composed of Iberdrola and Eole-RES, with a participation of Technip and STX. The turbines were to be built by a joint venture between Areva and Gamesa.[4]

In total, the four selected sites will have a capacity of 1928 MW. The fifth site (Le Tréport) was awarded at the end of the second tender, the winner being the

---

[4]Areva and Gamesa had established a joint venture named Adwen to build these turbines. Meanwhile, Gamesa has allied to Siemens. Given its financial difficulties, Areva wants to withdraw from the wind business. Before the end of 2016, Adwen could be either absorbed by Gamesa or sold to General Electric, which had already taken over the wind-related activity of Alstom in 2015.

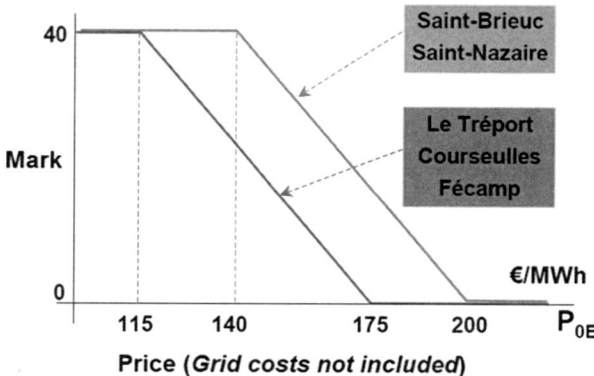

**Fig. 9** Mark according to required purchase price (CRE 2011)

consortium Engie and EDP-R, which also received the site Yeu-Noirmoutier, with Areva-Gamesa turbines. The total capacity will reach 992 MW.

Although prices are still provisional, the information available in the published results give an approximation (Fig. 10).

On 4 April 2016, the French government announced a third call for tenders for an offshore wind farm off the coast of Dunkirk.

Alongside these tenders, ADEME launched a call for projects for four offshore wind parks equipped with three to six floating wind turbines. Two projects seem well placed:

- Vertiwind, with a vertical axis wind turbine of 2 MW; a prototype should be tested in 2017 prior to subsequent construction of a farm equipped for 26 MW in the Mediterranean Sea;
- Winflo, innovative wind turbine; a 2.5 MW demonstrator is being tested in the UK before preproduction.

## 3.6    Hydropower

France is the second European producer of hydroelectricity, behind Norway. Generation in 2015 was 58.7 TWh (excluding generation from pumping), amounting to 10.8 % of the total power generation and 60 % of French electricity from renewable sources. Total hydropower capacity is 25.4 GW in mainland France, including 4.3 GW of pumped hydro storage (*stations de transfert d'energie par pompage*) (STEP). The country has just over 2000 small plants (with a capacity under 10 MW), totalling 2 GW of installed capacity and generating about 7.5 TWh/year (FHE 2016; RTE 2016).

The development of hydroelectric production in France is hampered by several considerations:

**Fig. 10** Purchase price (approximate)—tendering procedure for offshore wind (CRE 2012, 2014) (price includes grid costs)

|               | Expected Capacity MW | Selected Capacity MW | Purchase Price €/MWh |
|---------------|---------------------|---------------------|---------------------|
| First Tender  | 3,000               | 1,928               | 212                 |
| Second Tender | 1,000               | 992                 | 232                 |

1. The best sites are already equipped,
2. Uncertainty persists on the renewal of concessions currently granted to EDF or Engie,
3. The European Water Framework Directive especially affected France because of its variety of natural environments.
4. The various water users must share the resource in order to
   - Regulate the flow to prevent flooding,
   - Irrigate crops,
   - Supply drinking water,
   - Combat wildfires,
   - Sustain fishing and water-based activities.

To understand point 2, recall that in France, rivers belong to the public domain. Industrial facilities using water from rivers are only allowed to do this through temporary concessions. European directives governing public procurement impose competition when concessions are up for renewal. In France, 49 dams, representing 5.2 GW (20 % of French hydroelectric facilities), saw their concession expire at the end of 2015. However, local officials are worried by the possible arrival of foreign operators likely to relocate part of the activities and to be less sensitive to local concerns than existing recipients EDF (80 %) and Engie (17 %). Moreover, the dispersion of concessions could impede the upstream–downstream coordination of operations. Finally, no other European country is subject to an equivalent measure. In an attempt to defuse the hostility of local officials, the law of 17 August 2015 authorizes the creation of semi-public companies (*société d'economie mixte*) (SEM) or public–private partnerships (PPPs) allowing local authorities to remain involved. To date, this provision has not been used.

The European Water Framework Directive (2000/60/EC) was transposed into French law by the Law on Water and Aquatic Environments [*Loi sur l'eau et les milieux aquatiques* (LEMA) on 30 December 2006], which differentiates streams into two categories:

- Rivers of high environmental value will serve as a biological reservoir; they must ensure complete protection of species. New facilities are forbidden and existing facilities must be adapted to this purpose;
- On other rivers, the facilities shall allow the passage of sediments and migratory fish. Specifically, so-called fish ladders shall be built and the instream flow,

which provides water circulation in all circumstances, shall be increased from 1/40 (1984 act) to 1/10 of the average annual flow. This restriction leads to generation losses that can reach 10 % of annual volume.

Before the introduction of new constraints by LEMA, the additional hydropower potential was estimated at 7 TWh. The current target was reduced and now stands at 3 TWh by 2020. Given a loss of 2 TWh due to increased instream flow, new structures should provide 5 TWh. This figure can be achieved through the strengthening of existing structures by 2 TWh and through new installations in small hydropower by 3 TWh. The areas likely to receive new equipment are defined by the Regional Directorates of Environment, Planning and Housing (DREAL) and the National Office for Water and Aquatic Environments (ONEMA).

Until May 2016, all initiatives were encouraged by a FiT, with a guaranteed purchase price for 20 years, fixed since the order of 1 March 2007 at EUR 60.7/ MWh. This price could be increased by a bonus for small installations (EUR 5–25/ MWh) and increased again by another bonus favouring generation during peak periods. This mechanism will remain in place for new water plants with a capacity of less than 500 kW (continental France) as an open window. For installations with power between 500 kW and 1 MW, the guaranteed purchase price will be determined by a tender procedure (with limited quantities). Beyond 1 MW, the rule will be direct selling in the market, with additional compensation. The facilities already in service in May 2016 can sign a new 20-year contract providing access to FiTs (if their power is less than or equal to 1 MW) or additional compensation (for power greater than 1 MW) when they commit to a specific investment programme.

## 3.7 Other Renewable Sources of Electricity

**Ocean Energy** France hold enticing potential for power generation from the exploitation of marine currents, estimated between 5 and 14 TWh/year. Several industry players are interested in this energy:

- DCNS took control of Open Hydro and since early 2016 has been testing, together with EDF, a turbine off Paimpol-Bréhat (500 kW);
- The Engie group first considered installing HyTide turbines supplied by German manufacturer Voith Hydro; following the withdrawal of this actor, it has turned to Alstom—General Electric. Alstom, which integrated with GE Power on 1 October 2015, bought the British company Tidal Generation Ltd, which developed the model Oceade 18 (1.4 MW);
- Sabella has successfully tested a prototype off Benodet (2008–2012) and is now launching a range of turbines with three power levels. Its D10 model has been powering the island of Ouessant since 5 November 2015.

Many of these players must meet the conditions imposed by ADEME in its call for projects regarding two sites (Raz-Blanchard and Passage du Fromveur). Each project will receive a grant of EUR 30 million for 4 to 10 machines producing at least 2.5 GWh per year, benefiting from a guaranteed FiT of EUR 173/MWh. These first two pilot farms will test the turbine technology as well as the mode of installation, operation and maintenance. The ultimate goal is to lead to the establishment of commercial tidal farms and the creation of industries for the manufacture and export of machinery.

**Wave energy** Wave energy has significant potential around France, generally estimated at 40 TWh. The technology currently remains at the research stage. France will open a platform for experimentation near the town of Nantes (SEM-REV project). Among the various technologies, the CETO process is being tested on the island of La Réunion.

**High-temperature Geothermal** France operates a power plant fuelled by hot water from a volcanic zone in Bouillante (Guadeloupe). It consists of two units, B1 (4.5 MW) and B2 (11.5 MW). The plant sells its output at a price set by the CRE for Unit B1 at EUR 137/MWh and traded through an a local purchase agreement for Unit B2 at EUR 115/MWh (2013 values). Revenues proved to be lower than the cost of generation, which led to the payment of subsidies for the operation and delayed the construction of Unit B3 (20 MW planned) (CCE 2016). The government is now considering an ownership transfer of the entire plant to an independent operator.

An Enhanced Geothermal Systems (EGS) is being tested in Soultz sous Forêts in France. This technology includes increasing the permeability of rock by stimulation (hydraulic fracking), injecting cold water into deep underground hot rocks, collecting hot water and generating electricity on the surface. The first facility in the world operating on EGS technology, the plant has a capacity of 2.1 MWe.

### 3.8    Bioenergy That May Produce Heat or Electricity

France has the third largest forest area in Europe after Sweden and Finland, and wood is the first source of renewable energy in France (Fig. 1). However, the European Commission announced that it would draft a proposal for a directive to impose environmental constraints starting in 2020, in connection with solid biomass and biogas. According to the announcement, to be classified as renewable, these sectors will have to meet strict specifications for discarding products that require too much energy in the initial stages: collection, processing, routing. Only products resulting in a 70% reduction in greenhouse gas emissions would be deemed renewable. To date, no study has assessed the impact of such a rule on the development of these two sources.

In French regulations, the term *biomass* refers to products and by-products of the forest, wood industries, food and paper (such as black liquor) and agricultural residues and energy crops.

### 3.8.1 Electricity from Biomass

In France, the generation of electricity from biomass has been fostered by two mechanisms:

1. A FiT with a guaranteed purchase price for 20 years. It was updated by the order of 27 January 2011 at EUR 43.4/MWh. This price is increased by a premium of between EUR 77.1 and 125.3/MWh for plants with power greater than 5 MW whose energy efficiency is higher than 50 % and whose share of forest biomass in the supply exceeds 50 %. This price no longer applies to new installations (following the decree of 28 May 2016).
2. Tenders supervised by the CRE. Four series of tenders were launched in 2003, 2006, 2009 and 2010 (CRE 1 to 4) for a total of 1243 MW.

The purchase price appears to be insufficient; only facilities operating as cogenerators that provide good value for the heat achieve financial equilibrium. By the end of 2015, only seven plants had been completed according to the fixed FiT, for a total of 62.5 MWe. Tenders have also brought disappointments. According to the specifications, the projects were rated on the sustainability of supply, the relevance of the location and the overall energy efficiency. In a significant number of cases, the holders of the selected projects failed to follow through on their commitments, and these projects were abandoned. Figure 11 shows that of the four tenders, 1243 MWe projects were selected, but only 344 MW was eventually commissioned. Moreover, the average price seems very high, exceeding EUR 120/MWh at the end of the last three tenders.

At the end of the last tender (CRE 4), the selection of the Gardanne plant among the winners triggered a controversy. This plant, with a capacity of 150 MWe, will consume 2300 tons of wood per day, of which 55 % will be imported; this will exhaust the timber within a distance of 250 km, which may result in conflicts with other users. No cogeneration is being planned: the heat of the plant will be lost.

Faced with criticism over the fact that previous operations had favoured large plants, on 8 February 2016 the government launched a new call for tenders (CRE 5) restricted to projects with a capacity below 25 MW and accessible from 0.3 MW. It would run for 3 years (2016, 2017 and 2018) and cover 50 MWe/year, of which 10 MWe will be dedicated to projects with a capacity below 3 MW. This tender is open to biogas projects. Successful applicants will receive additional compensation designed to give an advantage to facilities that achieve a high efficiency, to those that meet strict emission limits for NOx and particulate matters, and, as regards biogas facilities, to those that incorporate livestock manure into their supply.

| | Average Price | Planned Commission | Minimum allowed Capacity | Selected Capacity | Commissionned Capacity by 2015 |
|---|---|---|---|---|---|
| | €/MWh | | MWe | MWe | MWe |
| CRE 1 | 85.5 | 2006 | 12 | 232 | 77 |
| CRE 2 | 128.3 | 2010 | 5 | 330 | 114 |
| CRE 3 | 145 | 2012 | 3 | 261 | 95 |
| CRE 4 | 137 | 2014 | 12 | 420 | 58 |
| Total | | | | 1243 | 344 |

**Fig. 11** Results of four tenders on biomass generation (AN 2013)

## 3.8.2 Electricity from Biogas

Despite the importance of the agricultural sector, suggesting a potential for biogas production assessed at approximately 4800 ktoe, France has only belatedly been interested in anaerobic digestion. In 2014, France ranked fifth in Europe in biogas production, with 421 ktoe of primary energy, far behind Germany (7434 ktoe) or even the Czech Republic (608 ktoe) (ADEME 2015; Eurobserv'ER 2015). The regulatory provisions being implemented since 2011 nevertheless have begun to bear fruit. They offer two methods of carrying out a project: either through the use of biogas for power generation or its purification into biomethane and injection into the natural gas grid.

Contrary to the rules in force in some countries, notably Germany, French support mechanisms do not apply to biogas from dedicated crop plants, such as corn. The rules cap at 25 % the incorporation of inputs from intermediate crops (*cultures intercalaires à vocation energétique*) (CIVE).

The development of biogas may benefit from EU Directive 2008/98/EC on waste, which calls for separate collection and recycling of bio-waste. Eventually, this collection could become mandatory. The increased use of biogas could then reduce resources for facilities generating electricity from renewable urban waste (Sect. 3.3).

There are several types of biogas plants:

1. Treatment of household or industrial waste and agricultural residues ("biogas farm");
2. Treatment of sewage sludge from industry or urban water;
3. Capture of landfill gas (the official name for a landfill is a storage facility for non-hazardous waste, or *installations de stockage de déchets non dangereux* (ISDND).

In 2011, a first support mechanism introduced the FiT, with a guaranteed fixed price for 15 years. Biogas production then took off. At the request of project holders, the government simplified the mechanism in 2015. It continues to apply to plants with a capacity of less than 500 kW. The new facilities in mainland France, whose power output is between 500 kW and 12 MW, are now eligible to the

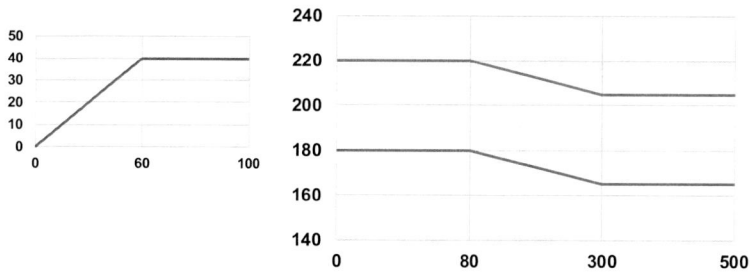

**Fig. 12** FiT price calculation for power from biogas (Order 10/30/2015). (In this example, the purchase price for power generated by a plant with a capacity of 100 kW incorporating 45 % of manure would be EUR 209/MWh)

additional compensation with a price determined by tender. The aforementioned CRE 5 tender involves biogas plants.

Figure 12 shows how to calculate the FiT price that applies to plant types 1 and 2 as of 1 December 2015. This price includes a bonus ranging between EUR 0 and 40/MWh proportional to the degree of incorporation of manure from 0 % up to 60 % of inputs. The price ranges between EUR 220/MWh for a small installation incorporating over 60 % of manure in its inputs and EUR 165/MWh for a 500 kW installation incorporating no manure.

As in the case of wind energy, the government included in the FiT a political concern: a bonus to help small farms and especially small breeding farms. With this support mechanism, about 50 new biogas plants are commissioned every year.

### 3.8.3    Electricity from the Incineration of Renewable Municipal Waste

Energy recovery from household waste remains underdeveloped in France: it achieved only 17.9 toe per 1000 inhabitants in 2014, vs. 87.1 in Sweden, for example. Electricity generation by incineration in waste processing plants amounted to 1.8 TWh in 2014 (Eurobserv'ER 2015). The revenue from the sale of electricity and heat represents on average a modest fraction of a plant's income; the bulk of resources stem from municipalities, through local taxation.

Currently, facilities that generate electricity from renewable municipal waste benefit from a FiT, the price of which includes a fixed sum rewarding availability and a variable sum proportional to the energy supplied. As of 1 June 2016, all new installations qualify for the additional compensation mechanism. Heat is sold through local agreements.

# 4 Heat

## 4.1 Overview

By the end of 2014, the share of heat from renewable energy appears to remain below the level needed to meet the 2020 target communicated to the European Commission in the National Action Plan; the production only achieves 78 % of the ideal trajectory. Nevertheless, as shown in Fig. 13, the new targets set in France after the vote on the law of 17 August 2015 do not mark a turning point.

To comply with these objectives, the system in place relies partly on facilities producing both heat and electricity. In this case, the support is based on the specific provisions that exist in favour of co-generation. There provisions were described earlier in Sect. 3.6. These facilities can also receive investment aid from local authorities.

For installations that do not generate electricity, support for renewable sources always takes the form of investment aid, distributed through two channels:

**Collective Uses or Professional Use** The promotion of heat from renewable sources is ensured by the Heat Fund, managed by ADEME:

- For large structures (>1000 toe/year), the support is based on tender procedures;
- For intermediate facilities (between 1000 and 100 toe/year), the aid is conditional on a series of criteria to fulfil;
- For small projects (<100 toe/year), the aid comes from the regions.

| HEAT (ktoe) | Former target 2020 | Achievement 2014 | New target 2018 | New target 2023 Low | New target 2023 High |
|---|---|---|---|---|---|
| Biomass & Renewable Waste | 12,600 | 8,846 | 12,000 | 13,000 | 14,000 |
| Geothermal | 250 | 129 | 200 | 400 | 550 |
| Heat pumps | 1,600 | 1,787 | 2,200 | 2,800 | 3,200 |
| Thermal Solar | 817 | 90 | 180 | 270 | 400 |
| Heat from Biogas | 900 | 111 | 300 | 700 | 900 |
| Biomethane injected into the grid of natural gas (2015) | | 7 | 146 | 688 | |
| Total | 16,167 | 10,970 | 15,026 | 17,858 | 19,738 |
| Of which supplied by District Heating | | 1,044 | 1,350 | 1,900 | 2,300 |

| TRANSPORTS | | Target 2018 | Target 2023 |
|---|---|---|---|
| Advanced Biofuels (%) | Gasoline | 1,6% | 3,4% |
| | Diesel | 1% | 2,3% |
| Biomethane for vehicles (ktoe) | | 60 | 172 |

**Fig. 13** Former targets, achievements, new targets for heat and biofuels (mainland France) [Sources: Former target 2020: Arrêtés du 15 décembre 2009 relatifs à la programmation pluriannuelle des investissements de production de chaleur; Achievement: SOeS Suivi Directive EnR 2014, SOeS Solaire thermique 2013, Syndicat des Energies renouvelables (SER) Panorama Biogaz 2015, Syndicat National du Chauffage Urbain Enquête 2014, tableau énergie produite en GWh; Target 2018 and 2023: Arrêté du 24 Avril 2016 relatif aux objectifs de développement des énergies renouvelables]

| | Number of projects | Total Investment | of which ADEME | RES per year | Support during 20 years |
|---|---|---|---|---|---|
| | | M€ | M€ | ktep | €/tep |
| Wood BCIAT | 147 | 867 | 334 | 808 | 20.6 |
| Wood non BCIAT | 762 | 1295 | 327 | 523 | 31 |
| Geothermal | 394 | 499 | 106 | 115 | 46 |
| Biogas | 51 | 200 | 31 | 68 | 22.9 |
| Thermal Solar | 1,590 | 154 | 73 | 7 | 521 |
| District Heating | 668 | 565 | 506 | 248 | 102 |

**Fig. 14** Results of Heat Fund (MEEM 2016)

The results of the Heat Fund over the period 2009–2015 appear excellent (Fig. 14). The first two lines separate professional uses—Heat from Biomass to Industry, Agriculture and Tertiary (BCIAT)—and uses in collective housing (non-BCIAT). The aid by tonne of oil equivalent of heat stemming from wood remains very modest, between EUR 21 and 31/toe or from EUR 1.8 to 2.8/MWh, compared with support for renewable electricity whose costs range between EUR 40/MWh (wind with FiT) and EUR 110/MWh (PVs on tender) (CEER 2015).

**Individual Uses** The promotion of heat from renewable sources is based on national or regional aid for citizens. Five types of aid are available, the first four being determined at the national and the latter at the local level:

1. Tax credit for the energy transition (*crédit d'impôt pour la transition energé tique*) (CITE),
2. Reduced value-added tax rate,
3. Eco-interest loan,
4. Aid from the National Housing Improvement Agency (*Agence nationale pour l'amélioration de l'habitat*) (ANAH),
5. Aid related to place of residence:
   - Local bonus paid by local authorities,
   - Temporary exemption from property tax.

Some aid is resource tested, others are intended for owners only. Installation work must be performed by a qualified installer who has been recognized as "safeguarding the environment" (*reconnu garant de l'environnement*) (RGE).

The Tax Credit for the Energy Transition is the largest form of aid; in 2016, the amount for this aid reached EUR 8000 for a single person and EUR 16,000 for a couple.

## 4.2    Comments on Renewable Heat

**Wood** In 2013, about seven million primary residences in France used wood for heating purpose. The aid measures are dedicated not only to new homes but also to the replacement of old appliances that are inefficient and polluting by new equipment with greater efficiency. In many areas, burning wood in open fires is now prohibited, and the use of modern appliances is encouraged.

The use of wood as energy competes with other forms of wood recovery: softwood lumber (carpentry, cabinet making) and industrial wood (chipboard, paper pulp). These latter forms currently absorb most of the wood sold from the French harvest, which is not enough to satisfy people's needs, so the country remains a major importer of wood. Competition is likely to increase in coming years owing to the opportunities offered by wood biochemistry, biomaterials and the growth expected for second-generation biofuels. Achieving the 2020 target will require increasing availability and stimulate forestry.

**Geothermal energy** This word refers to the heat of the Earth, available in certain areas over a hot aquifer, justifying deep geothermal operations. The water temperature ranges between 30 °C and 90 °C at a depth of 2000 m, which allows direct recovery of heat, optionally distributed by district heating if the field is large enough.

About 80 % of geothermal production in France is concentrated in the Paris Basin, located above the Dogger aquifer that extends over 15,000 km$^2$. The aquifer supplies 32 district heating networks and 5 isolated installations with water between 56 °C and 85 °C. Some 180,000 homes are connected (SER 2011). Support for geothermal energy consists in aid for investment and, when the geothermal source feeds a district heating network, a commercial incentive to encourage residents to connect their home to the network, in combination with regulatory constraints.

For collective domestic or professional use, the use of geothermal heat is promoted by both the Heat Fund, managed by ADEME, and the aid provided by local authorities. These subsidies also enable the development of smaller installations (heated greenhouses, farms). Preliminary studies (modelling of the underground and test drilling) are costly.

**Heat pumps** With almost 4.4 million heat pumps in service by the end of 2014, France ranks at the top among European countries. This superior result stems mainly from the success of reversible aerothermal heat pumps, which are easy to install and inexpensive and provide both heating in winter and cooling in summer (nearly 416,000 units sold in 2014). These devices, however, cannot claim the Tax Credit for the Energy Transition. This tax credit benefits air–water heat pumps (approximately 70,000 units sold in 2014) and so-called thermodynamic water heaters, which operate on the principle of the heat pump (72,500 units sold in 2014). (Eurobserv'ER 2015).

**Solar thermal** The simplicity of installing heat pumps and the advantages of reversibility won over the French public, who nearly abandoned solar thermal, despite favourable sunlight conditions in several regions. Only overseas departments offer high rates of development: they accounted for 81 % of new surfaces laid in France in 2013.

## 4.3    Biomethane from Biogas

Despite the advantageous guaranteed purchase price described in Sect. 3.8.2, generation of electricity from biogas proves to be profitable only for co-generation installations with good value for the heat. For this condition to be satisfied, the digester must be set up near an industrial business because there is hardly any district heating in France's rural areas. An alternative to electricity generation is to purify biogas and convert it into biomethane, which can then be injected into a natural gas grid.

Since 2011, biogas plants injecting biomethane into natural gas grid have received support in the form of a guaranteed purchase price. For plant types 1 and 2 (Sect. 3.8.2), the base price varies between EUR 95/MWh (gross calorific value) for injection capacities below 50 m$^3$/h and EUR 64/MWh for a capacity of at least 350 m$^3$/h. A bonus can be added to the base price, varying in proportion to the agricultural products incorporated into the inputs and another bonus varying in proportion to urban waste (Arrêté Biométhane 2011).

Biomethane producers sign a purchase contract with the supplier of their choice. The difference between the purchase price of biomethane and the natural gas price on the market is passed on to the end consumer via the so-called biomethane contribution. Biomethane producers can also receive a GO for each megawatt-hour they produce. When these GOs are sold, producers retain 25 % of the revenue, with the rest being used to reduce the biomethane contribution. This provision seems to suit producers since the GOs issued accounted for 99 % of the production of biomethane in 2014. However, only 30 % of all GOs issued could be sold.

The valuation of biomethane as motor fuel remains underdeveloped because the FiT for injection into gas grids seems more attractive. Several experiments have nevertheless been tried; the largest is located in Forbach on the site of the Méthavalor factory. It fuels in particular the fleet of commercial vehicles of the community.

## 5 Biofuels

### 5.1 The Situation in 2016

Although huge efforts are being devoted to the development of electric vehicles, it appears that compliance with the target set in the National Action Plan for 2020 in the field of transport relies primarily on biofuels. Given the investments already made, the first-generation biofuels will continue to play a leading role.

Following the adoption of the first European directive on biofuels (Directive EC 2003/30), France has implemented two support measures:

- A partial tax exemption for biofuels. For several years, biofuels have been totally exempt from the domestic tax on energy consumption of petroleum products (*taxe intérieure sur la consommation des produits pétroliers*) (TICPE); then it was gradually reintroduced and since 2016 the full rate applies again. To avoid overproduction, only plants that received ministerial approval benefited from the tax exemption.
- An incentive to purchase. The general tax on polluting activities (*taxe générale sur les activités polluantes*) (TGAP) hits operators who incorporate a biofuel share of less than a national floor, set at 7 % in 2013 for bioethanol and biodiesel, then raised to 7.7 % in 2015 for biodiesel (percentages are by volume). This tax applies to all operators (refiners' subsidiaries, supermarkets and independent sellers) and amounts to a quite high penalty.

To speed up the penetration of biofuels, France has gradually raised the ceiling for incorporating biodiesel in ordinary motor fuel: it was set at 8 % by volume as of 1 January 2015 (vs. 7 % in other European countries, sold under reference B7). The government also encouraged manufacturers to develop vehicles capable of running on a fuel containing up to 30 % biodiesel (called B30). These measures have borne fruit for biodiesel since it achieved a share of 8 % of volume sold since 2013, propelling France to the forefront of consumer countries, with 2541 ktoe in 2014. To date, the biodiesel that is consumed in France is derived from esterification, either with oils from rapeseed and sunflower grown in the country (approximately 50 %) or with imported vegetable oil (also approximately 50 %).

The ceiling for incorporating bioethanol has not changed (10 % by volume in the entire EU, sold under reference E10), but France requires distributors to set up pumps for motor fuel, including up to 85 % bioethanol (reference E85, only for so-called flexible-fuel vehicles). Despite these efforts, bioethanol only accounted for 6.1 % of sales by the end of 2014. The promotion of E85 failed: its market share is below 1 % (CCE 2016).

## 5.2 Biofuel Outlook in France

The outlook for biofuel remains unclear. In the short term, the collapse of oil prices has pushed biofuels into an unfavourable competitive position. In the longer term, uncertainty about public policies will constrain investment. Indeed, the new EU regulation restricts the share of biofuels from food crops (cereals and oilseeds) or energy crops on farmland to 7 % (bioethanol) or 7.7 % (biodiesel) and requires that at least 0.5 % of the energy consumed in transport in 2020 comes from advanced biofuels (Directive 2015/1513). However, no target is being considered for the following period. In its community framework proposal for 2030, the European Commission considers it unnecessary to provide for a specific objective in the transport sector (COM 2014-15 of 22 January 2014), and the European Council on 23 and 24 October 2014 endorsed this option.

Operators therefore lack visibility regarding requirements after 2020. The biodiesel sector appears particularly vulnerable because the French government is willing to reduce tax benefits that diesel fuel enjoyed until 2015. Such a move would favour gasoline-powered vehicles. In addition, current biodiesel producers are threatened by a new technology: the hydrogenation of vegetable oils. The product thus obtained, called hydrotreated vegetable oil, mixes more easily with fossil diesel than that resulting from esterification. It is referred to as "drop-in" biodiesel. Given its flexibility, this process could facilitate imports of oil produced outside Europe. Imports are currently hampered by duties introduced at the European level to fight the dumping practices of certain countries (Argentina, Indonesia, USA), but these duties should disappear in a few years.

As seen in Fig. 13, France has adopted a target on advanced biofuels (or second-generation biofuels) for 2020. Several research and demonstration projects give hope that the industrial stage will be reached soon. Examples:

- The Futurol project aims to develop and validate ethanol production through a biological process, from lignocellulose, stemming from agricultural, forestry by-products, dedicated residues or biomass;
- The BioTfueL project, led by the TOTAL group, aims to convert, through a thermochemical process, lignocellulosic biomass (e.g. straw, forest residues, dedicated crops) into biodiesel and bio-jet fuel. The project has reached the demonstration stage with a production launch of the Dunkirk plant scheduled for late 2016. It comprises three stages: pretreatment, gasification and synthesis;
- The Syndièse project aims to demonstrate the technical and economic feasibility of a complete chain of biofuel production on a single site, from the collection of biomass to synthesis gas with the introduction of hydrogen into the process to optimize performance.

The future of this work will depend on the price of fossil-based oil products, on the regulatory framework relating to the fight against the emission of greenhouse gases and on local measures against pollution from motor vehicles. Local governments that implement such measures in France, such as the city of Paris, currently prefer electric engines.

# 6 Conclusion

Foreign observers have often felt that the French policy on renewable energy was characterized by hesitation and groping. However, in retrospect, the French record of the last 10 years seems honourable. Certainly this long period of development, still unfinished at the time these lines are written, sometimes caused concern among investors, who want greater legal stability. Yet the reasons for optimism dominate. After the 2015 vote on a very comprehensive law, in March 2016 France adopted ambitious targets for renewable sources by 2023. The professionals concerned have largely expressed satisfaction. The rate at which the implementing texts are published shows a genuine will to succeed, and tender procedures prepared within the new regulatory framework have created a momentum in many sectors, as confirmed by this chapter. Other academic work could complement this one by analysing the impact of recent laws that strengthen the powers of local officials in the field of energy, the recent order that facilitates crowdfunding projects and the considerable effort being made to boost innovative technologies.

# References

ADEME. (2015). Agence de l'Environnement et de la Maîtrise de l'Energie, Estimation des gisements potentiels de substrats utilisables en méthanisation, Avril 2013.

AN. (2013). Rapport d'information de Mme Rohfritsch et M. Lambert pour la commission du développement durable et de l'aménagement du territoire de l'Assemblée Nationale (2013), page 56, except last column: Jean-Pierre Tachet Comité Interprofessionnel du Bois Energie (CIBE), Conference December 9, 2015 at Montpellier.

Arrêté Biométhane. (2011). Arrêté du 23 Novembre 2011 sur l'achat du biométhane injecté dans le réseau de gaz naturel.

CCE. (2014). Cour des Comptes, Rapport particulier 71058 sur le Bureau de Recherches Géologiques et Minières.

CCE. (2016). Cour des Comptes, Les biocarburants: des résultats en progrès, des adaptations nécessaires, 2016.

CEER. (2015). Council of European Energy Regulators, Status Review of Renewable and Energy Efficiency Support Schemes in Europe in 2012 and 2013, January 15, 2015.

CGDD. (2015). Commissariat Général au Développement Durable, Chiffres clés des énergies renouvelables, Edition Décembre 2015.

CRE. (2011). Commission de Régulation de l'Energie (CRE), Appel d'offres portant sur des installations éoliennes de production d'électricité en mer en France métropolitaine, 11 Juillet 2011, Présentation synthétique du cahier des charges.

CRE. (2012) & (2014). Author's calculation, based from CRE, Délibérations du 5 Avril 2012 et du 24 Avril 2014.

CRE. (2015). CRE, Délibération du 15 Octobre 2015.

EEA. (2009). *European Environment Agency, Europe's onshore and offshore wind energy potential*.

Eurobserv'ER. (2015). The state of renewable energies in Europe, Edition 2015.

FHE. (2016). Site Internet de l'association professionnelle France Hydro Electricité, rubrique « Les chiffres clés »: http://www.france-hydro-electricite.fr. Accessed 30 Jun 2016.

MEEM. (2016). Site Internet du Ministère de l'Environnement, de l'Energie et de la Mer, Le bilan du fonds chaleur. http://www.developpement-durable.gouv.fr/Le-bilan-du-Fonds-chaleur.html. Accessed 30 Jun 2016.

Order 10/30/2015: Arrêté du 19 Mai 2011 sur l'achat d'électricité produite à partir de biogaz, modifié par l'arrêté du 30 Octobre 2015.

Order 11/17/2008: Arrêté du 17 Novembre 2008 pour l'énergie éolienne.

Order 11/23/2011: Arrêté du 23 Novembre 2011 sur l'achat du biométhane injecté dans le réseau de gaz naturel.

PAN. (2012). Plan d'Action National de la France en faveur des énergies renouvelables, 2012.

PV Info. (2016, June). Site Internet « Photovoltaïque.info », Historique des tarifs d'achat garantis and Site Internet CRE.

RTE. (2016). Réseau de Transport d'Electricité (RTE), Bilan électrique 2015, édition de Janvier 2016.

SER. (2011). Syndicat des Energies Renouvelables (SER), La géothermie en France, Mars 2011.

SOeS. (2015). Service de l'Observation et des Statistiques (SOeS), Evolution des parcs éolien (2000-2014) et photovoltaïque (2005-2014), puissance cumulée en fin d'année.

**Michel Cruciani** is associated with the Centre de Géopolitique de l'Energie et des Matières Premières (CGEMP) and with Institut Français des Relations Internationales (IFRI). He teaches renewable energy in the master's program in energy, finance, and carbon at University Paris-Dauphine and is also an independent consultant working on energy issues. He has had long experience with Electricité de France, Gaz de France (before it became Engie) and the French Trade Union CFDT (Federation of Energy Workers). He was elected member of the board of Gaz de France. His interests relate to the actions of European institutions and questions related to the environment, energy efficiency and the development of renewables. He graduated from Ecole Nationale Supérieure d'Arts et Métiers.

# Marketing Renewable Energy in the United Kingdom

Catalina Spataru and Bruno Arcuri

**Abstract**

This chapter focuses on the renewable energy market in the UK. First we discuss the impact of privatization, then show what preconditions might be important. The main conclusion drawn from the analysis is that in the UK, as well as in other countries, new policy frameworks need to guide the transition from an energy system designed to achieve short-term efficiencies through market operation to a long-term approach that would embrace new uncertainties. Both market interests and environmental protection need to be secured in order to guarantee the levels of investment needed in the UK's renewable energy market.

**Keywords**

Markets • Renewables • Regions • UK • Solar • Wind

## 1 Introduction

The UK is producing most of its electricity from fossil fuels (coal and natural gas). Figure 1 shows the generation mix in the UK (2015), and Fig. 2 shows the electricity generation by source in the UK between 1998 and 2015.

In the UK, the high reliance on fossil fuels often provoked a price premium compared to continental Europe. In recent years we have seen a trade-off between gas- and coal-based power production in the UK. Indeed the rise in gas prices since the Fukushima disaster cause the share of natural gas to fall from 46 % in 2010 to 28 % in 2012 in the UK electricity mix. In contrast, the share of coal rose from 28 to 39 % during the same years as a result of falling coal and carbon prices (European Commission 2012).

C. Spataru (✉) • B. Arcuri
UCL Energy Institute, 14 Upper Woburn Place, WC1H 0NN London, UK
e-mail: c.spataru@ucl.ac.uk

© Springer International Publishing AG 2017
C. Herbes, C. Friege (eds.), *Marketing Renewable Energy*, Management for Professionals, DOI 10.1007/978-3-319-46427-5_17

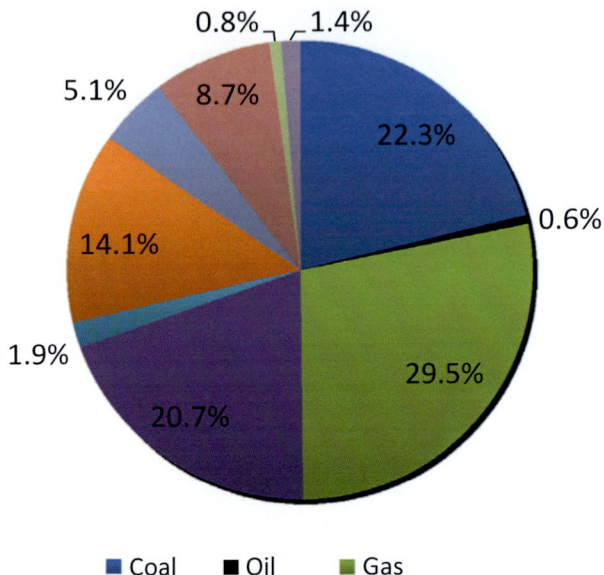

**Fig. 1**  Electricity mix in UK 2015 (Data source: DECC 2016)

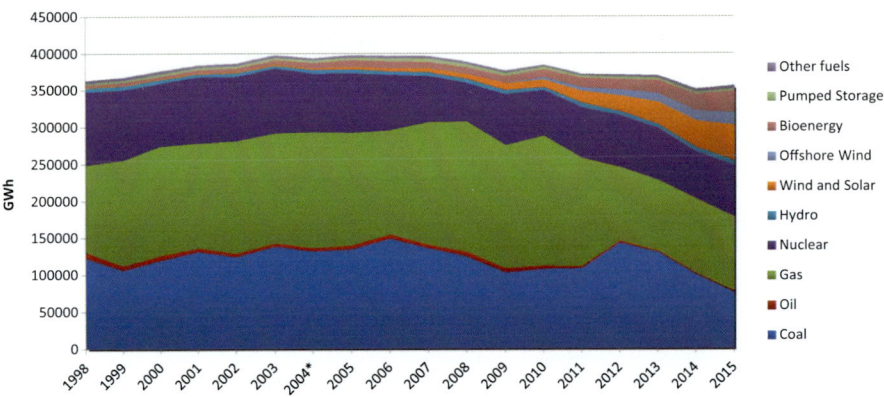

**Fig. 2**  Electricity generation by source in UK between 1998 and 2015 (Data source: DECC 2016)

## 2    Trade

The British Isles are historical net importers of electricity. There are currently four interconnections in service:

– IFA interconnector to France,
– BritNed interconnector to the Netherlands,

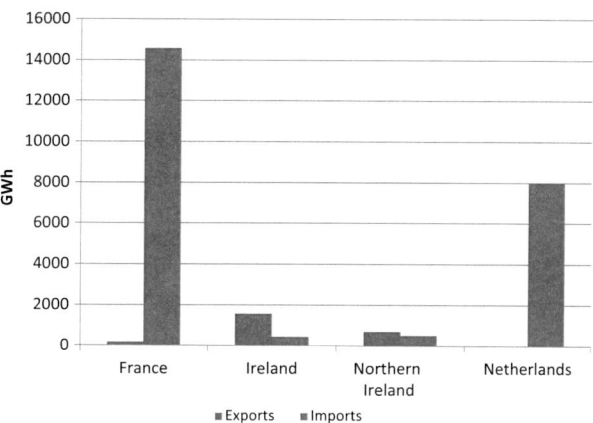

**Fig. 3** Physical electricity exchanges between Britain and its neighbours in 2015 (Data source: ENTSOE 2016)

– Moyle interconnector to Northern Ireland,
– East–West interconnector to the Republic of Ireland.

Most of the UK's imports come from France, followed by the Netherlands. Ireland and Northern Ireland import their electricity from Britain (Fig. 3).

## 3    Market Structure

### 3.1    Transmission and Distribution Network

England and Wales have an ownership unbundled transmission system operator (TSO). An independent TSO, National Grid, owns and operates the transmission network (National Grid 2016) (Fig. 4).

Scotland follows a legal ownership unbundled model. Scottish Power and Scottish and Southern Energy are vertically integrated companies that are involved in the whole electricity chain, from generation to retail, and manage the transmission network through their subsidiaries (respectively Scottish Power Energy Networks and Scottish Hydro Electric Transmission).

The Office of Gas and Electricity Markets (Ofgem) is in charge of the regulation of the energy market in Great Britain. It is a non-ministerial governmental department created in 2000 by the merging of Offer and Ofgas (Birchall and Dunstan 2010). On the distribution side, the UK counts seven DSOs (Fig. 5).

Private companies supply energy to consumers and consumers can choose which companies they buy energy from.

### 3.2    Generation and Retail

Six main companies, often called the Big Six, dominate the electricity supply in the UK. The Big Six are:

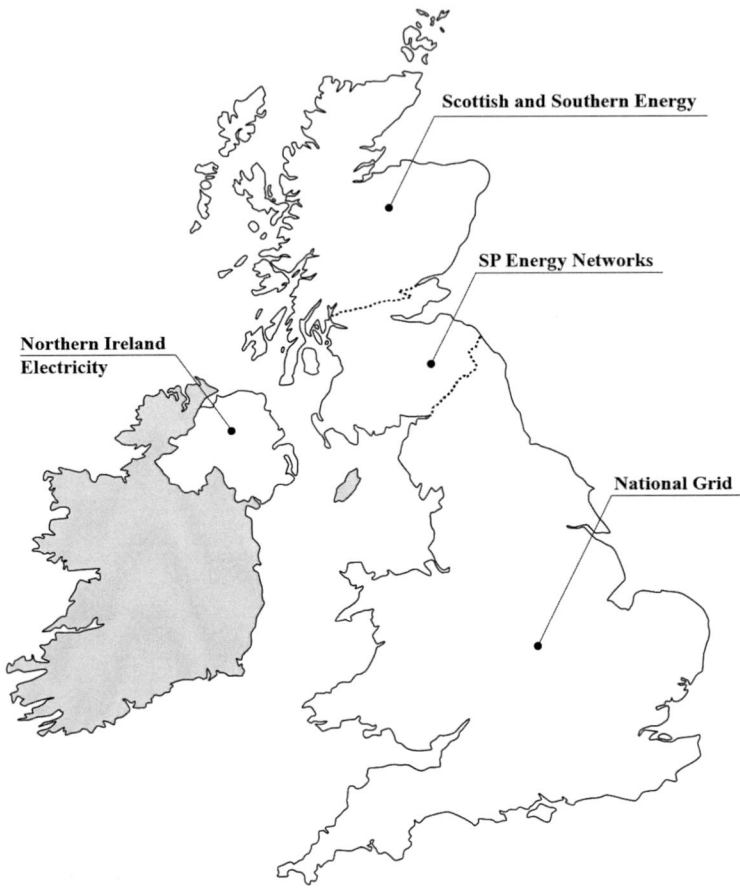

**Fig. 4** Electricity transmission network operators in UK (Adapted after National Grid 2016)

– British Energy (subsidiary of Centrica),
– Scottish and Southern Energy,
– ScottishPower (subsidiary of Iberdrola),
– RWE Npower,
– EDF energy,
– EON.

According to Sheffield Energy Resources Information Services (SERIS), in 2012 the Big Six controlled 96 % of the retail residential electricity market and 82 % of the non-residential electricity market. The rest of the supply is ensured by 21 smaller retailers (SERIS 2012). As the market opened to competition in the late 1990s, which is earlier than in most European countries, the switching rate between suppliers is one of the highest in Europe (15 % in 2011) (European Commission 2012). Regarding generation capacities, SERIS estimated that 74 companies owned

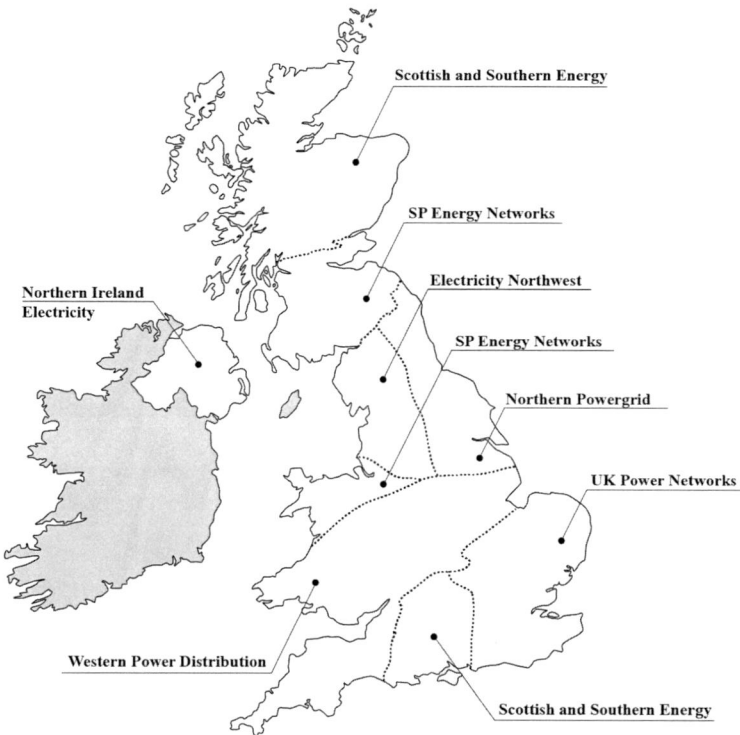

**Fig. 5** Electricity distribution network operators in UK (Adapted after National Grid 2016)

about 94 % of the generating capacity in 2012.[1] SERIS also determined that the Big Six owned 71.3 % of the total electricity generating capacities in 2012 (SERIS 2012).

## 4     History of the Market

From 1990 to 2001, in England and Wales, wholesale electricity was traded through an electricity pool mechanism. The pool mechanism was relatively simple. Each electricity producer was asked to inform the pool of the electricity prices and quantities that it could provide. National Grid planned the schedule of generation based on this information and a day-ahead estimate of the electricity demand and calculated the pool price. National Grid was also in charge of balancing the real-time demand and supply (Simmonds 2002). The New Electricity Trading Arrangements (NETA) replaced the pool in March 2001. NETA installed a classical bilateral electricity trading market composed of a forward and future market, short-

---

[1]Based on the last Seven Year Statement of National Grid combined with DUKES estimates.

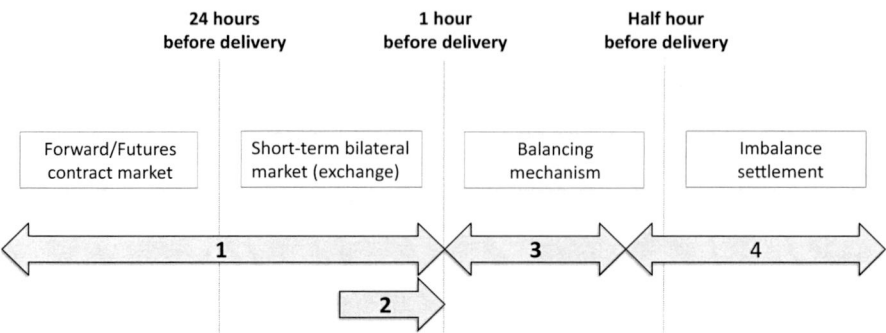

1. Generators, suppliers and traders buy and sell electricity;

2. Notification of contract volumes (to Settlement) and Final Physical Notification to National Grid
(as System Operator);

3. National Grid (as System Operator) accepts offers and bids for system and energy balancing;

4. Settlement of cash flows arising from the balancing process.

**Fig. 6** Organisation of electricity markets (Adapted after National Grid 2011)

term exchanges, a balancing mechanism and a settlement process (Simmonds 2002).

On 1 April 2005, the NETA were transformed to include Scotland in the scheme and became the British Electricity Trading and Transmission Arrangements (BETTA) (National Grid 2011). BETTA covers four types of electricity markets: the forward/future markets, the power exchange markets, the balancing mechanism market and the ancillary services market. Their organisation is described in Fig. 6. In terms of selling electricity, 90 % is sold through the forward/future market, 3 % through the power exchanges and 2–3 % through the balancing mechanism market (Wilson et al. 2011).

The markets operate on a half-hourly basis. The only mechanism mandatory for all companies is the imbalance settlement.

The UK has three power exchanges, which is rare in Europe:

– APX Power UK was created in 2000 and was first named UKPX (APX 2016),
– N2EX was launched in 2010 by NASDAQ OMX Commodities and Nord Pool Spot for UK contracts,
– ICE was formed in 2000 and is the first network of exchanges and clearinghouses in the world.

The energy markets are regulated by the Gas and Electricity Markets Authority (GEMA), which is the organisation responsible for setting strategy and policy priorities, making decisions on regulatory matter such as price control and enforcement. GEMA operates through the Office of Gas and Electricity Markets (Ofgem). Ofgem's principal objective is to protect electricity and gas consumers by promoting competition and regulating and delivering government schemes. According to Ofgem (2016), the scenario for renewable energy in the UK can be described by the following facts and figures:

- In 2015/16, 90,283,516 Renewable Obligation Certificates (ROCs) were issued for eligible renewable electricity generated under the Renewables Obligation;
- By June 2016, 771,998 solar panel installations were registered for the feed-in tariff;
- The generation capacity from Renewables Obligation–accredited UK offshore wind farms is 5 GW;
- The investment in offshore transmission projects totals £2.9 billion to date, connecting 4.4 GW of offshore wind.

## 5    Renewable Energy Source Integration: Challenges

Renewable energy sources (RESs) (wind and solar) are difficult to predict due to their variability. Wind speeds can vary from minutes to seconds and tend to be weakly correlated with high power demand: cold, windless winter evenings and hot, windless summer days (Grimston 2014). The Royal Academy of Engineering (2014) points out that considering consented and under construction wind projects, the UK has a total of 20.7 GW of wind capacity.

The electric system becomes exposed to weather risk when a significant proportion of the generating capacity comes from intermittent renewables. Darwall (2015) connects the weather risk and uncertainties inherent in farming to the reason the government heavily subsidises farmers, comparing this scenario to subsidies supplied to electricity generators. The report states that severe market distortions were introduced to the energy market due to government interventions to support investment in renewables, which transferred weather risks and system costs to the rest of the energy system. This would mean that renewables might increase the amount of subsidies and support of nuclear and combined-cycle gas turbines (CCGT) to keep the lights on.

Wind power substitutes the costs of fuel inputs, offering very low variable costs but making it a capital-intensive electricity generation source (Hughes 2012). When the wind blows at optimal speeds, wind farms force coal and gas power plants to reduce their output because they present higher variable costs. This scenario produces adverse impacts on prices and costs for thermal power plant investors. On the other hand, investors in wind and solar power are paid for the energy produced by the weather and, as stated by Darwall (2015), might even be paid without producing any energy in certain circumstances.

A study done by Ofgem (2009) assumed that onshore wind capital costs (£1.2 m/MW) are twice the size of that of CCGT (£0.6 m/MW), and for offshore wind this number is nearly five times that size (£2.8 m/MW). A report by UKERC (2010) also states that the capital costs of offshore wind have doubled in 5 years from approximately £1.5 m/MW in 2004 to over £3.0 m/MW in 2009, attributing this increase to factors such as commodity prices. Renewables also require transmission infrastructures to be built or reinforced. According to RenewableUK (2012), 10–20 % of the capital costs of developing an offshore wind farm are associated with the electricity transmission infrastructure. The UK government estimates a substantial investment in offshore networks worth up to £15 billion will be spent by 2020 to connect rounds 1, 2 and 3 of offshore wind (House of Commons 2010).

## 6    Incentive and Market-Distortion Effects

Policy interventions in the electricity market create unintended distortions that require further interventions. In this sense, subsidised intermittent electricity introduced to the market depresses the profitability of other generators and not only lowers the returns from investing in replacement capacity necessary to maintain continuity of supply but also makes it harder to predict (Darwall 2015). When there is low demand and high penetration of renewables, negative energy prices might appear, as observed in Denmark, Germany, Canada and the USA (California).

A report by OECD/NEA (2012) states that these distortions are likely to become more pronounced as wind and solar capacities are expanded. Two effects caused by the introduction of renewables on the market might lead to underinvestment in dispatchable technologies and, therefore, a decrease in security of supply at times of unfavourable meteorological conditions:

- **Compression effect:** Lower and more volatile wholesale energy prices impacting investment returns of conventional generating assets, which is amplified by favourable weather conditions for renewables;
- **Pecuniary effect:** Renewables investors are subsidised and, therefore, isolated from the effects of their output on the market, while conventional generators do not affect renewables generators.

## 7    Enhancing Competition and Protecting Consumers

Subsidies for intermittent renewables have damaged the functioning of the electricity market (Darwall 2015). An effective market would require removing subsidies and ensuring that renewables account for the risks they bring to the system. To accelerate the return to market pricing:

- Price support and incentives for planned renewable projects should be removed;
- Legal means to remove or reduce price support and obligations to purchase renewable power should be deployed;
- The costs of grid expansion and reinforcements should be allocated to renewables assets, taking them out of the National Grid's Asset Value;

– A revised and updated pool should be designed with international experience so that all generators submit bids to sell their energy production.

Reviving the pool would facilitate the entry of other generators into the market by reducing market barriers and the power of the Big Six. This right/obligation to sell at the pool's bid price would encourage renewable power producers to deal with conventional producers and internalise the intermittency costs of renewables, removing the need for a capacity market run by the government.

The structure of electricity prices is complex owing to the various tariffs. Green electricity suppliers have different rates for their electricity production, depending on the region. The predominant green tariffs on the market are green source (which buy electricity from suppliers marketing renewable generation) and green fund (customers voluntarily contribute money into a fund supporting new renewable initiatives).

## 8    Analysis of Proposed Electricity Market Reform

The electricity market is in need of wide-ranging reform (DECC 2011). The government's vision for the Electricity Market Reform (EMR), which is to establish a market that delivers secure power, an increasing share of renewables and carbon reduction simultaneously, will require a number of policy responses. According to Pollitt and Haney (2013), the increase in government intervention in the electricity market in recent years was motivated by sound reasons. The UK has set an 80 % carbon reduction target by 2050, compared to 1990 levels, as part of the 2008 Climate Change Act, and the electricity sector is key to the decarbonisation strategy.

There were four elements in the EMR proposed by DECC (2011):

– Contracts for difference (CFD),
– Carbon price support (CPS),
– Capacity market (CM),
– Emissions performance standard (EPS).

Achieving carbon and renewable targets put the electricity sector in line for large-scale decarbonisation. Pollitt (2012) describes the logic behind the four elements and questions whether this is good economics. Fixed prices for low carbon generation, or CFD, offer certainty and are high enough to support nuclear as well. CPS raises the price of carbon for fossil generation and encourages switching and benefits from reduced CfD payments and raised tax revenue. The CM allows fossil generation to provide back up for intermittent renewables via an availability payment, even though fossil generation is pushed to the margin and has low plant utilisation. EPSs then ensure that fossil generation plants are not built in the event that price-based incentives are not right. Pollitt (2012) also highlights that the motivation for EMR clearly lies with the Committee on Climate Change, 5-year carbon budgeting and the 2008 Climate Change Act.

The key objective of the EMR is to guarantee the level of investment needed in new low-carbon generation capacity and infrastructure in the most cost-effective way possible (DECC 2011). The white paper estimates investments of up to £110 billion in electricity generation (£75 billion) and transmission and distribution (£35 billion) by 2020. Studies prepared by Cambridge Economic Policy Associates (CEPA) and presented in the white paper suggests that using CfD would lead to an overall saving of around £2.5 billion over the period up to 2030.

The EMR key dates are as follows:

– **November 2008**: The 2008 Climate Change Act is introduced. The Committee on Climate Change is established as an independent body to advise the government on meeting carbon budgets;
– **December 2008**: The Committee on Climate Change publishes the first report, setting the electricity sector as key to the decarbonisation strategy, including heat and transport;
– **October 2009**: The Committee on Climate Change First Progress Report details key EMR elements;
– **May 2010**: Coalition agreement specifies four elements of EMR;
– **Dec 2010**: DECC publishes EMR proposals;
– **November 2012**: Energy bill introduced to House of Commons;
– **December 2013**: Energy Act 2013 introduces CFD and a CM

The "2010 to 2015 government policy: UK energy security" policy paper, issued by DECC (2015), states that the EMR currently operates two key mechanisms: CFD and CM.

## 8.1    Contracts for Difference

The EMR proposes a system whereby the government contracts electricity at fixed prices for a long period to be supplied by low-carbon generators. The government would pay the difference between the electricity average wholesale price and the price established in the contract. The EMR white paper indicates that the EU Emissions Trading System (EU ETS) carbon price has been volatile or too low to encourage investment in low-carbon electricity generation in the UK.

## 8.2    Capacity Market

A CM is a mechanism that introduces payments to generators that maintain availability and supply electricity to the market when required, thereby guaranteeing security of supply. At high levels of renewables, a CM might encourage small intermittent generators that do not contract back-up generation directly with fossil generators. The practical problem with CMs is that it is not clear who determines the level of capacity and how they determine it.

## 9    UK Electricity Market Reform and the EU

The UK EMR has been designed for the country's electricity market and targets; however, the national electricity market operates in a European context. Keay (2013) points out that there is a tension, and possibly an incompatibility, in the idea of separating the national energy and emission goals from the single European electricity market. As described in the last section, the UK EMR proposes a system in which liberalisation and environmental concerns transform the electricity sector into a public/private partnership, whereby the government (not the markets) defines the country's generating mix.

The EMR white paper claims that without the EMR, the electricity sector would have emissions intensity in 2030 of over three times the level advised by the Committee on Climate Change. Although this intervention in the market might be needed to support the development of low-carbon power generation, it goes against the concept of a single market in which the sources with the lowest costs, independent of country of origin, should be able to compete across the European market. Other EU countries also have energy and emission goals, but the UK's renewables targets are still seen as among the most ambitious. This might result in a complicated or compromised operation of the European single market.

Legal issues might arise from the reforms since they are designed to support specific sources of electricity generation. State aids such as subsidies or other forms of support of member states of the EU like the UK are bound to the EU State Aid rule. The commission can reject or modify proposed measures for state aids under EU law. According to DECC (2012), the UK government is designing the EMR to be consistent with European legislation. The policy review document also highlights that the UK government is working closely with the EU energy regulatory authorities group ACER and the EU transmission system operators group ENTSO-E to implement both the CFD and CMs.

Keay (2013) also raises questions related to specific elements of the EMR. As the UK approach points to a permanent involvement of the government in the electricity market, the longer the duration of the aid, the more likely it will generate distortions in a competitive marketplace. In the case of imports, it might be more difficult to maintain a certain UK scheme when contracting output from plants in other countries in Europe. It is also difficult to assess the contract of CMs outside the UK as in principle these auctions can be extended to other places in Europe; however, the UK's system might not be prepared to rely on non-domestic CMs. National CMs are more likely to serve national needs and create two separate income streams for generators (capacity and energy payments), lowering the average energy price and creating potential distortions when markets with and without CMs are coupled. A European solution would need to address the various issues related to the operation and specifics of national markets and power exchanges.

## 10    Discussions and Conclusions: Lessons for the UK

The UK implemented the CFD and CMs to address environmental and energy challenges, employing the market system to engage the private sector in investing in renewable energy. Unfortunately, CFD and CM can attract investors who are more focused on guaranteed rewards than on business innovations that could eventually reduce costs (Whitmill 2012). This guaranteed remuneration might undermine the idea that businesses need to think outside the box to ensure profitability (Onifade 2016). As a solution, the government might adjust the policy to encourage innovation, as has been discussed by Bolton and Foxon (2015), Finon (2013) and Kozlov (2014).

There are concerns about the structure of this hybrid system where the government, as the administrator of a market system, would be transferring the burden of financing the currently unstable renewable energy economy to the private sector (Darwall 2015). The author defends the idea that the EMR is the market without its discipline, combined with the inefficiency of the state without financial control and accountability. The electricity sector becomes a public/private partnership analogue to the Private Finance Initiative (PFI), existing in a zone where the state controls but is not financially accountable for the costs, which are paid by consumers, not taxpayers.

Onifade (2016) supports these concerns but highlights that, compared to previous regimes, the government's role in the CFD/CM system appears to be minimal. The central issue surrounding the influence of neoclassical economics on energy policy thinking is therefore profit maximisation versus public interest. It is acceptable throughout the world that governments should protect the public interest by performing regulatory and monitoring functions within the energy sector. In this sense, although the criticism is plausible in terms of profitability of the investment in the energy sector, the EMR in the form of the CFD and CM policies clearly considers environmental protection the priority. Bolton and Foxon (2015) argue that in the UK, as well as in other countries, new policy frameworks need to guide the transition from an energy system designed to achieve short-term efficiencies through market operation to a long-term approach that would embrace new uncertainties.

Both market interests and environmental protection need to be protected to guarantee the levels of investment needed in the UK's renewable energy market. Scholars (Blyth et al. 2015; Bolton and Foxon 2015; Finon 2013; Kozlov 2014; Kannan 2009; Levi and Pollitt 2015; Pye et al. 2015) have addressed some aspects of this issue.

# References

APX. (2016). *APX Power UK.* Accessed July 25, 2016, from http://www.apxgroup.com/trading-clearing/apx-power-uk

Birchall, D., & Dunstan, A. (2010). *The evolving role of Ofgem.* Accessed July 25, 2016, from http://www.utilityweek.co.uk/news/the-evolving-role-of-ofgem/766472#.U2icjvl_s_Y

Blyth, W., McCarthy, R., & Gross, R. (2015). Financing the UK Power Sector: Is the money available? *Energy Policy, 87*, 607–622.

Bolton, R., & Foxon, T. (2015). A socio-technical perspective on low carbon investment challenges—Insights for UK energy policy. *Environmental Innovation and Societal Transitions, 14*, 165–181.

Darwall, R. (2015). *Central planning with market features.* Surrey: Centre for Policy Studies.

DECC. (2011). *Planning our electric future: A white paper for secure, affordable and low-carbon electricity.* Accessed July 25, 2016, from https://www.gov.uk/government/uploads/system/uploads/attachment_data/file/48129/2176-emr-white-paper.pdf

DECC. (2012). *Electricity market reform: Policy overview.* Accessed July 25, 2016, from https://www.gov.uk/government/uploads/system/uploads/attachment_data/file/65634/7090-electricity-market-reform-policy-overview-.pdf

DECC. (2015). *2010 to 2015 Government policy: UK energy security.* Accessed July 25, 2016, from https://www.gov.uk/government/publications/2010-to-2015-government-policy-uk-energy-security/2010-to-2015-government-policy-uk-energy-security#appendix-5-electricity-market-reform-emr.

DECC. (2016). *National statistics energy trends: Electricity.* Accessed July 25, 2016, from https://www.gov.uk/government/statistics/electricity-section-5-energy-trends

ENTSOE. (2016). *Statistical database.* Accessed July 25, 2016, from https://www.entsoe.eu/data/data-portal/Pages/default.aspx

European Commission. (2012). *Energy markets in the European Union in 2011.* Luxembourg: Publications Office of the European Union.

Finon, D. (2013). The transition of the electricity system towards decarbonisation: The need for change in the market regime. *Climate Policy, 13*, 130–147.

Grimston, M. (2014). The full costs of generating electricity. *Journal of Power and Energy, 228* (3), 357–367.

House of Commons. (2010). *The future of Britain's electricity network.* London: House of Commons.

Hughes, G. (2012). *The performance of wind farms in the United Kingdom and Denmark.* London: Renewable Energy Foundation.

Kannan, R. (2009). Uncertainties in key low carbon power generation technologies—Implication for UK decarbonisation targets. *Applied Energy, 86*, 1873–1886.

Keay, M. (2013). *UK Electricity market reform and the EU.* Accessed July 25, 2016, from https://www.oxfordenergy.org/publications/uk-electricity-market-reform-and-the-eu/

Kozlov, N. (2014). Contracts for difference: Risks faced by generators under the new renewables support scheme in the UK. *Journal of World Energy Law and Business, 7*(3), 282–286.

Levi, P., & Pollitt, M. (2015). Cost trajectories of low carbon electricity generation technologies in the UK: A study of cost uncertainty. *Energy Policy, 87*, 48–59.

National Grid. (2011). *National Electricity transmission system seven year statement.* Accessed July 20, 2016, from https://www2.nationalgrid.com/WorkArea/DownloadAsset.aspx?id=43281

National Grid. (2016). *What we do: Electricity.* Accessed July 25, 2016, http://www2.nationalgrid.com/About-us/What-we-do/Electricity

OECD/NEA. (2012). *Nuclear energy and renewables: System effects in low-carbon electricity systems.* Accessed July 25, 2016, from https://www.oecd-nea.org/ndd/pubs/2012/7056-system-effects.pdf

Ofgem. (2009). *Project discovery: Energy market scenarios*. Ofgem. Accessed July 25, 2016, from https://www.ofgem.gov.uk/ofgem-publications/. . ./discoveryscenarioscondocfinal.pdf

Ofgem. (2016). *Infographic: Promoting a sustainable energy future*. Accessed July 25, 2016, from https://www.ofgem.gov.uk/publications-and-updates/infographic-promoting-sustainable-energy-future

Onifade, T. T. (2016). Hybrid renewable energy support policy in the power sector: The contracts for difference and capacity market case study. *Energy Policy, 95*, 390–401.

Pollitt, M. (2012). *Electricity market reform: Will it work and if so how?* University of Cambridge, Energy Policy Research Group. Accessed July 25, 2016, from http://www.eprg.group.cam.ac.uk/wp-content/uploads/2014/01/BeesleyLecturePollitt081112.pdf

Pollitt, M., & Haney, A. (2013). Dismantling a competitive electricity sector: The UK's Electricity Market reform. *The Electricity Journal, 26*(10), 8–16.

Pye, S., Sabio, N., & Strachan, N. (2015). An integrated systematic analysis of uncertainties in UK energy transition pathways. *Energy Policy, 87*, 673–684.

RenewableUK. (2012). *Potential for offshore transmission cost reductions. A report to The Crown Estate*. Accessed July 25, 2016, from https://www.thecrownestate.co.uk/media/5709/RenewableUK%20Potential%20for%20offshore%20transmission%20cost%20reductions.pdf

Royal Academy of Engineering. (2014). *WIND ENERGY implications of large-scale deployment on the GB electricity system*. London: Royal Academy of Engineering.

SERIS. (2012). *Who owns the UK Electricity Generating Industry—and does it matter?* Chesterfield: SERIS.

Simmonds, G. (2002). *Regulation of the UK Electricity Industry*. Bath: The University of Bath.

UKERC. (2010). *Great Expectations: The cost of offshore wind in UK waters—understanding the past and projecting the future*. UKERC. Accessed July 25, 2016, from http://www.ukerc.ac.uk/asset/967F73E3-E952-4CF7-B6C04BC2E978B016/

Whitmill, C. (2012). Is UK Energy policy driving energy innovation or stifling it? *Energy & Environment, 23*, 993–1004.

Wilson, I. G., McGregor, P. G., Infield, D. G., & Hall, P. J. (2011). Grid-connected renewables, storage and the UK electricity market. *Renewable Energy, 36*, 2166–2170.

**Catalina Spataru** is a Lecturer in Energy Systems and Networks at UCL Energy Institute in London. Her main research interests are in energy systems integration and operation, whole energy system modeling with forecasting for future years, impact of renewable energy sources in systems, coupling energy markets, and resource nexus. She has been involved in and led projects funded by industry, research councils, and other funding bodies. In addition, she has supervised PhD students, teaching a smart energy systems module for MSc students at UCL, and is currently Course Director for MRes in energy demand studies.

# Marketing Renewable Energy in Italy

Simona Bigerna, Carlo Andrea Bollino, and Paolo Polinori

**Abstract**

Following the abandonment of nuclear power with a post-Chernobyl referendum, Italian energy policy has created a culture in favor of renewables developed by entrepreneurs and paid for by end users in the form of subsidies. This has led to the success in Italy of incentives for the use of renewables, which in 2015 already met targets set for 2020. Focusing on the Italian case, this chapter initially describes the legal framework and, in particular, the incentive mechanisms and then analyzes the impact of renewables in the vertically integrated Italian electricity market with policy implications. The main results highlight that the massive spread of renewable energy sources (RES) has changed the attitude of policymakers from a command-and-control system to a more simplified and market-oriented approach. In particular, given the past intensive financial efforts, new legislation started to curb new RES investment by setting clear caps on the total financing allotments to the incentive policy. Furthermore, the massive injection of RESs has highlighted the inadequacy of the current electric market design. Finally, the large-scale penetration of RES into everyday life in Italy has increased consumer awareness of green electricity, stimulating a new quest for green electricity and better climate conditions.

**Keywords**

Support schemes • Electricity market • Incentive mechanism • Willingness to pay • Green electricity products

S. Bigerna (✉)
Department of Economics, University of Perugia, Via A. Pascoli, 20, 06123 Perugia, Italy
e-mail: simona.bigerna@unipg.it

© Springer International Publishing AG 2017
C. Herbes, C. Friege (eds.), *Marketing Renewable Energy*, Management for Professionals, DOI 10.1007/978-3-319-46427-5_18

# 1    Introduction

The story of renewables in Italy starts with the radical change in energy strategy following the abandonment of nuclear power with the post-Chernobyl referendum. Given the growing demand for electricity in the 1990s, policymakers have focused their efforts on the development of renewable energy sources (RES) for generating electricity from private companies that are independent of the state monopoly.

The policy was designed to encourage investment incentives in the capital account (capital subsidy) granted for a long enough period of time to individuals and to compel the public monopoly to purchase electricity produced from RES (RES-E). The Italian policy has favoured renewables' development through subsidy schemes paid by end users. This has allowed for the success in Italy of incentives for the use of renewables, following EU directives of 1997 and 2003, that led to Italy's energy policy.

The development of renewables has led to the setting of new ambitious targets for the electricity system worldwide. In Italy, RES have been growing consistently at a fast rate over the last 5 years, producing both negative and positive consequences. More expensive bills, grid problems, the potential losses from competition due to the crowding out of combined-cycle power plants set off against the positive effects of RES deployment related to climate change mitigation. RES have caused a new scenario to emerge in which a novel strategy is required to increase sustainability, the integration of RES into energy systems, promote innovation, and improve competition in the electricity market. This new strategy requires a deep knowledge of the institutional context in which RES have been developed and a constant monitoring of the energy scenario in relation to national and European targets. Furthermore, given the more challenging European targets, it is crucial to assess the feasibility of such targets in a scenario where governments reduce RES subsidies, thereby raising the degree of liberalization in the renewable sector.

In the light of the new European scenario, in this chapter we deeply analyze the Italian institutional context in the primary and retail markets, highlighting the main consequences of the deployment of RES. In addition, the supply and demand sides are analyzed to evaluate the financial sustainability of the challenging European environmental policies. The chapter is organized as follows. Section 2 describes the legal framework, focusing on the incentive mechanisms. Section 3 analyzes the impact of renewables in the vertically integrated Italian electricity market. The chapter concludes with Sect. 4, which presents policy implications.

# 2    Legal Framework

Italy implemented EU Directive 96/92/EC only in 1999, with the launch of Legislative Decree 79/98, which came into force on 1 April 1999 (known as the Bersani Decree, after the minister of industry at the time). The Bersani Decree was important for the Italian electricity system because it imposed an obligation beyond the minimum required by the EU directive to vertically integrated companies. In particular, it forced the monopolist ENEL to create independent companies for production, transmission, distribution, and sales to eligible customers. In addition, the Bersani Decree provided for the establishment of two agencies: (1) a RES operator (GSE, Gestore dei Servizi Energetici), which manages the payment of incentives to RES producers and is authorized to sell RES directly to the power exchange market, and (2) an electricity market operator (GME, Gestore dei Mercati Energetici), which manages the electricity market (the so-called power exchange) by organizing an auction system to meet the demand and supply of electricity in order to determine which production facilities will satisfy demand and how much will be paid; it also manages the market for energy efficiency certificates and that of emissions trading, that is, trading in $CO_2$ emission certificates.

The strong push for ownership unbundling and the obligation to promote RES are the driving forces behind RES deployment. In fact, new independent producers have started to become profitable in RES generation, and this has created a mass diffusion of entrepreneurial spirit. The growth of traditional energy source consumption together with the growth in demand and simultaneous increased attention on the sustainability of the energy system in Italy have kicked off a search for alternative sources of energy. RES have problems, such as greater land use, higher costs compared to plants powered by fossil fuels, and intermittency, which require storage systems, active management of networks, and integration with other sources.

On the retail market, the large-scale development of RES-E production places some new burdens in terms of transparency and competition in the electricity market, as highlighted by the Italian Regulatory Authority for Electricity Gas and Water, henceforth referred to as the Regulatory Authority (2010). In particular, the Regulatory Authority has stressed the need for a classification system that is comprehensible with respect to the several offers/bids of green contracts available in the Italian market. Such a classification system should be able to deal with two main problems: the negative perceptions of RES caused by a lack of knowledge of RES and RES-E by consumers and the low level of transparency of offers/bids.

## 2.1 Incentive Regulation

In Italy several regulatory frameworks have been created to promote RES, such as green certificates (GC), all-inclusive feed-in tariffs, simplified purchase and resale arrangements, net metering, and feed-in premiums.

### 2.1.1 The Mechanism of the Green Certificates

The mechanism of the GC (Fig. 1) is a rather complex one of "obligations" imposed on the producers of energy from fossil fuels and "benefits" offered to producers from RES. The GC system was introduced by the Bersani Decree, which ordered companies that produce or import electricity from non-RES to feed in, starting from

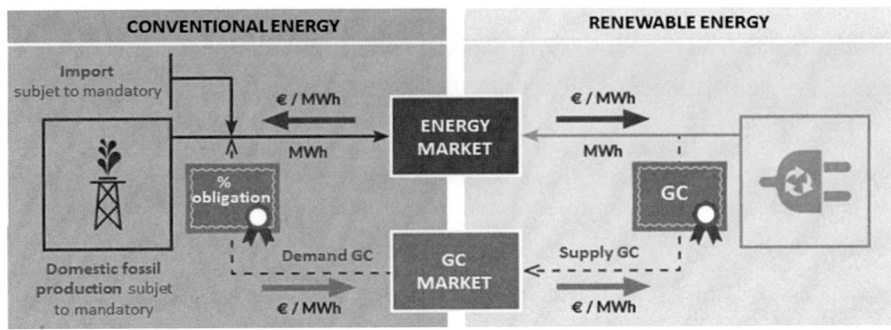

**Fig. 1** Scheme of mechanism of green certificates (GSE)

2001, a share of electricity generated by new or repowered plants, supplied by RES, and entered into service after 1 April 1999.

Producers of fossil fuels are required every year to inject into the energy system a certain percentage of their production as renewable energy. If they do not, they must buy a GC from a RES supplier. In this way, RES producers are granted every year a GC for each megawatt-hour produced, which they may trade, that is, sell to producers of fossil fuels who did not fulfill their obligation. In Italy, the government determines the percentage of obligation and the reference price. The share of RES, initially set at 2 % of energy in excess of 100 GWh, net of cogeneration, of plants' self-consumption and exports, was increased after 2002 with specific decrees, as required by the Bersani Decree, to limit greenhouse gas emissions and meet the country's international commitments to the Kyoto Protocol.

Legislative Decree 387/03, which implements Directive 2001/77/EC, fixed the increase at 0.35 % per year for the period 2004–2006, while the Regulation 244/07 (2008 Budget Regulation) raised the annual increase to 0.75 % for 2007–2012.[1]

The production of RES-E in plants that started operating or were repowered after 1 January 2008, however, is entitled to the certification of renewable generation for the first 15 years of operation. As for all RES, with the exception of photovoltaics (PV), the 2008 Annual Budget Law makes a distinction between plants with an average nominal power greater than 1 MW and plants of an average nominal power not exceeding 1 MW, for which it is possible, upon request, to opt for the all-inclusive feed-in tariff. In the case of plants with an average nominal power greater than 1 MW, the new incentive system is based on the issuance of GC by GSE.

The quantity of GC is calculated by multiplication coefficients differentiated for each specific type of RES for 15 years. To ensure the absorption of the GC by the

---

[1] According to the provisions of Regulation 244/07, the RES-E in plants that started operating or were repowered between 1 April 1999 and 31 December 2007 is entitled to certification of production from RES for the first 12 years of operation.

**Table 1** Evolution of green certificates (1 GC = 1 MWh) (calculated based on GSE data)

| Year | Energy subject to mandatory quota (TWh) | Mandatory rate (%) | Number of GC |
|------|------------------------------------------|---------------------|--------------|
| 2002 | 161.62 | 2.00 | 3,232,400 |
| 2003 | 182.03 | 2.00 | 3,643,200 |
| 2004 | 208.45 | 2.00 | 4,145,800 |
| 2005 | 193.75 | 2.35 | 4,553,073 |
| 2006 | 201.97 | 2.70 | 5,456,337 |
| 2007 | 190.11 | 3.05 | 5,798,350 |
| 2008 | 187.00 | 3.80 | 7,106,189 |
| 2009 | 187.22 | 4.55 | 8,518,286 |
| 2010 | 155.48 | 5.30 | 8,204,370 |
| 2011 | 147.84 | 6.05 | 8,944,202 |
| 2012 | 170.69 | 6.80 | 11,607,230 |
| 2013 | 168.11 | 7.55 | 12,692,129 |

market, the legislation determined that, starting from 2008 until the minimum coverage target of 25 % of RES-E is reached, GSE, at the request of the producer, should withdraw the GC in excess of demand and leave only those necessary to fulfill the obligation for the minimum rate of the previous year.

The GC mechanism was in operation from 2002 to 2012 (Table 1) and was abandoned for plants entering into operation after 31 December 2012.

## 2.1.2 Mechanism of All-Inclusive Feed-In Tariff

The all-inclusive feed-in tariff is a national scheme applicable to RES-E plants (excluding solar plants) commissioned after 31 December 2007 and with a nominal real power of less than 1 MW (200 MW for onshore wind plants). The all-inclusive tariff was introduced by the 2008 financial reform and then completed by the ministerial decree of 18 December 2008 (Renewables Decree) and by the resolution ARG/elt 1/09 of the Regulatory Authority. The tariff benefit is designed to promote small plants and is granted over a period of 15 years, during which its rate remains fixed and based on the amount of electricity fed into the grid, for all plants commissioned by 31 December 2012. It is the first mechanism in Italy that is differentiated by type of RES (Table 2).

According to Legislative Decree 28/11, the all-inclusive feed-in tariff remains constant for the entire entitlement period for all plants (Fig. 2) entering into operation by 31 December 2012; the tariff applies "also to biogas plants owned by farms or managed in connection with agricultural, food, farming and forestry that have entered into commercial operation before 1 January, 2008".

This represents an alternative to the GC scheme, so the right of option between GC and the all-inclusive tariff is exercised upon submitting an application for RES-E qualification. Before the end of the support period, only one transition is allowed from one support scheme to the other; in this case, the duration of the

**Table 2** All-inclusive feed-in rate (calculation based on GSE data)

| Renewable energy sources | All-inclusive feed-in rate (Eurocent/kWh) |
|---|---|
| Wind ($P < 200$ kW) | 30 |
| Geothermal | 20 |
| Wave and tidal | 34 |
| Hydro (other than the one indicated in previous point) | 22 |
| Biomass, biogases, and bioliquids complying with (EC) Regulation 73/09 | 28 |
| Landfill gas, sewage treatment plant gas, biogases, and bioliquids complying with (EC) Regulation 73/09 | 18 |

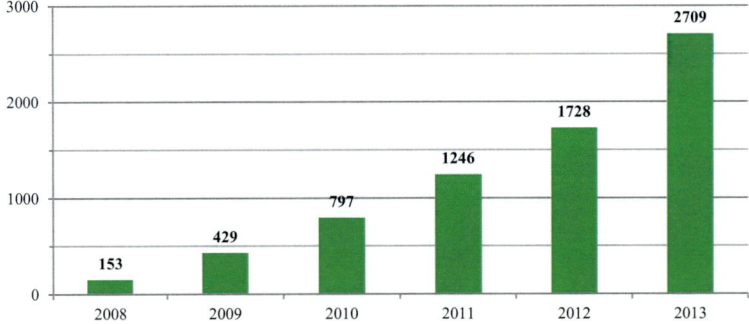

**Fig. 2** Number of plants in all-inclusive feed-in regime (calculation based on GSE data)

period of eligibility for the new support scheme is reduced by the period of eligibility that has already elapsed under the previous scheme.

### 2.1.3 Mechanism of Simplified Purchase and Resale Arrangements

The mechanism of simplified purchase and resale arrangements, introduced 1 January 2008, is not a subsidy but a facilitation of electricity sale to the grid. The mechanism of simplified purchase and resale arrangements for power production was established by Regulatory Authority Decision 280/07 and by the "Modalities and technical and economic conditions for the withdrawal of electricity in accordance with article 13, paragraphs 3 and 4, of Legislative Decree 29 December 2003 n. 387, and paragraph 41 of the Law of 23 August 2004 n. 239." This mechanism can be combined with GC and the feed-in tariff (only for the feed-in tariff must the plant be above 200 kW). Under these arrangements, producers sell the electricity generated to GSE instead of through bilateral contracts or directly on the Italian Power Exchange (IPEX). Eligible producers include plants having a nominal power of less than 10 megavolt ampere (MVA): RES plants or hybrid plants for the RES-E; plants of any capacity using wind, solar, geothermal, waves, tides, or hydro (run of river only); plants with a nominal power of less than 10 MVA: non-RES plants or hybrid plants for the portion of electricity generated from non-RES; plants having a nominal power greater than or equal to 10 MVA; plants using RES other than wind, solar, geothermal, waves, tides, or hydro (run of river only), provided that they are owned by a self-producer (as defined in article 2, paragraph 2, Legislative Decree 79/99). For access to the simplified purchase and resale arrangements, the producer remits to GSE a fee to cover the administrative costs up to a maximum of EUR 3,500.00 per year per plant. For plants of nominal power up to 50 kW, the producer remits to GSE an additional fee for meter aggregation. Thanks to this agreement between the producer and GSE, GSE (1) purchases and resells electricity to be fed into the grid at the zonal price or at a minimum guaranteed price, (2) transfers on behalf of the producer the fees for the use of the grid (dispatch and transmission fees) to distributors and to the transmission system operator. The price applied to the electricity purchased by GSE and injected into the grid is known as the average zonal price, that is, the average monthly price per hourly band, which is set on IPEX for the market area to which the plant is connected. Producers with small plants (nominal electrical capacity of up to 1 MW) benefit from guaranteed minimum prices for the first 2 million kWh per year, and they may receive more if the hourly zonal prices prove to be more advantageous. The guaranteed minimum prices are updated annually by the Regulatory Authority. At the end of each year, GSE makes adjustments for the plants, assessing whether the hourly zonal prices are higher than those resulting from the application of the minimum guaranteed prices.

### 2.1.4    Mechanism of Net Metering

Net metering was introduced by Regulatory Authority Decision ARG/elt 74/08 (subsequently amended and supplemented by Decision ARG/elt 186/09) and aimed at valuing the electricity injected into the grid according to predefined economic compensation criteria. The economic conditions of net metering are more convenient than those of simplified purchase and resale arrangements. Under the service, the electricity generated by a consumer/producer in an eligible onsite plant and injected into the grid can be used to offset the electricity withdrawn from the grid. This service has been activated at the request of interested parties since 1 January 2009. In particular, owners of one or more of the following plants may apply for the net metering service: (1) RES-E plants with a capacity of up to 20 kW, (2) RES-E plants with a capacity of up to 200 kW (commissioned after 31 December 2007), (3) high-efficiency combined heat and power plants with a capacity of up to 200 kW.

Consumers receive a monetary payment that offsets the value of the electricity produced and fed into the grid and the electricity withdrawn from the grid at different periods of time. The contribution is determined by GSE considering several issues, such as the characteristics of the plant and the type of contract signed by customer and supplier.

The related contribution paid by GSE to the customer provides (1) the minimum value of the energy cost and the equivalent in euros of electricity injected into the grid and (2) a reversal of service charges limited to the energy exchanged with the network.

For signing the agreement, the selling point and the point of entry must be the same, except in cases where the plants are powered by RES and (1) the user of net metering is a municipality, owner of the plant, with a population of up to 20,000 residents, and (2) the user of net metering is the Ministry of Defense, which is given unlimited power. Generally, the service of net metering produces a beneficial effect if, on an annual basis, the value of electricity injected into the grid is fully compensated for by the charges associated with the energy quantities of electricity withdrawn from the grid. Furthermore, for the total quantity of electricity exchanged with the network, the user of net metering is reimbursed by GSE for costs incurred to use the network in connection with transport services, dispatching users[2] who are owners of plants powered by RES, and general system charges. Net metering is not compatible with simplified purchase and resale arrangements or with the all-inclusive feed-in tariff.

### 2.1.5 Mechanism of Feed-In Tariff for PV

The feed-in tariff is an economic incentive for electricity generated from PV, thereby encouraging electricity production from RESs. Italy introduced the first feed-in tariff in 2005 (ministerial decree of 28 July 2005), followed by the introduction of several schemes. The fourth feed-in scheme, introduced with the ministerial decree of 5 May 2011, applies to PV with a capacity of at least 1 kW and commissioned between 1 June 2011 and 31 December 2016.

The impact of the five different schemes in terms of cumulative number of plants is shown in Fig. 3. The tariff differs depending on the capacity and type of plant and is granted over a period of 20 years. For plants commissioned by 31 December 2012, the feed-in premium scheme provides for a tariff for the electricity produced. The electricity fed into the grid may be purchased by GSE (simplified purchase and

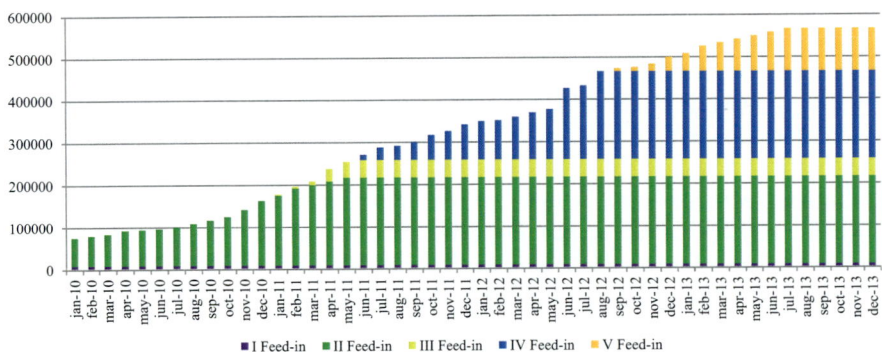

**Fig. 3** Cumulative number of plants under feed-in tariff (2010–2013) (calculation based on GSE data)

---

[2]According to the GME definition *Dispatching User* or *Dispatching Customer* is a party that has entered into a dispatching service contract with the TSO (Terna S.p.A.). For each Offer Point, it is the only party who/which is required to submit Offers/Bids into the Ancillary Services Market in respect of Offer Points authorised for this market.

resale arrangements) or economically offset by the value of electricity withdrawn from the grid net metering service. The fifth feed-in scheme introduced by the ministerial decree of 5 July 2012 has redefined the new tariff incentives for PV production starting in 2013. The novelty of the fifth feed-in scheme is that it sets a total limit on new incentives. This limit was reached already on 6 July 2013 at a cumulative annual cost of incentives of EUR 6.7 billion (Resolution 250/2013/R/EFR).

However, the fourth feed-in scheme is still in force for (1) small PV plants, PV plants integrated with innovative features, and concentrating solar power (CSP) coming into operation before 27 August 2012; (2) large PV plants listed in the relevant registers and producing certification of completion of work within 7 months (or 9 months for plants with a capacity greater than 1 MW) since the publication of the resulting ranking; (3) PV plants built on public buildings and public areas that went into operation by 31 December 2012.

## 2.1.6    Feed-In Tariff Mechanism for Concentrated Solar Power

The feed-in tariff support scheme for CSP is regulated by the ministerial decree of 11 April 2008, as subsequently amended by the ministerial decree of 6 July 2012. Plants eligible for support are new CSP plants (including hybrid ones) commissioned after 18 July 2008 (date of Regulatory Authority's Decision 95/08, implementing the aforementioned legislation) and meeting the following requirements: (1) they must be connected to the power grid (or to small isolated grids) and each plant must have a single point of connection; (2) they do not use substances and preparations classified as very toxic, toxic, or harmful under Directives 67/548/EEC and 1999/45/EC, as subsequently amended (plants located in industrial sites are not subject to this requirement); (3) feed-in tariffs for CSP plants vary, depending both on the nonsolar fraction, that is, the percentage of electricity from nonsolar sources generated every year, and the capturing surface, that is, the sum of the surface areas of all solar collectors that are part of the CSP (or hybrid) plant. The support period is equal to 25 years, beginning on the date of commissioning of the plant. Incentives are thus awarded only for electricity generated by the plant from the solar source; this electricity is measured by a metering system placed after the generators of the plant. Feed-in tariffs add to revenues from the sale of electricity generated and injected into the grid. Plants commissioned by 31 December 2012 are eligible for subsequent feed-in tariffs (Table 3).

Plants commissioned in the period from 31 December 2012 to 31 December 2015 and having a capturing surface of up to 2,500 m$^2$ are eligible for the following feed-in tariffs (Table 4).

Plants commissioned in the period from 31 December 2012 to 31 December 2015 and having a capturing surface exceeding 2,500 m$^2$ are eligible for subsequent feed-in tariffs (Table 5).

The values of the tariffs relate to plants commissioned in the period from 18 July 2008 (date of Decision 95/08 adopted by Regulatory Authority in compliance with the ministerial decree of 11 April 2008) to 31 December 2015. For plants

**Table 3** Feed-in tariffs for plants commissioned by 31 December 2012 (GSE 2016a, b)

| Non-solar fraction | Tariff (EUR/kWh) |
|---|---|
| Below 0.15 | 0.28 |
| 0.15–0.50 | 0.25 |
| Above 0.50 | 0.22 |

NB: The non-solar fraction (Fint) of a solar thermodynamic plant is defined as the share of net electricity generated from non-solar sources, expressed by the following relation: $Fint = 1 - Ps/Pne$, where Ps is the net electricity generated from the solar source and Pne is the net electricity generated by the plant

**Table 4** Feed-in tariffs for small plants commissioned from 31 December 2012 to 31 December 2015 (GSE 2016a, b)

| Non-solar fraction | Tariff (EUR/kWh) |
|---|---|
| Below 0.15 | 0.36 |
| 0.15–0.50 | 0.32 |
| Above 0.50 | 0.30 |

NB: The non-solar fraction (Fint) of a solar thermodynamic plant is defined as the share of net electricity generated from non-solar sources, expressed by the following relation: $Fint = 1 - Ps/Pne$, where Ps is the net electricity generated from the solar source and Pne is the net electricity generated by the plant

**Table 5** Feed-in tariffs for big plants commissioned from 31 December 2012 to 31 December 2015

| Non-solar fraction | Tariff (EUR/kWh) |
|---|---|
| Below 0.15 | 0.32 |
| 0.15–0.50 | 0.30 |
| Above 0.50 | 0.27 |

NB: The non-solar fraction (Fint) of a solar thermodynamic plant is defined as the share of net electricity generated from non-solar sources, expressed by the following relation: $Fint = 1 - Ps/Pne$, where Ps is the net electricity generated from the solar source and Pne is the net electricity generated by the plant

commissioned in the period from 1 January 2016 to 31 December 2016, the tariffs for the year 2015 will be cut by 5 %. For plants commissioned in the period from 1 January 2017 to 31 December 2017, the tariffs for the year 2015 will be cut by another 5 % (rounded to the third decimal place). In absence of further decrees, the tariffs set by the ministerial decree of 6 July 2012 for plants commissioned in 2017 will continue to apply for the years following 2017.

### 2.1.7   Feed-In Tariff Mechanism for Heating and Cooling

The ministerial decree of 28 December 2012, through the renewable energy for heating and cooling support scheme, introduced a specific support scheme to promote the use of thermal energy from RES and to improve energy efficiency. This scheme is one of the measures envisaged by the Legislative Decree 28 of 3 March 2011 to promote meeting the European mandatory targets assigned to Italy in terms of the RES share of energy consumption. Public administrations and private parties are eligible for incentives with reference to the following projects: energy efficiency improvements in existing building envelopes, replacement of existing systems for winter heating with more efficient ones, and replacement and eventual construction of new renewable-energy systems. There is a fund of EUR 900 million per year available, of which EUR 700 million is for private parties and EUR 200 million for public administrations. The incentive is spread out over a period of 2–5 years.

## 2.2    Incentive Schemes: Recent Developments

It should be stressed that the A3 tariff component is charged to end users for the promotion of RES-E production. The A3 tariff is finalized to cover the difference between the costs incurred by GSE for paying subsidies and purchasing electricity and the amount of revenue from selling energy on the electricity market. The amount needed to finance the A3 tariff has grown exponentially, increasing from EUR 3 billion in 2009 to almost EUR 15 billion in 2016.

From the second half of 2016 we are witnessing, instead, a reduction of A3 requirements, primarily owing to the end of the incentive period for some large plants. The calculations do not take into account the provisions of paragraphs 149 to 151 of article 1 of the law of 28 December 2015, no. 208 (Stability Law for 2016). According to this law, there are new incentives for electricity generation from biomass, biogas, and sustainable bio-liquids.

### 2.2.1    Incentive Mechanisms: Synthesis of Framework 2013–2018

**Number of contracts for incentives**  The increase in the number of contracts in recent years is mainly due to the strict adherence to the net metering mechanism. The end of the incentive period of large plants under the previous regime (CIP 6 and GC) reduces the A3 requirement, partially compensating for this increase. The resulting effect is, however, a significant decrease in power and energy managed by GSE.

**Incentive costs**  The end of the incentive period for the beneficiaries of the previous schemes (CIP-6[3] and GC) will lead to a gradual decrease in the share of energy benefitting from the incentive scheme and, therefore, a decrease in the energy sold by GSE on the electricity market. This phenomenon and the publication of specific measures aimed at reducing the electricity bill—including, for example, the so-called smoothing-photovoltaic incentives (Act 116/14) and smoothing RES incentives (Act 9/14)—led to a reduction of A3 requirements in 2015. It should be noted, however, that a new increase is expected in the estimated A3 requirements

---

[3]CIP-6 stands for the Inter-Ministerial Prices Committee Resolution 6 of 1992, which introduced an incentive mechanism for energy from RES and assimilated RES (e.g. cogeneration).

for 2016, primarily as a result of the GC mechanism. In addition to the costs associated with the buy-back of unsold certificates enacted by GSE of unsold titles, there are also costs related to the new incentive tariffs that will replace the GC scheme starting in 2016 (article 19 of the ministerial decree of 6 July 2012). Starting in 2017, the A3 economic requirements are expected to decline, partly as a result of the end of the incentive period for large plants.

**Feed-in tariff schemes** Law 11/15 has postponed until 31 December 2015 the deadline for the operational start-up of renewable energy plants in the earthquake zones of Emilia-Romagna, Veneto, and Lombardy. The number and size of the contracts may decrease as a result of inspection actions. Indeed, inspection actions in the period July–October 2015 yielded around 30 cases of incentive repeals. The incentive burden was reduced in 2015 due to the application of article 26.3 of Law 116/14 ("smoothing-incentives"), with a related savings of around EUR 400 million. Note that in the two-year period 2030–2031, the incentive period for about half of the existing plants will come to an end, leaving a total capacity of around 12 GW.

**Green certificates** The GCs are negotiable securities issued by GSE in proportion to the energy produced by RES qualified plants. During 2015 and 2016, the incentive period for many plants has ended or will end. The total capacity entitled to incentives can decrease as a result of inspection actions. The estimated financial A3 requirement increases in the year 2016. In addition to costs associated with the withdrawal of the remaining GC (relative to production in the second half of 2015), new incentives will be paid for 2016 production, according to article 19 of the ministerial decree of 6 July 2012. Starting in 2017, however, only the incentives provided for by article 19 of the ministerial decree 6 July 2012 will be paid. In addition, since 2015, Law 9/14 ("smoothing RES") has envisioned, on a voluntary basis, a reduction of the annual tariff and an extension of the incentive period.

**Feed-in tariff for small plants** The feed-in tariff for small plants is granted to qualified small RES plants (with a nominal capacity of up to 1 MW and 0.2 MW only for wind power plants) for a period of 15 years. The number and size of contracts may decrease as a result of inspection actions. The change in the cost of an energy buy-back is regulated by article 5, paragraph 7-bis of Legislative Decree 69/13. In addition, since 2015, Law 9/14 has envisioned, on a voluntary basis, a reduction of the annual tariff and an extension of the incentive period.

**CIP-6** The RES plants complying with the specific CIP-6 mechanism benefit from a contract enacting specific rates computed on the basis of the defrayed cost of installation and operation, the defrayed cost of fuel, and the incentive component. The Law 99/09 promoted voluntary resolution mechanisms early in CIP-6 contracts. Over the years there has been a decrease in the total CIP-6 power owing to early termination and natural expiration of the existing contracts. Overall, the amount of CIP-6 energy sales by GSE is decreasing in the electric market.

**Other RES** The incentives provided by ministerial decree 6 July 2012 apply to installations (not solar) that started operations as of 1 January 2013. The incentives are provided up to a cumulative cost relating to RES (non-PV) plants, other than PV, totaling EUR 5.8 billion per year. This value, on 30 September 2015, amounted to EUR 5.767 billion. The increase in the energy buy-back cost and incentive is estimated based on specific assumptions of entry into operation of the plants on the admissible register list. The ministerial decree of 6 July 2012 foresaw, in fact, a time limit for the operation start of 40 months since the closing of third admissible register list (August 2014). The Ministry of Economic Development is currently engaged in the publication of a new ministerial decree about RES incentives for plants (non-PV), starting in 2016.

**Energy buy-back scheme** The decreasing trend in energy buy-backs by GSE is due to the decrease in the number of existing contracts. The main causes are (1) the exclusion from guaranteed minimum prices since 2014 for some RESs (DL 145/13), (2) the application of balancing costs since 2013 (Decision 281/2012/R/efr), and (3) the conversion of some contracts to net metering since 2015 (due to the increase of the power limit to 500 kW to access qualification SEU—Law 116/14). The sharp reduction in the cost of buy-backs in 2014 depends firstly on the exclusion of certain energy plant from the benefit of the Guaranteed Minimum Price scheme (price effect of approximately EUR 490 million) and secondly on the end of a significant number of GSEs' dispatching contracts (quantity effect of approximately EUR 150 million).

**Net metering** There will be a gradual increase in the amount of power and the number of contracts, resulting in an increase in energy buy-backs by GSE (energy bought from individual RES plants and sold on the electric market) due to the increase to 500 kW of the power limit and direct access to qualification SEU (Law 116/14). The A3 requirements declined in 2014 as a result of the new method of calculation of the service fee, which provides for the introduction of a limit on the exemption from general system charges (Deliberations Regulatory Authority SI 570/2012/R/efr and 614/2013/R/efr).

## 2.3    Regulatory Framework: Recent Developments

Several new national legislative and regulatory laws were introduced in 2015, primarily focusing on the RES incentive mechanisms. Energy efficiency measures for public administration buildings were enacted by ministerial decree (application of Legislative Decree 28, 2011, article 40). In addition, new measures provided better coordination of the policies and actions envisioned by the National Fund for energy efficiency. GC measures for biomass controls were improved, amending the previous ministerial decree of 2010 requiring new tracing procedures for biomass. New measures for tax deductions were enacted by Law 208 (28 December 2015) starting in 2016, confirming tax reductions of 65 % for measures to improve energy efficiency and a 50 % deduction for building renovations. Incentives for biomass, biogas, and sustainable bioliquids were extended for 5 years. Environmental protection and the promotion of the green economy were the core of Law 221 (28 December 2015). It includes incentive arrangements for products derived from waste recovery and materials resulting from the dismantling of complex products, management of end of life of PV panels, construction requirements of thermal plants, and byproduct usage in biomass plants and biogas. New measures were enacted to improve the methodology for monitoring regional objectives in terms of the RES share of total consumption.

New, simplified rules for RES plant construction were enacted by ministerial decree (19 May 2015). Specifically, communication on the construction, connection, and operation of electricity production plants from RES and for the installation and exercise of micro-cogeneration units must use a single model. Procedures for the construction, connection, and operation of small PV systems integrated on the roofs of buildings were also simplified. Furthermore, new legislation in 2015 set guidelines for the National Statistical Program to monitor: heat derived from RESs; thermal energy produced by heat pumps, solar thermal collectors, and geothermal resources; and national consumption targets for RES. New legislation in 2015 concerning the incentive mechanism of white certificates has improved the monitoring process, updating the format for technical communications (21 T, 22 T, 36E, 40E, 47E data sheet).

In June 2016, a decree for renewables (excluding PV) was approved, providing EUR 435 million a year for new plants, fully reconstructed, reactivated, and upgraded and came into effect on 1 January 2013. The decree provides specific incentives for each source. In particular, to the most efficient "mature" technologies (such as wind) is assigned about half of the available resources. The remainder is equally divided among high-potential technologies that have strong growth prospects and the potential to penetrate foreign markets (such as solar thermal) and biological sources whose use is connected to the potential of the circular economy. Overall, the sector with a budget for the greatest incentives will be biomass, with EUR 105 million for energy recovery from waste and agricultural residues. The CSP will amount to EUR 98 million for the development of innovative technologies, and wind will be allocated EUR 85 million for onshore and EUR 10 million for offshore. Hydropower will be EUR61 million, and EUR 37 million will go to geothermal. However, EUR 29 million will be made available for so-called reworking, that is, to encourage the reclamation and redevelopment of old plants with the objective of ensuring adequate power efficiency (wind and water) with respect to costs and no new environmental impacts. The incentive period will last 20 years (25 for solar thermal). Finally, the development of green products and green contracts in the Italian retail market will require increased transparency of the regulatory framework, which will have a positive impact on consumer trust and confidence. In some countries there exists a mix of policy interventions ranging from private intervention by end user associations to regulatory acts of regulatory authorities.

In Italy, the Regulatory Authority approved a set of rules to ensure that electricity sold to individual customers is produced from RESs, also ensuring that it is not further commercialized. The tool used is the *garanzie di origine*, or guarantee of origin, that is, an electronic document used to certify that a given quantity of electricity has been produced from RES. Such a tool makes it possible to protect consumers in a context where there are lots of RES-E offers. In particular, the Regulatory Authority (2010) requires that the following information be indicated: (1) for *energy-based* contracts, the share of RES-E (or kilowatt-hour) or the share of electricity (or kilowatt-hour) produced by low-emission plants or the share of electricity (or kilowatt-hour) produced by high-efficiency plants; (2) for *fund- or project-based* contracts, the contribution for the project funding related to RES plants, low-emission plants or high-efficiency plants. In this way, the goal of the Regulatory Authority is twofold: to promote RES consumption through increased transparency and to develop tools for the promotion of a wide range of environmental services associated with electricity production.

## 3     Renewable Energy Sources: Looking Inside the Italian Electricity Market

Total RES-E generation has increased its penetration significantly in recent years in Italy, from around 54 TWh, that is, 18 % of gross electricity generation in 2004, to 112 TWh in 2013, that is, 39 % of gross electricity generation (Figs. 4 and 5). The contribution of hydropower saw some variability in the period 2004–2013, while geothermal energy was stable and PV and wind have contributed increasing shares to total RES-E generation (Table 6).

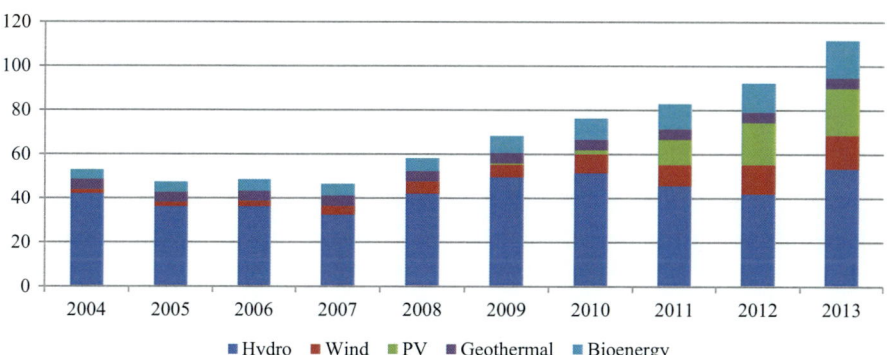

**Fig. 4** Trend of RES in Italian electricity mix, 2004–2013 (calculation based on TERNA data)

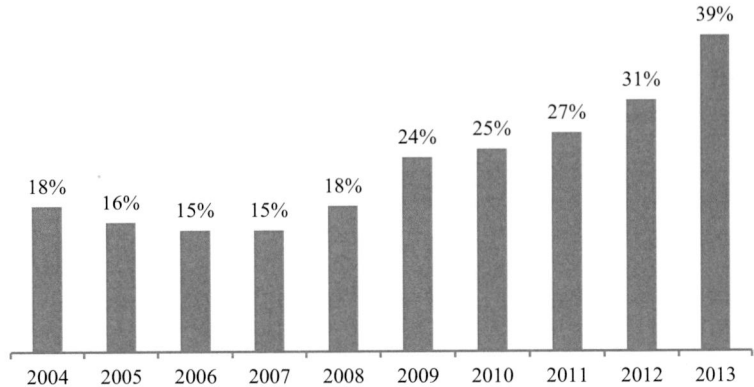

**Fig. 5** Share of RES in Italian gross electricity production, 2004–2013 (calculation based on TERNA data)

## 3.1    Supply Side

The fast-paced deployment of RES since 2010 has led to important changes in the Italian electricity market.

First, there has been a reduction in the number of plants operating in a competitive environment. Indeed, RES enjoy dispatching priority in the electricity market and then act as "price takers."

In addition, the solution mechanism of the market based on the marginal price implies that all plants are compensated at the most expensive price of the system selected in the day-ahead market (DAM). RES generators, which have very low or zero marginal costs, bid at (or close to) zero in the DAM and are compensated at the price of the more expensive thermal power plant without participating in the market. Therefore, competition occurs only among thermal power plants.

Second, RES might create congestion and security problems in the power network. Third, the capacity defined in the DAM is not easily executable because the plants selected based on economic merit do not guarantee a sufficient level of reserves (primary, secondary, and tertiary) that cannot be supplied by PV or wind power plants. Then, other markets have taken on the improper function of being "correct" on the DAM outcome to ensure the security of the system. Of course, such changes are expensive.

Fourth, the DAM does not distinguish between and valorize the energy supplied by thermal power plants and that offered by RES plants, which is more uncertain and therefore requires the availability of reserves, but may not offer reserves' services. This inequity generates a demand for a so-called capacity payment in favor of thermal power plants that allow for the operation of the electrical system (GME 2014). The strong growth of PV from 2010 to 2013 (Fig. 6) explains the important increase in RES total electricity consumption, from 3 to 4 % in 2010 to 30 % in 2013.

**Table 6** Trend of RES in Italian electricity mix and share of RES (calculation based on TERNA data)

| GWh | 2004 | 2005 | 2006 | 2007 | 2008 | 2009 | 2010 | 2011 | 2012 | 2013 |
|---|---|---|---|---|---|---|---|---|---|---|
| Hydro | 42,338 | 36,067 | 36,994.3 | 32,815.2 | 41,623 | 49,137.5 | 51,116.8 | 45,822.7 | 41,875 | 52,773 |
| Wind | 1,847 | 2,343 | 2,971 | 4,034 | 4,861 | 6,543 | 9,126 | 9,856 | 13,407 | 14,897 |
| PV | 4.0 | 4.0 | 2.3 | 39 | 193 | 677 | 1,906 | 10,796 | 18,862 | 21,589 |
| Geothermal energy | 5,437 | 5,325 | 5,527 | 5,569 | 5,520 | 5,342 | 5,376 | 5,654 | 5,592 | 5,659 |
| Bioenergy | 4,499 | 4,845 | 5,286 | 5,441 | 5,966 | 7,557 | 9,440 | 10,832 | 12,487 | 17,090 |
| Total | 54,125 | 48,584 | 50,781 | 47,899 | 58,164 | 69,255 | 76,964 | 82,962 | 92,222 | 112,008 |
| **Percentage** | **2004** | **2005** | **2006** | **2007** | **2008** | **2009** | **2010** | **2011** | **2012** | **2013** |
| Hydro (%) | 78 | 74 | 73 | 69 | 72 | 71 | 66 | 55 | 45 | 47 |
| Wind (%) | 3 | 5 | 6 | 8 | 8 | 9 | 12 | 12 | 15 | 13 |
| PV (%) | 0 | 0 | 0 | 0 | 0 | 1 | 2 | 13 | 20 | 19 |
| Geothermal energy (%) | 10 | 11 | 11 | 12 | 9 | 8 | 7 | 7 | 6 | 5 |
| Bioenergy (%) | 8 | 10 | 10 | 11 | 10 | 11 | 12 | 13 | 14 | 15 |
| Total (%) | 100 | 100 | 100 | 100 | 100 | 100 | 100 | 100 | 100 | 100 |

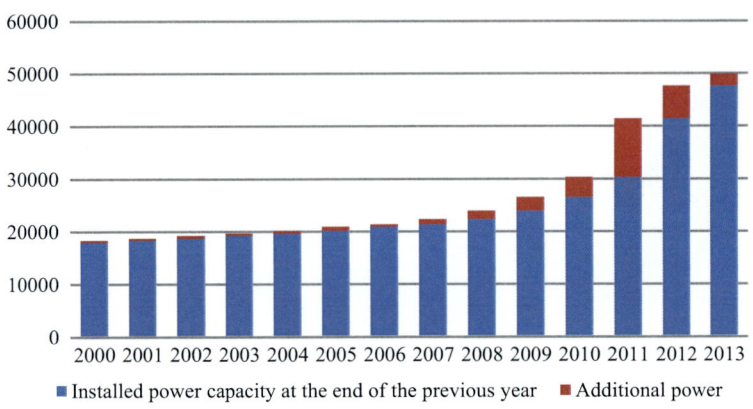

**Fig. 6** RES growth in Italy (MW) (calculation based on TERNA data)

**Table 7** IPEX main indicators (2010–2013) (GME annual reports, 2011–2014)

| Variable | Unit | 2010 | 2011 | 2012 | 2013 |
|---|---|---|---|---|---|
| Volumes trade on MGP | TWh | 318.56 | 311.49 | 298.67 | 289.15 |
| Liquidity | % | 63.1 | 58.2 | 60.1 | 72.0 |
| Price: Average | EUR/MWh | 64.12 | 72.23 | 75.48 | 62.99 |
| Minimum | | 10.00 | 10.00 | 12.14 | 0.00 |
| Maximum | | 174.62 | 164.80 | 324.20 | 151.88 |
| Number of participants, 31 December | Number | 207 | 192 | 200 | 223 |
| ENEL share | Percentage | 27.9 | 26.2 | 25.4 | 25.2 |
| Single-buyer share (Net CIP6) | | 21.11 | 26.67 | 22.35 | 13.04 |
| Sales by source | | | | | |
| Combined cycle | Percentage | 55.43 | 52.94 | 45.12 | 38.11 |
| Coal | | 9.04 | 11.20 | 12.81 | 10.80 |
| Other | | 11.34 | 11.51 | 11.50 | 12.07 |
| Hydropower | | 15.64 | 14.49 | 13.96 | 18.67 |
| Geothermal | | 1.89 | 2.06 | 2.10 | 2.18 |
| Wind | | 2.07 | 2.75 | 4.08 | 5.81 |
| Solar and other | | 2.45 | 3.48 | 9.24 | 11.00 |
| Pumped storage | | 2.15 | 1.57 | 1.19 | 1.36 |

Indeed, at the end of 2010, Italy had 156,000 PV plants with an installed capacity of 3500 MW. Installed capacity more than tripled over 2009 levels. As in 2010, the growth of these plants in 2011 was again outstanding, with 330,000 PV plants, that is, 12,700 MW of installed capacity.

The number and capacity of PV plants grew at a very fast pace in 2012, with 481,000 plants and 16,700 MW of installed capacity, and in 2013, with 591,000 plants and 18,000 MW of installed capacity (Table 7).

On the other hand, PV production increased by 13 % in the same period. The amount of electricity produced through the use of renewable incentives amounted

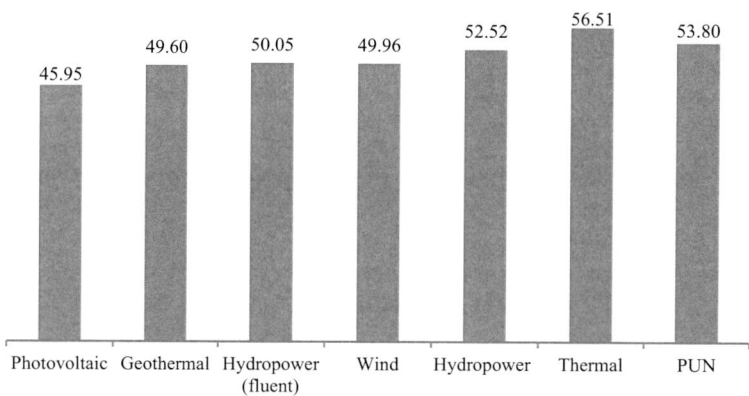

**Fig. 7** Day-ahead market: electricity price by RES and PUN - Prezzo Unico Nazionale, Single National Price- (our elaboration from GSE data; prices are in EUR/MWh, 2014)

to around 65 TWh in 2015, and the related aggregated cost has been approximately EUR 12.5 billion, of which around EUR 12.3 billion was covered by the A3 component of the electricity bill. Finally, information on electricity prices according to the different RES is shown in Fig. 7. The prices are quite similar, that is, the maximum difference is 20 % between thermal (most expensive) and PV (cheapest). On average, with the exception of thermal sources, all RES are cheaper than the average energy source mix. Indeed, the national average price in 2014 was EUR 53.80/MWh.

There has been an increase in the supply of green products in the Italian retail electricity markets in 2016. Both large and small operators offer different kinds of contracts for several types of customers. Operators involved in the renewable, green, or sustainable electricity supply are in many cases producers and retailers (e.g. vertically integrated operators). In 2015, the number of operators selling electricity on the free market was 487 vs. 450 in 2014. This result confirms the continued growth in the number of suppliers, especially of small operators that are not able to meet the needs of the entire country, which started in 2007, the year of liberalization. In the entire retail electricity market, overall sales grew by 2.2 % in 2015 compared to 2014; Enel and Edison are still the largest players. In 2015, Enel sold 85.4 TWh, equal to 33.7 % of the total, while Edison sold 17.1 TWh, approximately 7 % of the total. On the final sales market (end users), Enel controls 74 % of the electricity consumed by households, but Edison is the leading operator, followed by Enel, with respect to medium- and high-voltage end users (Regulatory Autority 2016a, b).

Focusing on RES-E production, two large players have a considerable share of green production (Nomisma 2012). Without taking into account nuclear power, Enel has a share of renewable equal to 38 %, while Edison's share is equal to 10 %. Other players with considerable shares of renewables in electricity production are A2A (39 %) and Iren (21 %). In Italy, around 70 % of utilities offer green products or options in the free market. This trend started in Europe in the late 1990s owing both to liberalization in the electricity markets and to environmental concerns. In this context, utilities understood the added value of green products and so started to offer them. It should be stressed that it is very difficult to compare all Italian green products data. However, using an institutional Website and the Regulatory Authority's Website (*Trova offerte*), we made a first comparison between green and conventional products for flat and time-in-use contracts for low-voltage clients, assuming a consumption of 2700 kWh and a power of 3 kW.[4] With reference to high- and medium-voltage consumers, we have estimated the number of contracts potentially provided by the main operators. In the Italian retail market, the differential between RES-E and normal electricity is quite small. If we consider flat products, the green ones have an average price of around EUR 0.08/kWh vs. EUR 0.07/kWh of non-green products, meaning that the premium is, on average, 11–12 %. With reference to the time-of-use product, the average premium price for peak consumption is negative and relatively close to zero. It amounts to EUR −0.14/kWh that is 2 % lower than the average flat tariff. In the off-peak period, the premium is positive and equal to EUR 0.018/kWh, that is, 30 % higher than the average conventional premium. These results could mean that on the free market the rates should induce consumers to shift from the peak to the off-peak period. However, for green products, the difference from peak and off-peak rates may not be so crucial given that customers should pay more attention to the environmental services associated with the rate. Finally, for medium- and high-voltage customers, there are no data on rates because they usually are negotiated directly between operators and customers. Nevertheless, several operators provide this type of product.

## 3.2   Demand Side

In Italy, RES play a central role in the national energy system. They are widespread in electricity production, in the thermal and the transport sectors. In 2014, the total consumption of energy from RESs amounted to 20.2 Mtoe, with a reduction of 2.4 % compared to 2013, mainly related to the thermal and transport sectors. Indeed, warmer weather in 2014 and a decline in fuel consumption were the main factors in the reduction. The reduction in gross total energy consumption has been considerable; in 2014, it decreased by 4.3 %, from 123.6 Mtoe to 118.6 Mtoe compared to 2013. The share of renewables in consumption was 17.1 % in 2014; this value is higher than the target assigned to Italy by Directive 2009/28/EC for 2020 (17 %) and is close to the objective set by the National Energy Plan (NEP). It should be stressed that this performance is also conditioned by the negative trend of

---

[4]Operators and offers take, into account are ENEL (Enel, 2016) (Energia pura casa, E-light, E-light verde, Semplice luce, Tutto compreso green); EDISON (Energia impatto zero); E.On (Luce verde più, Luce verde bioraria); Illumia (Energia sostenibile); Engie (Casa più verde Web); A2A (Prezzo sicuro verde); Hera Comm (Prezzo Fisso Hera Natura Luce); Sorgenia (Senza pensieri; Libero casa); Lifegate (Lifagate energy); Acea (Sostenibile più) (Acea Energia, 2016); ènostra (soloverde MONO, solo verde BIO, solo verde ALTRI USI); Enegan (Green domestico).

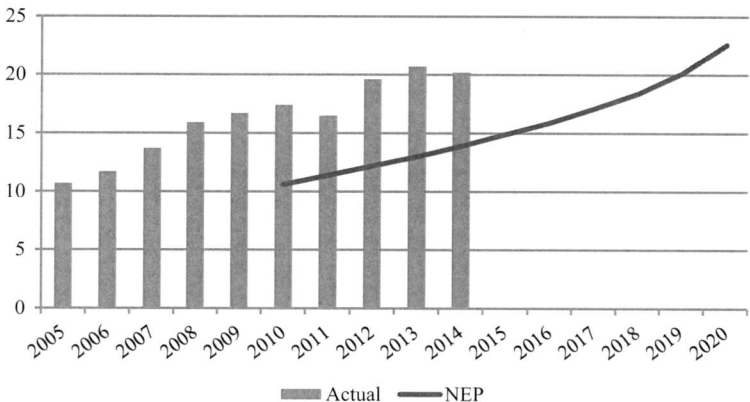

**Fig. 8** Gross final RES consumption (our elaboration based on TERNA data)

**Table 8** RES share of gross final energy consumption

| Gross final energy consumption from RESs (A=A1+A2+A3) | 2010 | 2011 | 2012 | 2013 | 2014 | 2015 |
|---|---|---|---|---|---|---|
| | 17.35 | 16.51 | 19.63 | 20.74 | 20.25 | 21.13 |
| Electricity sector (A1) | | | | | | |
| Hydropower | 3.73 | 3.78 | 3.80 | 3.87 | 3.94 | 3.94 |
| Wind | 0.76 | 0.88 | 1.07 | 1.21 | 1.28 | 1.31 |
| Solar | 0.16 | 0.93 | 1.62 | 1.86 | 1.92 | 1.96 |
| Geothermal | 0.46 | 0.49 | 0.48 | 0.49 | 0.51 | 0.53 |
| Biomass | 0.81 | 0.93 | 1.06 | 1.46 | 1.61 | 1.62 |
| Heating sector (A2) | | | | | | |
| Geothermal | 0.14 | 0.14 | 0.13 | 0.13 | 0.13 | 0.13 |
| Solar-thermal | 0.13 | 0.14 | 0.16 | 0.17 | 0.18 | 0.19 |
| Biomass | 7.65 | 5.55 | 7.52 | 7.78 | 7.04 | 7.69 |
| Heat pumps | 2.09 | 2.27 | 2.42 | 2.52 | 2.58 | 2.58 |
| Transport sector (A3) | | | | | | |
| Biofuels | 1.42 | 1.4 | 1.37 | 1.25 | 1.06 | 1.18 |
| **Gross final consumption (B)** | **133.32** | **128.21** | **127.06** | **123.86** | **118.6** | **122.21** |
| Electricity | 28.48 | 28.7 | 28.31 | 27.48 | 26.8 | 27.11 |
| Oil and biofuels | 50.13 | 49.7 | 46.61 | 45.02 | 45.41 | 46.69 |
| Natural gas | 38.5 | 35.53 | 35.45 | 35.22 | 30.9 | 32.6 |
| Coal | 2.91 | 3.41 | 3.32 | 2.37 | 2.48 | 2.08 |
| RESs for heating and waste | 13.3 | 10.87 | 13.37 | 13.77 | 13.01 | 13.73 |
| **National target (A/B) (%)** | **13.01** | **12.88** | **15.45** | **16.74** | **17.07** | **17.29** |

Source: Calculation based on TERNA (2016) data. (In bold main indexes)

total energy consumption as a result of the economic crisis. However, the actual RES aggregated consumption in Italy is following a path that often exceeds the target set by NEP (Fig. 8 and Table 8).

Moving from an aggregate to disaggregate perspective, in Chapter "Global Markets and Trends for Renewables" we highlighted the importance of the attitudes of citizens to environmental considerations in promoting green product deployment. Attitude determines citizens' assessment of the positive externalities of RES-E production and, consequently, how much they are willing to pay to support RES-E and green services. Consequently, it seems crucial to have a better understanding of the determinants of consumer attitudes. Since 2007, we have investigated the attitudes and preferences of Italian citizens for RES-E production (Bollino 2009; Polinori 2009; Bigerna and Polinori 2014). National surveys were conducted in 2007 (July, November, and December), 2008, and 2011 and are representative of the national level. In the survey conducted in November 2007, 1091 Italians were interviewed, while the sample sizes of the other surveys were around 1600. The stratified samples are highly representative of Italy's 60 million residents. The surveys were conducted using a computer-aided telephone (CATI) and Web (CAWI) interviewing methods by two prominent Italian marketing and consulting companies. In the surveys, each respondent was questioned about (1) RES and their potential development, (2) the Italian energy system, and (3) amount of money (bids) needed to support RES development in Italy. Figure 9 shows the statistics on "Knowledge of RES," that is, whether or not respondents

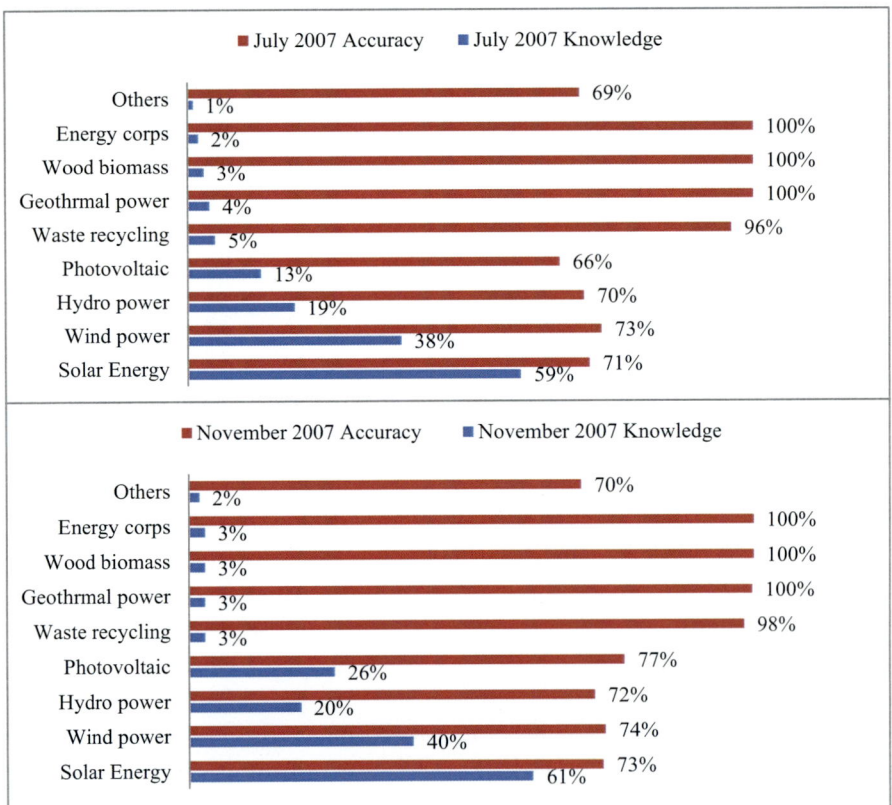

**Fig. 9** Knowledge of RES (our elaboration on original data)

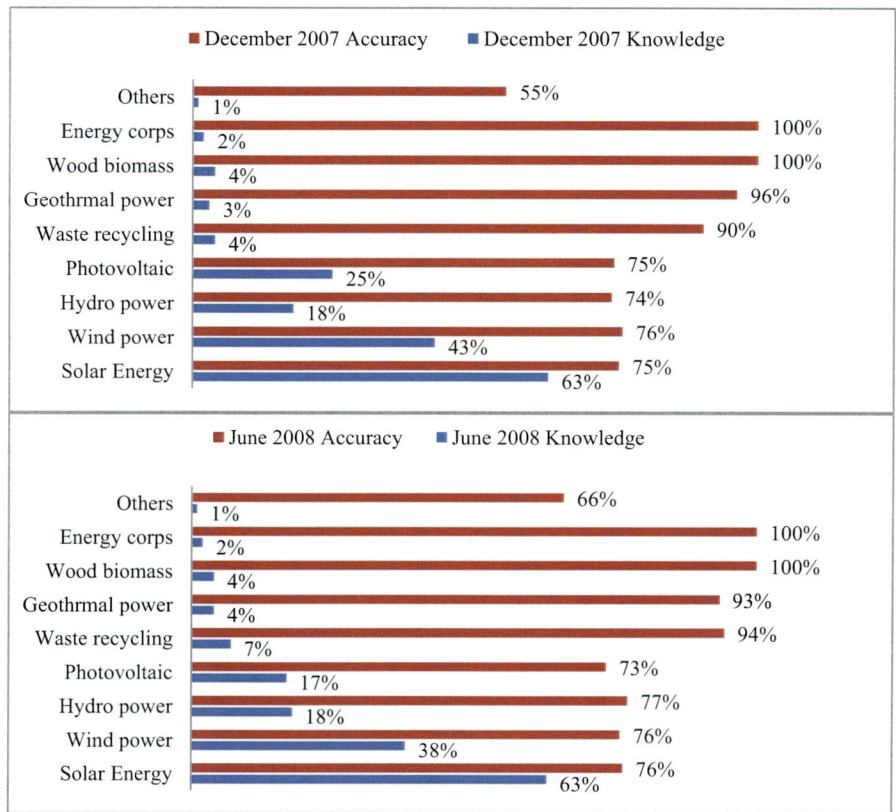

**Fig. 9** (continued)

have sound knowledge of RES. On average, 77.4 % of the total respondents answered that they knew about RES. In particular, respondents had good knowledge of solar power, hydropower, and wind power, whereas they knew little about biomass, energy crops, and geothermal power. Furthermore, respondents were asked to identify RES from among a set of energy sources. In this way we highlight the importance of the description and understanding of the valuation scenario (Soderqvist and Soutukorva 2006). We noted that, on average, more than 75 % of the respondents were able to correctly identify different types of RES (Fig. 9).

The quality of data, that is, their accuracy, shows that respondents' knowledge about RES-E production increased in the investigated timeframe, and they were able to understand the evaluated scenario proposed. This is an important verification survey. Indeed, if respondents did not take the decision process seriously or understand the questions, their responses would not reveal their true preferences. An important aspect to understanding refers to how people were affected by the environmental change proposed, particularly if Italians mainly focused on the

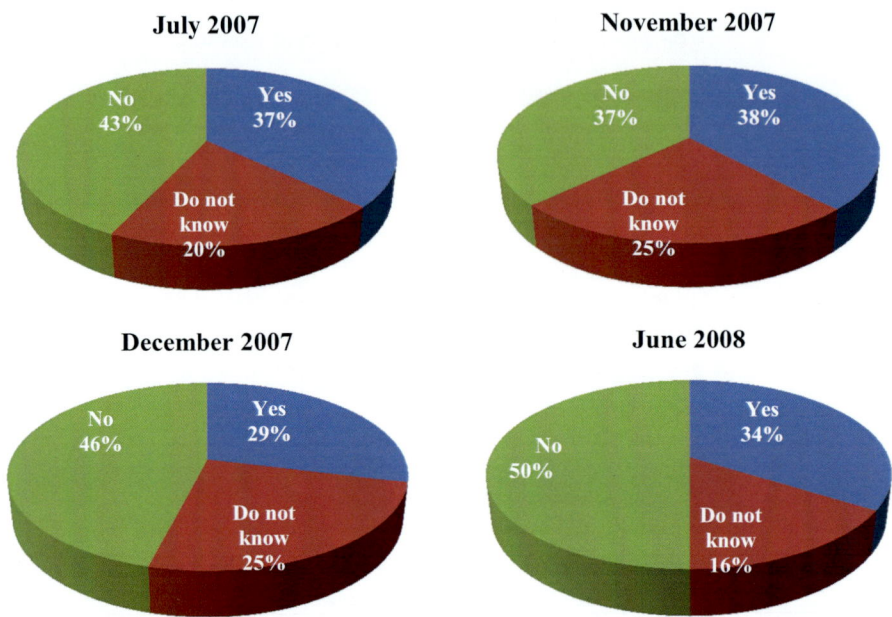

**Fig. 10** Extra premium for RES-E production: preferences (our elaboration on original data)

environmental benefits of RES-E or if they mainly perceived the higher costs and technical problems associated with RES-E production.

If environmental benefits compensate for the costs associated with RES-E production, it is expected that citizens will be willing to pay an extra premium for RES-E production. Respondents' preferences for the extra premium in each survey are shown in Fig. 10. Respondents who are unwilling to pay an extra premium are in the majority in all investigated cases. However, combining the favorable (more than 30 %) and indecisive responses (around 20 %), we obtain an absolute majority.

In the last section of each survey, consumers' willingness to pay (WTP) was elicited using different formats (payment card, bidding game, contingent evaluation multiple-bound dichotomous choice). The aggregate descriptive results are shown in Fig. 11. Among respondents with a positive WTP, the average amount respondents are willing to pay increases from EUR 6.70 in July 2007 to EUR 8.40 in June 2008. Another important result is that, on average, 80 % of citizens are willing to pay from EUR 0.1 to EUR 10, while only 5 % are willing to pay EUR 20 or more.

Econometric results (Bollino 2009; Bigerna and Polinori 2014) confirm these WTP amounts. In particular, the highest mean WTP amount obtained is EUR 9.39, with a confidence interval of (EUR 9.24, EUR 9.50), while the lowest is EUR 3.74 with a confidence interval of (EUR 3.45, EUR 3.91), based on the different degrees of uncertainty.

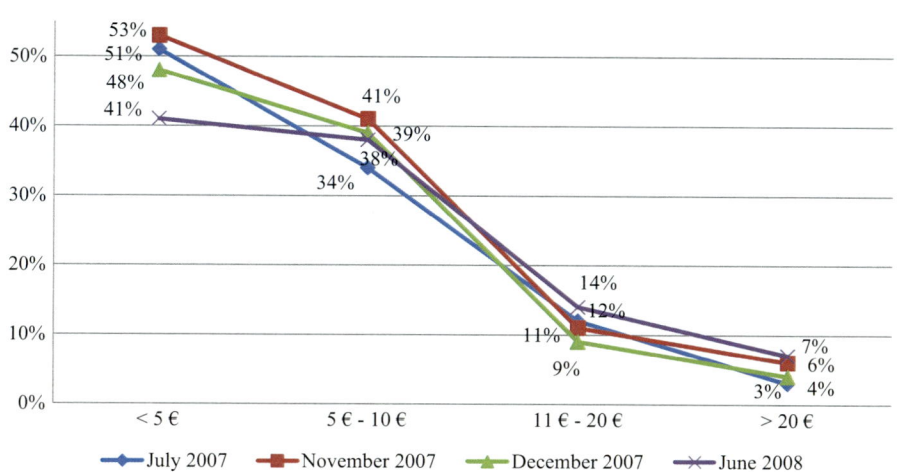

**Fig. 11** Extra premium for RES-E production: WTP (our elaboration on original data)

The importance of understanding the impact of uncertainty on the WTP in contingent valuation studies has been frequently treated in the literature [see Bigerna and Polinori (2014) for a brief review]. Then, in November 2007, a specific section of the questionnaire focused on this topic (Bigerna and Polinori 2014). Results confirm that Italians' preferences for RES-E are characterized by heterogeneity. According to different models estimated, the mean WTP range from EUR 12.76, with a confidence interval of (EUR 11.36, EUR 14.80), to EUR 15.09, with a confidence interval of (EUR 14.02, EUR 16.47). Obviously, estimated median WTPs are more conservative ranging from EUR 4.62, with a confidence interval of (EUR 4.11, EUR 5.13), to EUR 8.05, with a confidence interval of (EUR 7.32, EUR 8.93), and they are assumed to be paid by more than half the respondents.

More recently, other authors have investigated Italians' attitudes toward several types of RES. However, only Cicia et al. (2012) used a representative sample at the national level. They provided information about Italians' preferences for various energy sources, highlighting that all respondents had a negative attitude toward nuclear power, even if they might be willing to buy electricity produced from nuclear power if it were offered at a lower price than fossil energy. Also in their study, Italians demonstrated better knowledge of solar energy and wind energy compared to other RESs, and these two sources are the only ones supported by the respondents in monetary terms. The WTPs estimated by the authors is comparable with those of previous results. Indeed, to consume electricity from wind energy, respondents indicated a WTP of EUR 7.05 to EUR 47.32 bimonthly in values 2007, while the premium ranged from EUR 12.02 to EUR 46.63 for electricity from solar energy. With reference to the other sources considered, Italians stated that they would buy electricity generated from these sources only if it were cheaper. Assuming the robustness of the previous results, it seems that Italians are largely willing to support RES development.

### 3.3    RES Development in Italian Electricity Market: Recent Developments Based on European Targets

Preliminary estimates show a capacity increase from 2014 to 2015 of around 1000 MW (+2 %), mainly concentrated in wind turbines (+5 %) and PV (+2 %). Overall production, however, is estimated to decrease by around 14 TWh (−12 %). The estimated contraction is associated with hydroelectric production, not so much because 2015 was a particularly bad year (according to preliminary estimates, hydropower would amount to 43,902 GWh, in line with the average value of hydraulic production in the last 15 years) but because in 2014 weather conditions were extremely favorable and production reached an unprecedented record (see following chart). Based on these assumptions, in 2015 the share of RES generation of gross domestic consumption would reach 32.8 %, down sharply compared to 2014 (37.5 %).

Monitoring the 2020 target set by RES Directive 2009/28/EC, Directive 2009/28/EC establishes for each EU member state binding targets for RES development in 2020 expressed in terms of the RES share of gross final consumption of energy. The target assigned to Italy in 2020 is 17 %. In accordance with Legislative Decree 28/11, progress toward this objective is monitored annually by GSE, according to the methodology approved by the decree of 14 January 2012 of the Ministry of Economic Development. Table 8 provides a general framework of the variables considered for target monitoring (RES uses in Italy in the electrical, heating, and transport sectors) as well as the macro components of gross final consumption of energy in the country. Data for 2010–2014 are official historical series, while for 2015 they are preliminary estimates. On the basis of these assumptions, the RES share of gross final consumption of energy can be preliminarily estimated at about 17.3 %, a slightly higher value than that for 2014 (17.1 %). Accordingly, the RES increase, approximately 900 ktoe (of which over 70 % is attributable to the use of biomass in the heating sector), would then be proportionally higher than the increase in total gross final consumption (3.6 Mtoe).

### 4    Conclusions and Policy Implications

During 2015, RES grew at their strongest pace ever, shortening the economic gap from fossil fuels. Among RES, wind, thermal, and PV have developed dramatically, as has green energy used in transport. RES are already competitive with fossil fuels in many markets. In addition, the pressure of climate change is forcing a growing number of countries to incorporate a "green component" into their national policies.

Even if there are still many challenges that need to be addressed, such as the effective integration of large shares of RES-E in the electrical system, political instability, and regulatory and fiscal constraints, we can derive four main positive lessons from the recent Italian case.

First, the massive widespread use of RES in everyday life has gradually changed the attitudes of policymakers. From a very strong trend toward a command-and-control approach in which all technical and economic issues related to RES deployment in the system are tightly regulated, recent legislation has proven to be more simplified and more market oriented. Second, the surprising success of RES investment by both residential consumers and small and medium-sized commercial and industrial enterprises made it possible to reach the target set for 2020 already in 2015. The financial effort has been significant, and the political net benefit of high RES deployment started to be eroded by an onerous increase in household electric bills. To this new scenario Italian policymaker reacted with a bold new approach involving incentive cuts. All new legislation quickly started to curb new RES investment by setting clear caps on total financing allotments to incentive policies. Third, there has been growing awareness of the inadequacy of the electric market design, demonstrated by the inversion of the peak and off-peak price profiles, with a massive injection of RES (especially PV) during the day (Bigerna and Bollino 2016). As a consequence, conventional thermal plants' utilization rates dropped (from an optimal 6500 h to less than 2000 h in 2015), financial distress hit some relevant generators, and the exercise of market power increased during the evening hours, from 7:00 to 10:00 p.m. This last fact can be viewed as an attempt by generators to recover fixed costs within a short time frame, when solar is nonfunctioning. Fourth and last, but not least, on the positive side consumers have become increasingly aware of green electricity. This has stimulated two courses of action. On the one hand, regulators have enacted better rules for transparency and information dissemination in the retail market. On the other hand, retailers have evolved their marketing attitude, taking into consideration the end consumer of electricity and treating them as customers who want value and not only users who receive a state-monopoly regulated service. Thus, throughout Italy, new, high-quality products have started to be offered on the market to satisfy the new demand for green electricity and better environmental conditions.

## References

Acea Energia. (2016). *Offerte Casa e Business*. Accessed June 03, 2016, from http://www.aceaenergia.it/

Bigerna, S., & Bollino, C. A. (2016). Optimal price design in the wholesale electricity market. *Energy Journal*. doi:10.5547/01956574.37.SI2.sbig

Bigerna, S., & Polinori, P. (2014). Italian households' willingness to pay for green electricity. *Renewable and Sustainable Energy Review*. doi:10.1016/j.rser.2014.03.002

Bollino, C. A. (2009). The willingness to pay for renewable energy sources: the case of Italy with socio-demographic determinants. *Energy Journal*. doi:10.5547/ISSN0195-6574-EJ-Vol30-No2-4

Cicia, G., Cembalo, L., Del Giudice, T., & Palladino, A. (2012). Fossil energy versus nuclear, wind, solar and agricultural biomass: Insights from an Italian national survey. *Energy Policy*. doi:10.1016/j.enpol.2011.11.030.

Enel. (2016). *Mercato libero*. Accessed June 04, 2016, from https://www.enelenergia.it/mercato/libero/it-IT

GME–Gestore del Mercato Elettrico. (2014). *Relazione annuale 2014*. Accessed June 09, 2016, from https://www.mercatoelettrico.org/it/MenuBiblioteca/documenti/20150720RelazioneAnnuale2014.pdf

GSE –Gestore dei Servizi Elettrici. (2016a). *Open data*. Accessed June 1, 2016, from http://opendata.gse.it/opendata/

GSE –Gestore dei Servizi Elettrici. (2016b). *Data set*. Accessed June 01, 2016, from http://opendata.gse.it/dataset

Nomisma. (2012). *I vantaggi del mercato libero dell'elettricità e del gas*. Bologna: Nomisma Energia srl.

Polinori, P. (2009). *Italians, renewable energy sources and EU climate vision*. IAEE Newsletter. Accessed June 15, 2016, from http://www.iaee.org/en/publications/newsletterdl.aspx?id=64

Regulatory Autority -AEEGSI. (2010). *Controllo della vendita ai clienti finali di energia elettrica da fonti rinnovabili*. Accessed June 10, 2016, from http://www.autorita.energia.it/it/docs/dc/10/023-10dco.jsp

Regulatory Autority -AEEGSI. (2016a). *Relazione annuale sullo stato dei servizi e sull'attività svolta*. Accessed June 22, 2016, from http://www.autorita.energia.it/allegati/relaz_ann/16/RAVolumeI_2016.pdf

Regulatory Autority -AEEGSI. (2016b). *Trova Offerte*. Accessed June 02, 2016, from http://www.autorita.energia.it/it/trovaofferte.htm

Soderqvist, T., & Soutukorva, A. (2006). *An instrument for assessing the quality of environmental valuation studies*. Stockholm: Swedish Environmental Protection Agency. Accessed June 18, 2016, from https://www.naturvardsverket.se/Documents/publikationer/620-1252-5.pdf

TERNA. (2016). *Dati statistici*. Accessed May 25, 2016, from https://www.terna.it/it-it/sistemaelettrico/statisticheeprevisioni/datistatistici.aspx

**Simona Bigerna (Ph.D.)** is Assistant Professor of Economics at the Department of Economics, Professor of International Economics, Economics of the Third Sector, and Economics of Growth and Innovation at the University of Perugia. She is a Faculty member in the Ph.D. program in Internationalization of Small and Medium Enterprises and visiting professor at several European Universities. Her research focuses on energy economics, renewables, environmental policy, and the transport industry, at the theoretical and empirical levels. She has published in leading international journals, including Energy Policy, Journal of Cleaner Production, Renewable Energy, Renewable and Sustainable Energy Reviews, and The Energy Journal. She has served as referee for many international scientific journals. She is a member of the editorial board of the Review of Economics and Institutions. She is a member of the Interuniversity Research Centre on Pollution and the Environment (CIRIAF) and the International Association of Energy Economics (IAEE). She has been a member of the Spin-offs and Startups research group at the University of Perugia.

**Carlo Andrea Bollino** is Professor of Economics at the University of Perugia, Professor of Energy Economics at University LUISS in Rome, and president of the Italian Association for Energy Economics (AIEE). He has served as Chief Economist at Eni, an economist in the Bank of Italy, and a research associate in Project Link for the United Nations. Prof. Bollino has also taught economics at the University of Pennsylvania (USA) and at universities in Campobasso, Sassari, and Urbino (Italy). He graduated with a degree in economics from Bocconi University and earned his Ph.D. in Economics from the University of Pennsylvania in 1983, under Nobel laureate Prof. Lawrence Klein. His research interests include econometric modeling, consumer behavior, energy markets, sustainable and renewable energy, liberalization, and regulation policy. He is the author of over 200 scientific articles.

**Paolo Polinori (Ph.D.)** is Associate Professor of Economics in the Department of Economics at the University of Perugia, professor of economics (undergraduate), regulatory economics (graduate), and energy and environmental economics (Ph.D.) at the University of Perugia. He is a faculty member of the Ph.D. program in economics and a visiting professor at several European universities. His research focuses on energy and environmental economics, transport economics, industrial economics, and agrifood systems. He has published in leading international journals, including *Energy Policy, Renewable and Sustainable Energy Reviews, Sustainability, British Food Journal,* and *The Energy Journal,* and served as a referee for many international scientific journals. Prof. Polinori holds membership in the Interuniversity Research Centre on Pollution and the Environment (CIRIAF), the International Association of Energy Economics (IAEE), and the Global Trade Analysis Project at Purdue University, as well as in several economic associations. He is an economic consultant for Monitoring Economics and Territory and has won important national awards for his research activities. Finally, he is managing editor of the *Review of Economics and Institutions.*

# Marketing Renewable Energy in Japan

Jörg Raupach-Sumiya

**Abstract**

This chapter provides a comprehensive analysis of Japan's policy framework and regulatory context for marketers of renewable energy (RE) and describes recent developments in Japan's market for RE, its competitive landscape, and emerging RE-based business models. Fundamental regulatory reform of Japan's energy system, the complete liberalization of its electricity and gas markets, and policies to promote the deployment of RE have expanded consumer choice and created new business opportunities for RE in Japan. In the aftermath of the Fukushima nuclear disaster, the interest of Japanese consumers in RE is growing. However, market transparency and disclosure rules regarding power sources are still wanting, and constraints in the supply capacity for RE still limit the sourcing opportunities for RE marketers. Furthermore, uncertainties regarding the outcome of the market liberalization process and the future course of RE-related policies remain high, while the commercial viability of differentiation strategies based on environmental value added must still be tested.

**Keywords**

Japan's energy policy • Regulatory reform • Japan's renewable energy market

J. Raupach-Sumiya (✉)
College of Business Administration, Ritsumeikan University, Osaka, Japan
e-mail: raupach@fc.ritsumei.ac.jp

© Springer International Publishing AG 2017                                          375
C. Herbes, C. Friege (eds.), *Marketing Renewable Energy*, Management for Professionals, DOI 10.1007/978-3-319-46427-5_19

# 1    The Regulatory Context for RE in Japan

## 1.1    Japan's Energy Policy

Energy-related policies and the regulatory framework for energy markets determine supply- and demand-side conditions for the market integration of renewable energy (RE) and shape the fundamental context for RE providers and marketers. For Japan, the fundamental direction of energy policy is spelled out in the *Basic Energy Policy*, which assesses the long-term outlook for Japan's energy supply and demand, determines the targeted mix of energy sources, and provides the overall policy framework (METI 2015a). It is reviewed every 3 years. The latest plan was adopted in April 2014 following controversial debate about the future role of nuclear energy. It provides a comprehensive analysis of Japan's post-Fukushima energy situation and raises fundamental concerns about Japan's high and growing dependency on imports for fossil fuels, as well as the associated rise in energy costs and greenhouse gas emissions following the shutdown of almost all nuclear power plants. Because of these concerns, the Japanese government is calling for accelerated investment in RE and is promoting its deployment. At the same time, it still considers nuclear energy an indispensable, clean "baseload energy source" while pointing to the relatively high cost and unreliability of RE.

This basic direction of Japan's energy policy has been translated into a targeted energy mix with a share of 20–22 % of electricity supply for nuclear energy and a share of 22–24 % for RE by 2030.[1] Following the adoption of the Paris Agreement under the United Nations Framework Convention on Climate Change on 12 December 2015, the Japanese government announced a new Energy Reform Strategy in April 2016, which calls for JPY28 trillion (approx. USD 300 billion) public and private investment by 2030 in the promotion of energy efficiency, the cost-efficient expansion of RE, and the development of new energy technologies (METI 2016a).

A cornerstone of Japan's energy policy is the strengthening of competition in the energy sector and the deregulation of Japan's electricity and gas markets, which until 1995 had been dominated by a limited number of vertically integrated utilities that enjoyed regional monopolies in generation, transmission, and distribution. In 1995, the electricity market was first opened to independent power providers (IPPs),

---

[1]Though the promotion and expansion of RE is a declared important policy objective, the Japanese government stresses the continued importance of large-scale "baseload technologies" such as nuclear, hydro, and coal, aiming for a 60 % share of the electricity supply for these baseload technologies. These targets are, however, highly controversial. On the one hand, the targets for nuclear power are deemed unrealistic because they would require the extension of the lifetime of existing nuclear facilities beyond the formerly restricted lifetime of 40 years or the construction of new nuclear facilities (Bloomberg 2015a). On the other hand, the Ministry of the Environment, leading think tanks, and nongovernmental organizations call the targets for RE unambitious and too low. Furthermore, the targeted expansion of RE relies to a large degree on solar and biomass, while wind and geothermal continue to play a minor role (Nikkei Technology Online 2015).

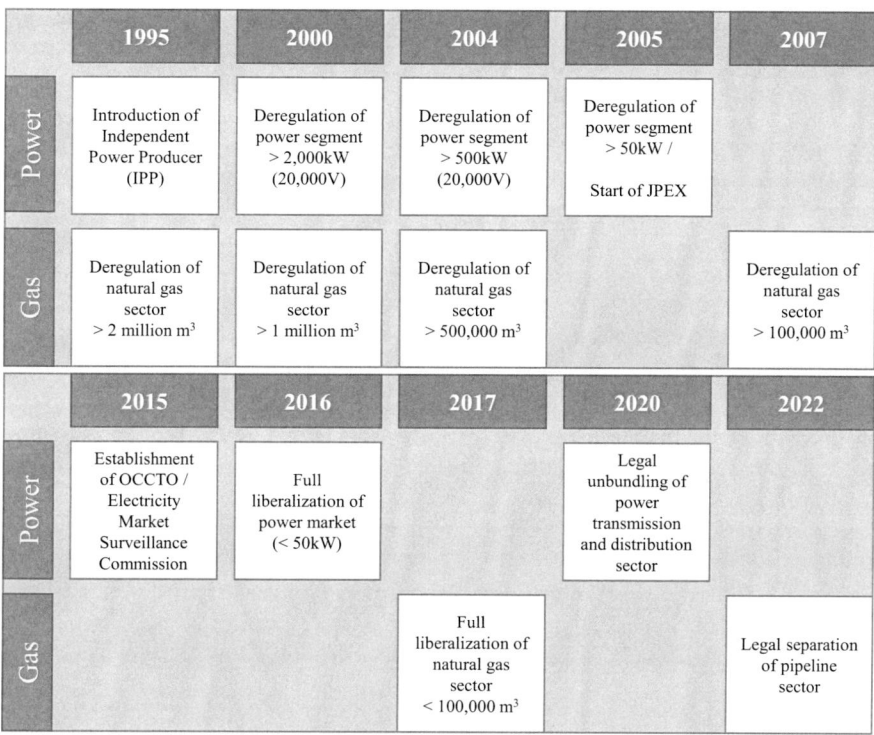

**Fig. 1** Deregulation of Japan's power and gas market (author's illustration based on METI 2014, 2015b)

followed by the gradual liberalization of the retail market from 2000 onwards (Fig. 1) (METI 2014). In 2003, a power wholesale market was established (Japan Electric Power eXchange/JEPX) that began operation in 2005. Since 2005, the power retail market segments have been deregulated for large- and mid-scale commercial users; however, the regional utilities still monopolize the important retail segments of small-scale commercial users and private households.[2] Since 2013, the government has enacted a series of laws and amendments that will fully liberalize Japan's electricity and gas markets in three steps by 2020 for electricity and by 2022 for gas. The policy aims to secure a stable supply of electricity and gas,

---

[2]The electricity market is subdivided into several segments based on connected power and voltage: very large-scale users with a capacity exceeding 2000 kW and above 20,000 V, large-scale users with a capacity of 500–2000 kW and up to 20,000 V, mid-scale users with a capacity of 50–500 kW and 20,000 V, small-scale users (6000 V), and households (100~200 V) consuming less than 50 kW (METI 2014). Similar reforms were enacted in the gas market between 1995 and 2007, in principle opening the market up to competition except for small commercial customers and private households (consumption of less than 100,000 m³).

to reduce tariffs, and to expand business opportunities and choices for consumers (METI 2015b).

In Step 1, the Organization for Cross-Regional Coordination of Transmission Operators (OCCTO) was inaugurated in July 2014 as an independent regulatory authority sponsored by more than 500 power providers, operators, and retailers. It started operations on 1 April 2015 with the objective of improving and monitoring a more efficient, cross-regional management and nationwide balancing of electricity supply and demand, coordinating among regional transmission operators, and establishing rules for grid access and grid management (Smart Japan 2015a).

In Step 2, the low-voltage retail segment ($<$50 kW) was opened to full competition on 1 April 2016 by abolishing regional monopolies, introducing price competition (though applying transition rules), and allowing retail customers to freely choose their electricity provider.[3] The Amended Electricity Business Act replaced the prevailing licensing categories for power providers and suppliers with three general categories for participants in the electricity markets: (1) generation, (2) transmission/distribution, and (3) retail. Full competition is induced in the power generation and retail segments, while transmission and distribution remain regulated regional monopolies (METI 2015c). Power generation companies with more than 10 MW capacity (including RE capacity) must notify the authorities and regularly submit power supply plans. Retailers are subject to registration and approval by the authorities, must demonstrate sufficient and secure capacity of power supply, and must comply with so-called balancing rules, which require them to strictly observe their supply plans and otherwise impose penalties for deviations.

In the final step, Step 3, the government will aim for an unbundling of the transmission/distribution business from the generation and retail business and the legal and organizational separation of the three business segments (METI 2014). The objective is to strengthen the neutrality of the transmission/distribution sector and to secure open and fair third-party access. For electricity, unbundling will be implemented starting 1 April 2020 and for gas in 2022. At the same time, gas and electricity rates will become fully deregulated.

The general direction of Japan's energy policy promises a far-reaching overhaul of the country's energy system. The RE business in Japan will be profoundly affected by these fundamental reforms. With regard to power generation and transmission, RE investment is influenced by how promotional policies for RE are aligned with the reform of Japan's energy system, in particular in relation to market access, grid connection, and system integration. With respect to the retail segment, labeling and marketing rules for RE, balancing requirements for RE suppliers and retailers, and the development of wholesale markets will determine the extent to which consumers can enjoy transparency and true choice.

---

[3]With about 84 million customer contracts valued at JPY 75trillion (approx. USD 825 billion), this segment accounts for around 40 % of Japan's electricity market.

## 1.2    Promotional Policies for Renewable Energy

Policies promoting RE in Japan have a rather long history, starting with the Sunshine Project in 1973, the Energy Efficiency Law in 1979, and the Non-Fossil Fuel Energy Law in 1980, which promoted energy conservation as well as research and development for alternative energies as a reaction to the two oil crises (Jordan-Korte 2011). Since the mid-1990s, the Japanese government has enacted a series of laws to promote RE as part of Japan's international commitments to climate change mitigation. In 1992, a policy was initiated that aimed for RE deployment by means of voluntary purchase agreements between electric utilities and RE providers. In 1994 it launched the 70,000 Solar Roofs subsidy program to promote the installation of photovoltaic (PV) systems in residential areas. Most noteworthy is the 1997 Special Measures Law Concerning the Use of New Energy by Electric Utilities, which was amended in 2002. The law, usually referred to as the Renewable Portfolio Standard (RPS) law, set annual targets for the use of RE and obligated electric utilities and retailers to generate RE by themselves, purchase the required amounts from RE providers, or acquire a Tradable Renewable Certificate of renewable electricity production from other suppliers. The law for the first time established an institutional framework for integrating RE into the electricity market, thereby stimulating RE-related investment. PV power received a further boost from the New Buyback Program for Photovoltaic Generation, which was launched in 2009. Instead of setting targets for RE purchases at market prices, the buyback scheme guaranteed purchases of surplus electricity from PV power systems at a fixed price for 10 years. As a result of these various policies, Japan emerged as the world's leading producer of solar energy in the early years of the millennium (Jordan-Korte 2011; METI 2011).

The RPS law and PV buyback program was replaced by the Special Law to Promote Renewable Energy, which came into effect in July 2012 and finally introduced a full-fledged feed-in tariff (FIT) scheme for RE. The FIT law calls for a 3-year period of accelerated growth of RE development and requires (in principle) electric utilities to connect RE power plants to the grid and purchase the generated electricity at rather generous FITs guaranteed for 20 years (METI 2012).[4] It certifies and promotes five categories of RE that are further subdivided into subsegments: PV power (below 10 kW, 10 kW and above), wind (less than 20 kW, 20 kW and above), small- and medium-scale hydropower (less than 200 kW, 200 to less than 1 MW, 1 MW to 3 MW), geothermal power (less than 15 MW, 15 MW and above), and certain forms of biomass and biogas.[5]

---

[4]The exception is PV power installations below 10 kW, where purchases have been limited to surplus energy at tariffs guaranteed for only 10 years. The FITs are subject to an annual review and have been revised four times since 2012, most recently on April 2016, mainly lowering the rates for solar power.

[5]In the case of biomass, only the use of biomass fuel that does not harm other industries that use the material are subject to a FIT. The energy producer must provide documentation on the source of the certified biomass material and regularly calculate a biomass ratio (METI 2012). The reform of

The adoption of the FIT system has triggered strong growth of RE investment in Japan, making it the world's third largest market for RE facilities (Bloomberg 2015b). As of December 2015, the total registered RE capacity eligible for compensation under the Japanese FIT program amounts to around 95 GW or about 37 % of Japan's total capacity of 258 GW for electricity generation (Table 1). However, only about 35 % of the capacity (33 GW) is actually under operation,[6] still an increase of approximately 25 GW (or more than three times) since the introduction of the FIT system (METI 2015d). The rapid growth of RE since 2011 has almost entirely been driven by solar power, where newly installed capacity amounted to 23.7 GW (or 93 % of the total registered RE capacity), while wind (377 MW, 5 %), biomass/biogas (344 W, 4 %), small-scale hydro (125 MW), and geothermal (9 MW) have played a noteworthy minor role.

The recent boom in RE investment and the bias toward solar power has triggered a controversial debate about the rise in the associated surcharge costs for consumers and industry and about their impact on grid stability. The Japanese government responded with a set of emergency measures in January 2015 that increased the flexibility of the power utilities to curtail the output of RE providers, expanded the scope of facilities as possible targets for curtailment, softened the related compensation rules for curtailing RE output, and tightened the certification procedures for new RE facilities (METI 2015e). These events have triggered a fundamental reform of the FIT program with the intention of correcting the bias toward solar power, reducing the cost burden, and enhancing the efficiency of RE deployment and integration into power trading and distribution (METI 2015f). The reform measures were approved by the Japanese Cabinet in February 2016 and will be implemented in February 2017 (METI 2016b). The new law introduces various new schemes (including auctions) for determining tariffs, tightens the rules for RE providers with respect to facility certification, interconnection, and operation, and shifts the obligation for grid connection and purchase of RE from the power retailer to the local transmission operator. The transmission operator is required to sell the purchased RE to the power wholesale market as a matter of principle or to register the supply contract with specific retailers with the METI minister.[7]

---

the FIT law in 2015 further separated wood-fired power plants using timer from forest thinning into smaller plants consuming less than 2000 kW and plants consuming 2000 kW to promote smaller-scale installations. It also introduced a tariff for offshore wind.

[6]Under the original scheme, the FIT was granted as soon as the planned facility had been certified. Eager to secure higher tariffs and to increase profits, investors often certified projects with questionable economic feasibility or postponed the start of operations, hoping for lower investment costs. As a result, many certified facilities have not yet become operational.

[7]As of April 2016, METI also revised the method to calculate the compensation of purchasing costs of RE under the FIT system: instead of calculating the so-called avoidable cost by referring to the production costs of conventional power sources, they are now linked to trading prices on the wholesale market. While the measure will not come into force for most existing contracts until 2020, it is applied to all new RE installations as well as for retailers who sell the purchased RE on the wholesale power market (METI 2015h).

**Table 1** RE installations in Japan under FIT (as of December 2015)

| | Newly registered RE installations[a] | | RE installations under operation[b] | | | Total registered capacity | |
|---|---|---|---|---|---|---|---|
| | Capacity (MW) | % of capacity | Capacity (MW) | % of capacity | No. of installations | MW | Operative (%) |
| Solar (<10 kW) | 4256 | 5 | 8296 | 25 | 2,010,644 | 8955 | 93 |
| Solar (>10–50 kW) | 25,839 | 30 | 8120 | 24 | 347,066 | 25,985 | 31 |
| Solar (>50–500 kW) | 4058 | 5 | 2377 | 7 | 10,659 | 4144 | 57 |
| Solar (>500–1000 kW) | 4559 | 5 | 2576 | 8 | 3673 | 4566 | 56 |
| Solar (>1000–2000 kW) | 12,956 | 15 | 5057 | 15 | 3373 | 12,970 | 39 |
| Solar (>2000kW) | 28,078 | 33 | 2228 | 7 | 187 | 28,087 | 8 |
| Wind | 2339 | 3 | 2906 | 9 | 403 | 4868 | 60 |
| Small-scale hydro | 714 | 1 | 334 | 1 | 333 | 922 | 36 |
| Geo-thermal | 73 | 0 | 10 | 0 | 17 | 74 | 14 |
| Biogas/Bio-mass[c] | 2709 | 3 | 1472 | 4 | 360 | 4107 | 36 |
| Total | 85,581 | 100 | 33,375 | 100 | 2,376,715 | 94,678 | 35 |
| of which solar | 79,746 | 93 | 28,653 | 86 | 2,375,602 | 84,706 | 34 |

Source: Agency for Natural Resources and Energy http://www.fit.go.jp/statistics/public_sp.html (access on May 3rd, 2016)
[a]Installations subject to FIT excluding transfer of existing installations into FIT system
[b]Installations subject to FIT incl. transfer of existing installations into FIT system
[c]Capacity with consideration of biomass ratio

## 1.3    Regulations for Marketing RE

In the context of the reform of Japan's energy system and FIT scheme, METI has issued a set of regulations and guidelines that also directly affect RE marketers. These regulations and guidelines refer, on the one hand, to balancing obligations of RE retailers and, on the other hand, to marketing methods that seek to differentiate RE from other energy sources by claiming specific environmental benefits.

In the liberalized power market, the registered power providers and licensed power retailers are obliged to submit their final, daily supply and demand plans to the transmission operator 1 h before the daily closing gate (METI 2015g). The respective plans are divided into 30-min segments and matched accordingly. Should supply or demand deviate from the original plan, the transmission operator assumes the balancing responsibility and imposes imbalance fees on the party that violated its plan. In the case of solar and wind power, which often are generated by distributed, smaller-scale facilities and which, by nature, are unstable and depend on weather conditions, the regulation allows for the formation of a so-called balancing group between several power providers and a power retailer that engages in a joint supply and purchase agreement. The balancing group is based on the notion that demand and supply can be better matched through the combination of several supply locations, thereby levelling the fluctuations in solar and wind power supply. There are two cases. In case one, the transmission operator assumes the responsibility and balancing risk for the supply plan, and the cost for the imbalance fee are included in the FIT surcharge. In case two, the power retailer assumes these responsibilities as well as associated costs and agrees to purchase all of the actually produced power volume supplied by the group. These special measures reduce the commercial risks for providers of solar and wind power and offer marketers of RE opportunities for new business models.

During the process of deliberations about guidelines for power retailers in the liberalized power market, two issues related to the marketing of RE received special attention: first, the disclosure of information about the sources of the purchased power, and second, the handling of marketing claims that attempt to differentiate their offerings through certain qualitative attributes (e.g. environmental benefits, regional sourcing). The latter particularly also refers to power produced under the FIT scheme (METI 2016c). Both issues are deeply related to the broader discussion about consumer protection, consumer choice, and anticompetitive behavior in the liberalized power market.

Despite requests by consumers and environmental protection groups for a mandatory, detailed disclosure of information that indicates the type of energy source (e.g. gas, nuclear power, RE), the Japanese government decided to recommend a voluntary disclosure of power sources. At the same time, it developed detailed rules for the disclosure of sourcing information (METI 2016c).

–  In case of disclosure, the guidelines prescribe eight general categories that
    classify the different power sources (e.g. fossil-fuel-based power plant based

on coal/gas/oil, nuclear power, RE sources except FIT power, FIT power). Within these categories, further specifications are possible.

- Disclosed information on power sources must be related to actual (in some cases planned) purchases based on data from the previous year (or the actual plan). Vague claims such as "procured power contains a high ratio of RE" or "power is procured from stable sources" are considered inappropriate.
- The indication of RE as a power source is considered appropriate only if it is procured outside the FIT scheme at negotiated prices or if the RE is procured under the FIT scheme, but reimbursement for the surcharge is foregone (METI 2015h). Only in these cases is the claim of environmental benefits as an additional value added considered legitimate because the costs are borne by the marketer, who thereby provides incentives to invest in RE.
- FIT power refers to RE procured under the FIT scheme, and disclosure must meet three requirements: (1) clear reference as "FIT power" without the use of other expressions like "solar power" or "renewable power, (2) clear indication of its volume share of the total purchases, and (3) explanation of the meaning of "FIT power" by adding the following statement: *The costs of FIT power are borne by the general public by means of a surcharge, thereby differing from ordinary renewable energy sources with respect to the handling of the cost burden and greenhouse gas emissions. It is considered similar to the nation's average energy mix that also contains power from fossil-fuel-based sources and its average greenhouse gas emissions.* In the case of FIT power, the claim of environmental benefits as a differentiating, value-adding quality attribute is considered to be inappropriate because costs are not borne by the marketer and no additional RE investment is initiated. Therefore, claims such as "green power" or "clean power" are not appropriate for FIT power (METI 2016c).
- The claim "produced locally, consumed locally" (*chisan chishō*) is considered a differentiating attribute that, in principle, offers additional value added to consumers. The claim may also be associated with environmental benefits, as the transportation losses are reduced and as often regional power is generated by clean, renewable sources such as hydro or biomass. The claim also suggests regional economic benefits such as employment creation or generation of regional income. The guideline is kept rather general but requires marketers to clearly indicate the location of the power suppliers, the regional source of the used fuel material, and the region to which the power is mainly supplied. The location of the suppliers and consumers should be in the same clearly specified region.

A different scheme for the marketing of RE is the Green Energy Certification system that was first established in 2000 with the objective of stimulating investment in RE sources (UNESCAP 2012). The mechanism splits the generated volume of RE into two components, physical energy to be sold on the market and a tradable certificate that represents the environmental value added of the energy generated from the certified RE sources. By purchasing these certificates, commercial firms and private households without their own source of renewable power or heat are enabled to claim that they are using energy from RE sources. Importantly, FIT power is not eligible for green energy certification except for the portion that is consumed by the FIT power producer and not sold under the FIT scheme. Presently, the market for green certificates is very small and has limited relevance for marketing, although some RE providers and marketers offer the issuance of Green Power Certificates as part of their services.[8]

The balancing rules and marketing guidelines for RE aim at the development of a level, competitive, and transparent playing field that offers RE marketers new opportunities for innovative business models based on environmental differentiation. Marketers, who also invest in commercially viable RE projects, who engage in the formation and leadership of balancing groups based on RE, who act as an intermediary of Green Energy Certificates, or who concentrate on regional supply and distribution of RE power may well be capable of developing competitive, RE-based business models. However, marketers who have not secured access to their own RE sources and who rely mainly on the procurement of FIT power face challenging balancing requirements and strict limitations on claiming environmental benefits. Furthermore, marketers of FIT power will be exposed to additional commercial risks because the compensation for FIT power purchases will no longer be linked to the predefined, rather stable additional cost for the generation of conventional power. Instead, the compensation will be tied to wholesale power prices, which are expected to be rather volatile in the short term. Some experts, therefore, fear that highly volatile wholesale power prices on Japan's still rather underdeveloped power exchange may squeeze margins of FIT power marketers (Smart Japan 2015b).[9] However, generally, the opportunities and challenges for RE marketers depend to a large degree on how the Japanese market for RE will develop, on consumer attitudes, and on the overall competitive landscape in the newly liberalized retail power and gas markets.

## 2    The Market for RE in Japan

### 2.1    Consumer Attitudes

The Japanese market for renewable power is still rather small despite the recent rapid growth in RE investment. In fiscal year 2014, purchases of RE under the FIT scheme amounted to 28.6 TWh, or approximately 3 % of the total power consumption (962T Wh) (METI website 2016a). The Institute for Sustainable Energy Policies (ISEP), a nongovernmental think tank, estimates that in 2014, RE met

---

[8]As of 31 March 2016, 1092 facilities with a capacity of 422 MW have been certified, and 37 partners have been signed up as intermediaries (Green Energy Certification Center website).

[9]Note: Although the Japanese government aims to stimulate the development of a liquid and deep wholesale power market, the market is still rather small. For example, the volume that was traded on the JPeX spot market in the first half of 2015 amounts to around 1.5 % of Japan's total power consumption (METI 2015i).

around 12.5 % of Japan's electricity demand, of which around 50 % represents large-scale hydropower, while distributed forms of RE, such as solar (2.2 %), biomass (1.5 %), small-scale hydro with less than 10 MW capacity (1.5 %), wind (0.5 %), and geothermal (0.2 %), amount to slightly more than 6 % (ISEP 2015).[10] In short, distributed forms of RE are in rather short supply, and most of it is provided under the FIT scheme. From a supply-side point of view, this makes Japan a seller's market for RE and should place suppliers of RE in a favorable competitive position. An essential question, therefore, is whether there is demand for RE in Japan and whether Japanese consumers are willing to pay for a perceived, additional environmental value added?

In the course of the full liberalization of the Japanese retail power market, various surveys by the Japanese government (Smart Japan 2014, 2015c) and private institutions (JCCU 2014a, 2015; Mizuho Research 2015; Tokyo Gas 2015; Dentsu 2016; PWC 2016) provide insights about consumer attitudes with regard to RE and prospective purchasing patterns. While the surveys differ with respect to specific focus and design details, the overall results can be summarized as follows:

- The Japanese public generally favors the promotion of RE (JCCU 2014a, 2015);
- Stability of supply and lower electricity rates enjoy absolute priority (>80 %) when consumers are choosing power suppliers, ranking high above other criteria;
- However, only about 20 % of the respondents decided or would seriously consider switching suppliers (Smart Japan 2015c; Dentsu 2016);
- A significant number of respondents (between 23 and 61 %, depending on the source) emphasize that they consider the share of RE as the first or second most important criterion for supplier selection (Smart Japan 2014; Mizuho 2015) and believe that disclosure should be made mandatory (>90 %) (JCCU 2015);
- A majority of respondents express a general willingness to pay higher prices for the use of RE (JCCU 2014a); however, only a minority (5–22 %) intend to actually switch at higher rates, while most will do so only at the same or lower rates (Mizuho 2015);
- The willingness to pay for RE at somewhat higher rates has even decreased from 30.8 % in 2013 to 22 % in 2016 (PWC 2016), and people above the age of 60 are more inclined to use RE even at higher rates than average-age and especially younger Japanese (Mizuho 2015);
- Interestingly, a small number of Japanese consumers (11 %) expressed a preference for local suppliers in their region (Smart Japan 2014), and a majority would choose their local municipality or the local city gas supplier, valuing their reliability, trustworthiness, and contribution to the local economy (Mizuho 2015).

---

[10]The figures include RE that is not covered by the FIT system, in particular hydropower plants with a 10–30 MW capacity. Large-scale hydropower is a long-established pillar of Japan's energy system, serving also as a power storage reserve for its nuclear power plants.

The results underline the paramount importance of lower electricity rates for Japanese consumers in their choice of power suppliers. At the same time, Japanese consumers appear rather conservative and assume a "wait-and-see" posture when reflecting about their supplier choice. Furthermore, the surveys reveal that Japanese consumers also value other so-called quality attributes of power suppliers, such as reliability, stability of supply, various value-added services, and the type and location of the power source. These attitudes, therefore, indicate that opportunities for differentiation strategies by power providers exist in general and that there is room for an ecology-based positioning approach by RE marketers. Although the apparent number of consumers ready to switch to RE still seems to be quite small, and despite a significant resistance to change itself, the high level of environmental awareness and interest in RE, as well as a strong critical attitude toward nuclear power, suggests a significant latent market potential for RE in Japan. Further, even just 5 % of Japan's 84 million retail power contracts constitute a sizable and commercially attractive market potential of more than 4 million contracts. This leads to the question of the actual choices presently being offered to consumers in Japan's liberalized power market.

## 2.2    The Competitive Landscape

Japan's power market has undergone profound changes in recent years, yet the 10 large, vertically integrated utilities, like Tokyo Electric Power or Kansai Electric Power, still enjoy an overwhelming market share in power generation and retailing while owning the transmission and distribution grids. The Big 10 generated 68 % (744 TWh) of Japan's total power output of 1090 TWh in 2013, mostly based on fossil fuels (90 %) and large-scale hydro (8 %) (FEPC 2015). They commanded a 96 % share of the retail power market in 2014 (METI Website 2016a). However, power sales and the market share of the Big 10 have recently seen a sharp decline in the higher-voltage market (>50 kW). The share of independent power producers and suppliers (PPSs) has doubled within two years from 4.3 % in April 2014 to 8.6 % in February 2016, as many industrial and commercial customers have moved away from the Big 10 owing to rising electricity rates. At the same time, the number of registered PPS firms has increased from 55 to 127, indicating a growing intensity of competition (Smart Japan 2016a).

Similar developments are expected in the low-voltage retail power segment that was opened to competition on 1 April 2016. Since registration started in January 2016, the number of participants in this new market has surged to 286 (as of 18 April 2016) covering a broad range of very diverse players (Smart Japan 2016a). Besides the Big 10 and their subsidiaries (19 companies) and the long-established PPS firms (22 firms), there are many new entrants coming from the gas industry (51 firms), from the telecom/broadcasting/railway industry (34 firms), from the oil industry (9 firms), and various other business fields such as retailing, housing/construction, engineering/facility management, trading, finance, and electronics. Forty-four firms are explicitly categorized as RE suppliers, but the other categories also have quite a number of companies with a focus on RE. Noteworthy is also the number of regionally active firms (approx. 20 firms), many of which are (partially) owned by municipalities (at least 13 firms), consumer cooperatives (around 5 firms), or local citizen groups who position themselves on the "regionally produced, regionally consumed" platform and actively promote regional RE investment.

As of April 2016, more than 810,000 contracts have been switched to new suppliers, most of them in the large metropolitan areas around Tokyo and Osaka. Gas companies like Tokyo Gas or Osaka Gas have apparently grabbed the lion's share of new contracts (Kankyō Business online 2016; Smart Japan 2016b) by offering attractive discount packages that combine gas and power. A similar approach is being taken by cable TV firms like J:Com or mobile phone providers such as Softbank or KDDI, who attract customers with discount packages that combine power with mobile phone or TV services and often also offer the accumulation of bonus points from popular bonus and e-money systems like Rakuten or T-Points. These powerful firms also benefit from their extensive sales networks. The liberalization of the retail power market also offers new business opportunities for the Big 10 by invading each other's so far protected regional markets. Their preferred strategy is to form strategic alliances to strengthen their sales networks, for example, Tokyo Electric Power with Softbank, Kansai Electric Power with KDDI, the oil companies Tonen General and JX Energy, Osaka Gas with NTT Docomo, or Chubu Electric Power with the retailer Edion and the gas firm INPEX.

A key issue from the perspective of environmentally conscious consumers is the lack of transparency about the power source structure, which most companies are not (yet) disclosing. Most comparison Websites, like power hikaku.com, kakaku. com, or enechange.jp, focus on price comparisons and reveal very little information on quality attributes like the power source structure. Environmental nongovernmental organizations like Greenpeace (Greenpeace Website), Friends of the Earth (Powershift Website), or the Green Purchasing Network (GNP Website) advocate mandatory disclosure, run campaigns under slogans such as power shift or iSwitch, and have launched Websites that rate and promote power suppliers on the basis of disclosure of their power source structure and carbon footprint, active promotion of RE and support for regional or citizen-initiated RE investment, refusal to use nuclear- or coal-fueled power, or their independence from the large utilities. At present, the number of listed power suppliers with a focus on RE is rather limited; for example, the Powershift Website endorses just 14 RE suppliers, most of which are rather small with limited territorial reach.

One of the main reasons for the rather limited choice of RE suppliers is their constrained ability to source sufficient and reliable volumes of RE because the Japanese RE market is still comparatively small. It is also to an overwhelming degree based on widely distributed, smaller-scale solar power, which is cumbersome to contract and whose owners may find it more convenient to keep their FIT-based contracts with the established Big 10. In addition, large-scale solar and wind power facilities, as well as hydro or biomass plants, which are often favored by RE suppliers as a stable, backup power source, are usually owned by a few large companies that are frequently associated with the large power utilities (EU-Japan Centre 2014). The large-scale hydro plants are owned by the Big 10 or by municipalities who have engaged in long-term supply contracts with them (Smart Japan 2015d). This makes it difficult to form strong RE-based balancing groups. As a result, many suppliers that focus on RE are forced to rely on backup agreements with suppliers of conventional power and depend to a substantial degree on purchases from the wholesale market, which exposes them to greater commercial risks. Their claim of environmental benefits, therefore, remains rather weak. These constraints, on the other hand, serve as an opportunity for suppliers that secure reliable RE sources. The following section shines a light on some interesting emerging business models for RE marketers.

## 2.3    Emerging Business Models

Various types of actors entering the retail power market are positioning themselves as RE marketers. A number of long-established PPS firms, like the market leader Ennet, a joint venture by NTT Facilities, Tokyo Gas and Osaka Gas (Ennet Website), Summit Energy, a 100 % subsidiary of the general trading house Sumitomo Trading (Summit Energy Website), or Orix, one of Japan's leading financial service conglomerates (Orix website), emphasize their investments in RE facilities and comparatively high share of RE power as part of their general corporate social responsibility program. Idemitsu, one of Japan's leading oil and gasoline companies, has established Idemitsu Green Power and claims that 78 % of the supplied power comes from recycling and RE facilities (e.g., wind parks, biomass, geothermal, solar) with a combined capacity of 123 MW (Idemitsu Green Power Website). And one of Japan's leading IT and mobile phone providers, Softbank, has invested heavily in RE and entered the retail power market in a tie-up with Tokyo Electric Power. It offers a wide range of power service packages, frequently based on combined mobile and power services, but also markets a "FIT Power" package claiming a 67 % RE share (Softbank Power Website). While these companies can be expected to play an important role in the emerging RE market segment, four emerging RE-based business models from newcomers are particularly noteworthy:

- Aggregation model based on FIT power,
- Integrated RE service providers,
- Municipality-centered balancing groups,
- Regional power retailing by consumer cooperatives.

**Aggregation Model Based on FIT Power**  The business model of these actors is based on the nationwide aggregation of a large number of smaller-scale solar power installations from private households and companies, thereby securing access to regionally diversified RE power sources. The regional diversification and aggregation make it possible to balance fluctuations in the supply of solar power during the day, allowing these providers to offer their customers attractive rate packages for FIT power. At the same time, they try to generate profit by selling excess RE power to the wholesale market. Furthermore, they provide various other products and services to their customers.

An example of the implementation of this strategy is NTT Smile Energy (NTTSE Website). The company was founded in June 2011 as a joint venture between Japan's leader telecom provider NTT West (66%) and the electronics company Omron (34%). It initially specialized in the marketing of the so-called Eco Megane (Eco glasses) solar power monitoring system. The system is installed together with solar panels and allows for real-time Internet-based and mobile monitoring of generated solar power and the provision of service and maintenance services for solar panels. Backed by an installation base of 40,000 units and an aggregated capacity of 750 MW of solar power, NTTSE is expanding into trading and retail sales of solar power in a partnership with Ennet. In 2015, it introduced a product, Eco Megane Plus, to customers in the low-voltage segment (<50 kW) who install solar panels with the monitoring system and allowed their customers to purchase the generated RE power at a premium to the FIT. In this way, NTTSE is able to aggregate regionally diversified RE power, provide forecasts on RE availability, and assure its partner Ennet a comparatively stable supply of solar power to sell on the wholesale or Over-The-Counter (OTC) market (Smart Japan 2015e, f). In April 2016, NTTSE entered the liberalized retail power market with a service package that also targets customers who install new solar panels with the Eco Megane monitoring system (Smart Japan 2016c). These customers are encouraged to sell their excess solar power to other utilities under the prevailing FIT scheme and to buy FIT power from NTTSE at the same rate as other utilities. The scheme assures customers that during the day, between 8:00 a.m. and 4:00 p.m., they receive 100% solar power regardless of weather conditions and the season. Outside this time period, NTTSE assures supply by relatively environmentally friendly gas-generated power through its partnership with Ennet. NTTSE offers additional rate discounts for their customers' energy-saving efforts and promises that 1% of revenue from FIT power sales will be invested in new RE facilities. The scheme is based on supply from 8000 solar power installations with a capacity of 270 MW to service 30,000 customers nationwide.

**Integrated RE Service Provider Model** Another group of RE marketers that position themselves as integrated RE service providers is composed of companies that started in the construction business then entered the engineering, procurement, and construction (EPC) business for RE and now seek an entry into the retail power market. The leader in this group is Japan's top EPC company, NTT Facilities, but there are also younger, venturelike firms such as the West Group (West Website). The company started in 1984 as a home builder and construction service company. In 2007, it expanded into the sales, planning, and installation of solar systems, first for residential houses, then for commercial customers and municipalities. Today the West Group is one of Japan's leading EPC firms in the solar business. It has grown rapidly into a company with approximately JPY 46 billion (around USD 420 million) and around 480 employees in fiscal year 2015 and is listed on the JASDAQ stock market. Backed by its strong base in the EPC business for solar systems, the West Group provides a wide range of RE-related services such as operation and maintenance of RE facilities, energy management for commercial buildings, and fund management and leasing for larger-scale solar projects. A strategic focus is placed on providing energy consulting and management services for municipalities and local governments in their efforts to develop integrated, local energy concepts based on the utilization of local RE resources. To this end, in 2014 it established West Denryoku as its retail power subsidiary in 2014 positioning itself as a one-stop partner for regional governments with broad competencies in power generation, power management, and conservation, as well as power trading.

Nihon Eco Systems is another example of a company seeking to leverage its broad expertise in the EPC solar business through an entry into the power retail market and position itself as a RE provider (Nihon Eco System Website). The company started in 1997 as a sales company for energy-saving devices and entered the solar business in 2000 by becoming an agent for Sharp solar systems on residential buildings. In 2010, it expanded into the commercial segment and established itself as an EPC provider for medium-scale solar systems for industrial users. In 2016, it registered as a power retail company intending to generate synergies with its residential solar system business that boasts more than 36,000 installations. For this purpose, it launched a rather unique service package under the name *Jibun Denki* (my own power) (Smart Japan 2016d). Under the scheme, the company installs solar systems on residential homes free of charge. The generated power is owned by the company and sold under the FIT program. In exchange, the company sells power to the homeowner, offering a competitive set of two types of rates. The first rate relates to the self-generated solar power that is consumed by the residential owner; the homeowner repurchases the power from the company at a competitive rate, implying that the homeowner uses its own power. In times when the self-generated solar power is insufficient, the company provides power it procures from Ennet at a competitive market price. The contract runs for 20 years, after which the solar system is handed over to the homeowner. With this scheme Nihon Eco System, which now belongs to the information technology and telecom engineering service group Nippon Consyst, is aiming for 100,000 free-of-charge installations within 5 years.

Yet another innovative company is Looop (Looop Website). Established in 2011, Looop started its sales business in solar power and battery systems. Its unique approach rests on the concept of do-it-yourself product kits that are easily assembled and, therefore, are also sold via the Internet. Under the My-Kit brand, Looop first targeted the consumer segment but rapidly expanded its product range to applications for farmers, land owners, and commercial users. At the same time, Looop invested in its own solar facilities, often as showcases for its unique product concepts and technologies. Since 2011, Looop has sold around 1500 kits with a capacity of 120 MW, and its own RE investments amount to 9 MW. Looop has also entered the retail power market and positions itself as an integrated RE provider that gets 26 % of its power from RE sources and is endorsed by the Power Shift campaign (Power Shift Website).

**Model of Municipality-Centered Balancing Groups** A number of municipalities have founded their own local power companies aimed at regional economic development and the utilization of local RE sources, with more expected to follow (Asahi Digital 2016). Typically, these power companies are set up as joint ventures with industrial partners to minimize commercial risks and gain access to professional management know-how. Some, usually smaller, firms, like Izumisano Denryoku in Osaka Prefecture or Ōta Denryoku in Gumma Prefecture, are majority owned by the local city, while larger ones, often located in bigger regional cities like Kitakyūshū in Fukuoka Prefecture (960,000 inhabitants), Hamamatsu in Shizuoka Prefecture (810,000 inhabitants), Yamagata in Yamagata Prefecture (250,000 inhabitants), or Tottori in Tottori Prefecture (190,000 inhabitants), usually hold a minority share. The key managerial challenge of municipality-centered power companies is gaining secure access to RE power sources and the stabilization of power supply. For this purpose, most schemes are created as regional balancing groups that integrate various power sources (e.g. public sewage facilities, waste incineration, large-scale RE facilities like solar parks, hydro, or biomass plants, and PV systems on residential homes) with energy management and trading capabilities. Industrial partners usually contribute larger-scale power sources like biomass plants or solar parks and handle the operation, energy management, and trading sides of the business. Many municipality-centered power firms have started out by supplying power to the higher-voltage market, preferably public buildings (e.g. city office buildings, schools, hospitals) or selected industrial facilities, but are eager to expand into the low-voltage segment for local citizens and smaller businesses. For example, Miyama, a city of 38,000 inhabitants on the southern island of Kyushu, started selling power to its citizens in April 2016 (Smart Japan 2016e). The city established the local power company Miyama Smart Energy in February 2015 as a joint venture of the city (55 %), a local energy service company (Kitakyushu Smart Community, 40 %), and a local bank (5 %) (Miyama Smart Energy Website). For power sourcing it relies on a 5 MW solar park and offers the purchase of solar power at a premium over the prevailing FIT rate to those citizens with a solar system on their roof. This serves also as an incentive to those citizens who have not yet installed a solar system on their homes. At the same time, the city is promoting a home energy management system (HEMS) that monitors power consumption. The scheme helps to reduce power consumption, predict power supply, and design rate packages in line with consumption patterns. Backed by 1200 homes with solar systems and 2000 HEMS-equipped households, Miyama Smart Energy intends to supply power to 70 % of citizens within 3 years. It is also considering investing in its own local power grid.

A central message of these municipalities to their local communities is their reliability and regional roots, which commits them to regional economic and social development. For many the entry into energy management is an essential part of a comprehensive strategy to reduce their energy dependence and to create a regional, circular economy that keeps income within the regions and creates new funding sources for community services (e.g. education, child care, health care). The mayor of Miyama points to the annual power bill of JPY 2 billion (USD 18 million) that leaves the regional economy. Their scheme of "local generation, local consumption" aims to keep these funds within the region and to use them for the support of the local economy.

**Regional Power Retailing by Consumer Cooperatives**  A fourth group of actors with a promising business model as RE marketers are Japan's powerful consumer cooperatives, in particular the 130 community-based retail coops. With their extensive retail stores and home delivery service network, more than 20 million members, and revenues of more than USD 24 billion in 2014, retail coops are the backbone of the Japanese Consumers' Cooperative Union (JCCU), which itself is the world's largest coop organization (JCCU 2014b). As of April 2016, at least six community-based retail coops have registered as power retailers, among them some of Japan's largest, like Co-op Kobe (1.7 million members), Osaka Izumi Co-op (480,000 members), and Pal System Tokyo (440,000 members), as well as the consumer cooperative union Seikatsu Club (300,000 members) (Smart Japan 2016f). The huge membership base, the close connection with consumers who frequently share concerns about food safety and environmental protection, and their strong relations with rural communities and farming establish them as convincing RE marketers and makes them potentially very powerful competitors in the regional power markets. Their main strategic challenge, however, is the lack of sufficient and stable access to RE power sources and the associated commercial risks of energy sourcing, trading, and management. Careful not to overstretch the business, the consumer coops apparently take a long-term approach to developing energy services as a third pillar next to food and health care. In a first step, many of them have invested in their own RE power plants to provide power for their own facilities (e.g. stores, distribution centers) and launched tie-ups with other, larger-scale RE providers to secure a power supply. Some of them, like Seikatsu Club Union, mobilize their member base to invest in joint RE projects like wind power (Seikatsu Club Website) or, like Pal System Tokyo, sponsor citizen- or farmer-based initiatives to invest in RE projects. As the first community-based retail coop, Osaka Izumi Coop started marketing power to its members as 70 % of 70,000 respondents to an internal survey expressed an interest in procuring power from the cooperative. More than 11,000 customers have already signed up for the new service (Osaka Izumi Coop presentation). The cooperative already covers more than 60 % of its own power consumption from RE sources based on its investment in solar power (approx. 11 MW) and a tie-up with a large biomass operator.

Community-based retail coops can be seen as a spearhead of a citizen-driven movement for regional power based on RE. Japan's so-called Community Power movement can look back to a long history of citizen-initiated investments in RE projects that has gained noteworthy momentum since the Fukushima nuclear disaster (Raupach-Sumiya and Tezuka 2017). Under the motto "Gotōchi Energy" (Local Energy) a grassroots movement is rapidly growing eager to invest in local RE projects and establish regional power retail operations.

## 3 Outlook

The Japanese market for RE is a seller's market, constrained by the still comparatively low share of RE in Japan's energy mix. Backed by growing consumer interest in RE and the substantial purchasing power of Japanese consumers, marketers of RE should face a promising future. However, the prospects for RE must be evaluated within the overall regulatory and political context in Japan. The full liberalization of Japan's power and gas retail markets opens the industry to full-fledged competition, and a large number of companies from various industries, many of them financial heavyweights, have lined up to enter the market. Some of them are visibly positioning themselves as RE marketers. Innovative business models based on differentiation through RE and clean, regional sourcing are emerging. Yet, uncertainties run high. On the one hand, it remains to be seen whether liberalization will lead to truly open and fair competition, whether third-party access to the grid will be unhindered, and whether a deep and liquid wholesale market will develop. On the other hand, the commercial viability of the various strategic approaches and business models is still to be proven, and consumer attitudes must still be tested. Most importantly, the supply of RE itself needs to grow requiring continued, high investment in RE and energy grids. The energy policy of Japan's government actively promotes the expansion of RE while aiming for a more orderly and cost-efficient process of RE deployment. But the indicated, substantial changes in the regulatory framework and promotional policies for RE seem to raise the bar even further for RE investors and marketers.

## References

Asahi Digital. (2016). *13 municipality-owned power companies are offering locally generated power for local consumption at a discount (Jichitaidenryoku 13-sha, aitsugu setsuritsu denki chisan chishō wariyasu)*, March 1, 2016. Accessed May 14, 2016, from http://digital.asahi.com/articles/ASJ2H457VJ2HTIPE016.html?_requesturl=articles%2FASJ2H457VJ2HTIPE016.html&rm=741

Bloomberg New Energy Finance. (2015a). *Tokyo too bullish on nuclear, too bearish on gas and solar.* Press Release on 2.6.2015. Accessed May 3, 2016, from http://about.bnef.com/press-releases/tokyo-bullish-nuclear-bearish-gas-solar/

Bloomberg New Energy Finance. (2015b). *Global trends in renewable energy investment 2015.* Frankfurt School of Finance & Management GmbH. Accessed May 3, 2016, from http://www.fs-unep-centre.org

Dentsu. (2016). *Survey on consumer attitudes in regard to energy liberalization (Enerugi- jiyūka ni kan-suru seikatsusha ishiki chōsa).* Press release on January 22, 2016. Accessed May 6, 2016, from http://www.dentsu.co.jp/news/release/pdf-cms/2016009-0122.pdf

EU-Japan Centre for Industrial Cooperation. (2014, February). *The clean energy sector in Japan—An analysis on investment and industrial cooperation opportunities for EU SMEs*, Tokyo.

Federation of Electrical Power Companies in Japan (FEPC). (2015). *Electricity in Japan 2015.* Accessed May 7, 2016, from http://www.fepc.or.jp/english/library/electricity_eview_japan/__icsFiles/afieldfile/2015/08/10/2015ERJ_full.pdf

Institute for Sustainable Energy Policies (ISEP). (2015). *Renewables 2015 Japan Status Report (Shizen enerugi- hakushō 2015), Tokyo, 2015.* Accessed May 3, 2016, from http://www.isep.or.jp/images/library/JSR2015all.pdf

Jordan-Korte, K. (2011). *Government promotion of renewable energy technologies—Policy approaches and market development in Germany, United States, and Japan.* Springer Gabler.

Japanese Consumers' Co-operative Union (JCCU). (2014a, July 2). *Survey on consumer attitudes regarding renewable energy and energy saving (Setsuden to saisei kanō enerugi- ni kan-suru shōhisha ishiki chōsa).*

Japanese Consumers' Co-operative Union (JCCU). (2014b). *Co-op 2014 facts and figures.* Accessed May 15, 2016, from http://jccu.coop/eng/public/pdf/ff_2014.pdf. Accessed May 6, 2016, from http://jccu.coop/info/press_140702_01_01.pdf

Japanese Consumers' Co-operative Union (JCCU). (2015). *Survey about consumer attitudes regarding the future electrical power system (kore kara no denryoku no arikata nit suite no shōhisha ishiki chōsa),* May 8, 2015. Accessed May 6, 2016, from http://jccu.coop/info/press_150508_01_01.pdf

Kankyō Business Online. (2016). *One month after liberalization of power market the number of power switching contracts exceeds 810,000—competition in the Tokyo area intensifies (Denryoku jiyūka kara ikkagetsu—denryokukeiyaku no henkō hachijūichiman ni—Shutoken de kyōsō gekka),* May 11, 2016. Accessed May 14, 2016, from https://www.kankyo-business.jp/news/012622.php?utm_source=mail&utm_medium=mail160512_d&utm_campaign=mail

Ministry of Economy, Trade and Industry (METI). (2011). *Feed-in tariff scheme for renewable energy,* October 2011. Accessed May 3, 2016, from http://www.meti.go.jp/english/policy/energy_environment/renewable/pdf/summary201209.pdf

Ministry of Economy, Trade and Industry (METI). (2012). *Feed-in tariff scheme in Japan,* updated on July 17, 2012. Accessed May 3, 2016, from http://www.meti.go.jp/english/policy/energy_environment/renewable/pdf/summary201207.pdf

Ministry of Economy, Trade and Industry (METI). (2014). *Energy White Paper 2014.* Accessed May 3, 2016, from http://www.enecho.meti.go.jp/about/whitepaper/2014pdf/whitepaper2014pdf_3_4.pdf

Ministry of Economy Trade and Industry (METI). (2015a). *Strategic Energy Plan, April 2014 (provisional translation).* Accessed May 3, 2016, from http://www.enecho.meti.go.jp/en/category/others/basic_plan/pdf/4th_strategic_energy_plan.pdf

Ministry of Economy Trade and Industry (METI). (2015b). *Japan's electricity market deregulation,* June 2015. Accessed May 3, 2016, from http://www.meti.go.jp/english/policy/energy_environment/electricity_system_reform/pdf/201506EMR_in_Japan.pdf

Ministry of Economy Trade and Industry (METI). (2015c), *Reform of electricity system (Denryoku shisutemu kaikaku-ni tsuite).* Accessed May 3, 2016, from http://www.enecho.meti.go.jp/category/electricity_and_gas/electric/system_reform.html

Ministry of Economics Trade and Industry (METI). (2015d). *Feed-in-tariff-system public information website (Koteikakakukaitorihō jōhōkōhyō webusaito).* Accessed May 3, 2016, from http://www.fit.go.jp/statistics/public_sp.html

Ministry of Economy Trade and Industry (METI). (2015e). *News release on January 22, 2015.* Accessed May 4, 2016, from http://www.meti.go.jp/english/press/2015/0122_02.html

Ministry of Economy Trade and Industry (METI). (2015f). *Status report on the reform of the system to promote RE deployment (saiseikanō enerugi-no dōnyūsokushin ni kan-suru seidokaikaku-no kentōjōkyō nit suite hōkoku),* December 27, 2015. Accessed May 4, 2016, from http://www.meti.go.jp/committee/sougouenergy/shoene_shinene/shin_ene/pdf/015_b02_00.pdf

Ministry of Economy Trade and Industry (METI). (2015g). *Changes in working rules in the application of the FIT system related to the full liberalization of the power retail market (Kouri zennmenjiyūka ni muketa koteikakakukaitoriseido no unyōminaoshi nit suite),* November 2015 (renewed on December 9, 2015). Accessed May 4, 2016, from http://www.enecho.meti.go.jp/category/saving_and_new/saiene/kaitori/dl/kouri_free/20151125.pdf

Ministry of Economy Trade and Industry (METI). (2015h). *About the detailed system design for the fully liberalized power retail market—Materials submitted by the Bureau of the System Design Working Group for the 14th meeting (Dai 14kai seido sekkei wa-kingu gru-pu jimukyoku teishutsu shiryō—kouri zenmen jiyūka ni kakawaru shōsai seido sekkei nit suite)*, July 28, 2015. Accessed May 5, 2016, from http://www.meti.go.jp/committee/sougouenergy/kihonseisaku/denryoku_system/seido_sekkei_wg/pdf/014_06_04.pdf

Ministry of Economy Trade and Industry (METI). (2016a). *Energy Reform Strategy (Enerugi-Kakushin Senryaku)*, April 2016. Accessed May 3, 2016, from http://www.meti.go.jp/press/2016/04/20160419002/20160419002-2.pdf

Ministry of Economy Trade and Industry (METI). (2016b). *Outline of the draft for the partial revision of the Special Law to Promote Renewable Energy (Denkijigyōsha ni yoru saiseikanō enerugi-denki no chōtatsu ni kan-suru tokubetsusochihō no ichibu wo kaisei-suru hōritsuan no gaiyō)*, February 2016. Accessed May 4, 2016, from http://www.meti.go.jp/press/2015/02/20160209002/20160209002-3.pdf

Ministry of Economy Trade and Industry (METI). (2016c). *Guidelines regarding retail power marketing (Denryoku no kourieigyō ni kan-suru shishin)*, January 2016. Accessed May 4, 2016, from http://www.meti.go.jp/press/2015/01/20160129007/20160129007-1.pdf

Mizuho Information and Research Center (Mizuho). (2015). *Survey report on the attitudes of consumers regarding the selection criteria for retail power suppliers and services in the face of power market liberalization (Denryoku jiyūka ni mukete no shōhisha no denryoku kouri kigyō sa-bisu sentaku kijun ni kan-suru ishiki chōsa chōsa ripo-to)*, June 8, 2015. Accessed May 6, 2016, from http://www.mizuho-ir.co.jp/publication/report/2015/pdf/e-jiyuka0608.pdf

Osaka Izumi Coop presentation material obtained at Symposium by the Miyako-no Agenda 21 Forum on April 9th, 2016 in Kyoto.

PricewaterhouseCoopers K.K. (PWC). (2016). *The 2016 survey on attitudes about the retail power market just before liberalization (Denryoku kouri shijō chōsa 2016 jiyūka chokuzen mae hen)*, January 26, 2016. Accessed May 6, 2016, from http://www.pwc.com/jp/ja/japan-press-room/press-release/2016/electricity-market-survey160226.html

Raupach-Sumiya, J., & Tezuka, T. (2017). Community power in Japan. In L. Holstenkamp & J. Radtke (Eds.), *Handbook of energy transition and participation*. Heidelberg: Springer (in press).

Smart Japan. (2014). *54% of households consider to switch their power supplier, lower electricity rates are the most important selection criteria (Denryoku no kōnyūsaki wo kentō-surukatei ha 54%, ryōkin no yasusa ga saidai no sentaku riyū)*, June 6, 2014. Accessed May 6, 2016, from http://www.itmedia.co.jp/smartjapan/articles/1406/30/news026.html

Smart Japan. (2015a). *OCCTO starts operation on April 1 with 7 tasks (Kōiki Kikan-ga 7shurui-no gyōmu-wo shigatsu tuitachi-ni kaishi)*, February 20, 2015. Accessed on May 3, 2016, from http://www.itmedia.co.jp/smartjapan/articles/1502/20/news039.html

Smart Japan. (2015b). *The FIT system changes with full liberalization of the retail power market–link of scheme to trading prices on wholesale market (Kouri zenmen jiyūka de sai ene no kaitori seido-ga kawaru, oroshidenryoku shijō no torihiki kakaku ni rendō)*, May 19, 2015. Accessed May 5, 2016, from http://www.itmedia.co.jp/smartjapan/articles/1505/19/news037.html

Smart Japan. (2015c). *80% consider switching their power supplier/fears about power quality (denryoku no kōnyūsaki henkō wo kentō-suru hito ga achiwari ni, denki no shitsu ni fuan mo)*, November 24, 2015. Accessed May 6, 2016, from http://www.itmedia.co.jp/smartjapan/articles/1511/24/news029.html

Smart Japan. (2015d). *Municipalities are urged to dissolve their supply contracts with power utilities and introduce competitive auctions (Denryoku kaisha to no baiden keiyaku wo kaishō-shiyasuku, jichitai ni kyōsōnyūsatsu wo onagasu)*, January 5, 2015. Accessed May 7, 2016, from http://www.itmedia.co.jp/smartjapan/articles/1501/05/news010.html

Smart Japan. (2015e). *NTT group purchases solar power starting operation in February in the Tokyo area (Taiyōkō no denryoku wo NTT group ga kaitori, Tōkyō Denryoku no kannai kara ni-gatsu ni kaishi)*, January 30, 2015. Accessed May 14, 2016, from http://www.itmedia.co.jp/smartjapan/articles/1501/30/news023.html

Smart Japan. (2015f). *NTT Smile Energy announces its business strategy—Raising the value of RE through visualization (Mieruka de saisei kanō energugi- no kachi wo takameru, NTT Smile ga jigyōsenryaku wo happyō)*, November 16, 2015. Accessed May 14, 2016, from http://www.itmedia.co.jp/smartjapan/articles/1511/16/news042.html

Smart Japan. (2016a). *The number of switching cases for power suppliers increases to 680,000, the share for commercial customer to more than 8% (Denryoku kaisha kara norikaere wa 68-manken ni, kigyō muke no shea wa 8% dai kōhan)*, April 27, 2016. Accessed May 7, 2016, from http://www.itmedia.co.jp/smartjapan/articles/1604/27/news048.html

Smart Japan. (2016b). *Tokyo Gas and Osaka Gas increase the number of new power contracts, appealing with combined discount rate packages and disclosure of power source structure (Denryoku no keiyaku kensū wo nobasu Tōkyō Gas and Ōsaka Gas, setto waribiki to dengenkōsei wo api-ru)*, April 7, 2016. Accessed May 7, 2016, from http://www.itmedia.co.jp/smartjapan/articles/1604/07/news048.html

Smart Japan. (2016c). *NTT group starts to offer FIT power at same rate as utilities starting in May (Denryokukaisha to dōitsuryōkin de FIT denki-wo kyōkyū, NTT group ga gogatsu ni kaishi)*, April 14, 2016. Accessed May 124, 2016, from http://www.itmedia.co.jp/smartjapan/articles/1604/14/news047.html

Smart Japan. (2016d). *Free-of-charge solar power installations for residential homes—The first nationwide model for retail power sales that leverages solar power installations (Jūtaku ni zero en de hatsudennsetsubi wo teikyō—taiyōkō wo katsuyō-shita zenkoku hajime no denryoku kōuri moderu)*, March 22, 2016. Accessed May 14, 2016, from http://www.itmedia.co.jp/smartjapan/articles/1603/22/news054.html

Smart Japan. (2016e). *Locally generated and consumed energy changes cities—municipalities enter power retailing (Enerugi- no chisan chishō de machi ga kawaru, jichitai ga denryoku no kōuri ni noridasu*, January 13, 2016. Accessed May 15, 2016, from http://www.itmedia.co.jp/smartjapan/articles/1601/13/news019.html

Smart Japan. (2016f). *Citizen Groups and Consumer Co-ops enter the power retailing business, offering renewable energy to households (Shimin Denryoku ya Seikyō ga zokuzoku to kouri denki jigyōsha ni, saisei kanō enerugi- wo katei he)*, February 5, 2016. http://www.itmedia.co.jp/smartjapan/articles/1602/05/news105.html

Tokyo Gas. (2015). *Survey on attitudes regarding energy market liberalization (Denryoku jiyūka ni kan-suru ishiki chōsa)*, December 8, 2015. Accessed May 6, 2016, from http://prtimes.jp/main/html/rd/p/000000001.000016718.html

United Nations Economic and Social Commission for Asia and the Pacific (UNESCAP). (2012). Low carbon green growth roadmap for Asia and the Pacific: Case Studies and Policy Papers—27. Case Study Japan Green Power Certification Scheme, January 2012. Accessed May 5, 2016, from http://www.unescap.org/sites/default/files/27.%20CS-Japan-Green-Power-Certificate-Scheme.pdf

## Websites

Enechange Japan. Accessed May 7, 2016, from https://enechange.jp/

Ennet website. Accessed May 14, 2016, from www.ennet.co.jp/about/clean.html

The Green Energy Certification Center website. Accessed May 5, 2016, from http://eneken.ieej.or.jp/greenpower/jp/

Greenpeace iSwitch campaign website. Accessed May 7, 2016, from https://act.greenpeace.org/ea-action/action?ea.client.id=1980&ea.campaign.id=47973

Green Purchasing Network (GNP) website. Accessed May 7, 2016, from http://www.gpn.jp/guideline/electric_power.html

Idemitsu Green Power website. Accessed May 15, 2016, from http://idemitsu.co.jp/igp/

Kakaku.Com. Accessed May 7, 2016, from http://kakaku.com/energy/

The Japan Natural Energy Company Limited website. Accessed May 5, 2016, from http://www.natural-e.co.jp/company/index.html

Looop website. Accessed May 14, 2016, from http://looop.co.jp

Ministry of Economy Trade and Industry (METI website). (2016a). Statistics from Electrical Power Survey (Denryoku Chōsa Tōkei), website. Accessed May 6, 2016, from http://www.enecho.meti.go.jp/statistics/electric_power/ep002/results.html#headline1

Ministry of Economy Trade and Industry (METI website). (2016b). Public information website on FIT system (Kotei kaitori kaku seido jōhōkōkaiyō webusaito). Accessed May 6, 2016, from http://www.fit.go.jp/statistics/public_sp.html

Miyama Smart Energy website. Accessed May 15, 2016, from http://www.miyama-se.com/company.html

Nihon Eco System website. Accessed May 14, 2016, from http://www.j-ecosystems.co.jp

NTT Smile Energy (NTTSE website). Accessed May 8, 2016, from http://nttse.com

Orix website. Accessed May 14, 2016, from http://www.orix.co.jp/grp/business/eco/

Power Hikaku website. Accessed May 7, 2016, from http://power-hikaku.info/

Powershift. Accessed May 7, 2016, from http://power-shift.org/

Seikatsu Club website. Accessed May 15, 2016, from http://www.seikatsuclub.coop/activity/20160122.html

Softbank Power website. Accessed May 15, from http://softbank.jp/energy/price/fit

Summit Energy website. Accessed May 14, 2016, from http://www.summit-energy.co.jp/about/index.html

West Group website. Accessed May 14, 2016, from http://www.west-gr.co.jp/en

**Jörg Raupach-Sumiya** received his Ph.D. at the Department for East Asian Economic Studies, Faculty for Economic Studies, University Duisburg/Essen, Germany. He has been living and working in Japan since 1990. After his business career as Senior Consultant at Roland Berger Strategy Consultants (Tokyo/Japan), Managing Director at Trumpf Corporation (Yokohama/Japan) and NEC SCHOTT Corporation (Minakuchi/Japan), he became Professor for International Management at the College of Business Administration of Ritsumeikan University in Osaka, Japan, in 2012. His main research topics include the energy industry, regional economic effects of renewable energy, energy transition, Japanese management, and the Japanese economy.

Zeitfracht Medien GmbH
Ferdinand-Jühlke-Straße 7
99095 Erfurt, Deutschland
produktsicherheit@kolibri360.de